T0282301

LONDON MATHEMATICAL SOCIETY LECTURE NOTE

Managing Editor: Professor J.W.S. Cassels, Department of Pure Mathematics and Mathematical Statistics, University of Cambridge, 16 Mill Lane, Cambridge CB2 1SB, England

The titles below are available from booksellers, or, in case of difficulty, from Cambridge University Press.

London Mathematical Society Lecture Note Series. 204

Combinatorial and Geometric Group Theory

Edinburgh 1993

Edited by

Andrew J. Duncan
University of Newcastle

N.D. Gilbert
University of Durham

James Howie
Heriot-Watt University

CAMBRIDGE
UNIVERSITY PRESS

CAMBRIDGE UNIVERSITY PRESS
Cambridge, New York, Melbourne, Madrid, Cape Town, Singapore,
São Paulo, Delhi, Dubai, Tokyo, Mexico City

Cambridge University Press
The Edinburgh Building, Cambridge CB2 8RU, UK

Published in the United States of America by Cambridge University Press, New York

www.cambridge.org
Information on this title: www.cambridge.org/9780521465953

© Cambridge University Press 1995

First published 1995

A catalogue record for this publication is available from the British Library

ISBN 978-0-521-46595-3 Paperback

Contents

Foreword

'This is a truly wonderful book, which unfortunately our library is too small to contain.' - *Pierre de Fromage*

In the spring of 1993, a 10-day workshop was held at Heriot-Watt University, Edinburgh, under the auspices of the International Centre for Mathematical Sciences, on Geometric and Combinatorial Methods in Group Theory. This volume contains papers contributed by participants at the workshop. Some report work presented by the authors in lectures at the conference, and all of them are on topics closely related to the central theme of the conference. Survey articles were kindly contributed by S M Gersten, R I Grigorchuk, P H Kropholler, A Lubotzky, A A Razborov and E Zelmanov, who were among the invited conference speakers.

The problem section at the end of the book is made up of problems presented at a problem session on the final day of the conference. The session was chaired by S J Pride, and the list was compiled using notes taken by S Wreth together with written comments from the presenters of the problems.

The editors are deeply indebted to V Metaftsis and S Wreth for their invaluable assistance in the running of the conference. We could not have succeeded without their help. In addition, we are grateful to the many other people who helped to make the conference a success. These include the Scientific Committee of Steve Gersten, Steve Pride and Sasha Razborov; John Ball, Frank Donald and Elmer Rees of the International Centre for Mathematical Sciences; and Patricia Hampton, Isobel Johnson, Barbara Kollmer, Markus Kreer, and Donald Smith of Heriot-Watt University. It is a pleasure to record our gratitude to them, as well as to those who gave conference lectures or merely took part. We are also grateful to the Science and Engineering Research Council, who provided the principal funding for the conference under grant number GR/H57219, and to the Royal Society and the Royal Society of Edinburgh for additional funding.

The production of the present volume could not have taken place without the valuable help of Roger Astley, David Tranah and their colleagues at CUP, not to mention an army of anonymous referees, and of course the contributors themselves. To all of them we offer our thanks.

The Editors
July 1994.

Participants

H Abels	Universität Bielefeld
A G Ahmad	University of Glasgow
R Alperin	San Jose State University
S Bachmuth	University of California at Santa Barbara
B Baumslag	Imperial College, London
G Baumslag	City College, CUNY
N Benakli	Princeton University
M Bestvina	U C L A
W Bogley	Oregon State University
B H Bowditch	University of Southampton
N Brady	University of California
M R Bridson	Université de Genève
C J B Brookes	Universität Frankfurt
P Brown	University of California, Berkeley
I M Chiswell	Queen Mary and Westfield College
A Clifford	SUNY at Albany
D E Cohen	Queen Mary and Westfield College
D J Collins	Queen Mary and Westfield College
L P Comerford	Eastern Illinois University
G Conner	Brigham Young University
A Deutsch	Université de Neuchâtel
W Dicks	Universitat Autònoma de Barcelona
A J Duncan	University of Newcastle
M J Dunwoody	University of Southampton
M Edjvet	University of Nottingham
B Farb	University of California, Berkeley
K Fujiwara	Keio University
S M Gersten	University of Utah
N D Gilbert	Heriot-Watt University
D Gildenhuys	McGill University
C McA Gordon	University of Texas at Austin
R I Grigorchuk	Moscow Institute of Railway Transportation
M Hagelberg	Ruhr-Universität Bochum
J Harlander	Universität Frankfurt
M Hartl	Universität Bielefeld
S Hermiller	MSRI
J Howie	Heriot-Watt University
O Kakimizu	Niigata University
O G Kharlampovich	McGill University
C Kilgour	University of Glasgow
A C Kim	Pusan National University
E Koudela	Queen Mary and Westfield College

D Krammer	University of Utrecht
P H Kropholler	Queen Mary and Westfield College
P Latiolais	Portland State University
T H Lenagan	University of Edinburgh
A Lubotzky	University of Chicago
M Lustig	Ruhr-Universität Bochum
I G Lysionok	Academy of Sciences, Moscow
A J MacIntyre	University of Oxford
C Maclachan	University of Aberdeen
T Maeda	Kansai University
H Meinert	Universität Frankfurt
V Metaftsis	Heriot-Watt University
A G Myasnikov	University of Omsk
G A Niblo	University of Southampton
D B Nikolova	Bulgarian Academy of Sciences
A Yu Ol'shanskii	Moscow State University
Ch Pittet	Université de Genève
S J Pride	University of Glasgow
A A Razborov	Academy of Sciences, Moscow
E G Rees	University of Edinburgh
S Rees	University of Newcastle
A W Reid	University of Texas at Austin
V N Remeslennikov	University of Omsk
S Rosebrock	Universität Frankfurt
G Rosenberger	Universität Dortmund
A Rosenmann	Universität Essen
Z Sela	Princeton University
M Shapiro	City College, CUNY
P Shawcroft	Brigham Young University
A M Sinclair	University of Edinburgh
G C Smith	University of Bath
J R Stallings	University of California, Berkeley
R Stohr	UMIST
M Stoll	Universität Bonn
E Swenson	Brigham Young University
O Tabachnikova	University of Bath
O Talelli	University of Athens
R M Thomas	University of Leicester
E C Turner	SUNY at Albany
E Ventura	Escola Universitaria Politecnica de Manresa
K Vogtmann	Cornell University
S Wreth	Heriot-Watt University
E Zelmanov	University of Wisconsin
P-H Zieschang	Universität Kiel

On bounded languages and the geometry of nilpotent groups

MARTIN R. BRIDSON AND ROBERT H. GILMAN [1]

Abstract

Bounded languages are a class of formal languages which includes all context free languages of polynomial growth. We prove that if a finitely generated group G admits a combing by a bounded language and this combing satisfies the asynchronous fellow traveller property, then either G is virtually abelian, or else G contains an element g of infinite order such that g^n and g^m are conjugate for some $0 < n < m$.

The introduction of automatic groups [5] has precipitated a host of questions about the roles which formal language theory and geometry play in the study of normal forms for finitely generated groups, particularly groups which arise in geometric settings. For example, when a group G is given as the fundamental group of a compact Riemannian manifold, words in a fixed set of generators for G have a natural interpretation as paths in the universal cover of the manifold; it is natural to ask how the geometry of the manifold is reflected in the linguistic complexity of normal forms for elements of G. The results presented here and in [3] can be interpreted as providing a partial answer to this question in the case where the manifold under consideration is a quotient of a nilpotent Lie group.

It has become customary in geometric group theory to refer to a set of normal forms for elements in a finitely generated group as a *combing* of the group. There is much work to be done on the problem of determining how various geometric and linguistic constraints on the type of combings which a group admits are reflected in the structure of the group. The results presented here contribute to this task. These results arose in the course of our work on the structure of normal forms for elements in 3-manifold groups [3].

Theorem A *If a finitely generated group G admits a combing by a bounded language, and if this combing satisfies the asynchronous fellow traveller property, then either*

1. *G is virtually abelian, or*

2. *there is an element $g \in G$ of infinite order such that for some m, n with $0 < m < n$, g^m and g^n are conjugate in G.*

---

[1] The first author was supported in part by NSF grant DMS-9203500 and FNRS (Suisse). The second author thanks the Institute for Advanced Study for its hospitality while this paper was being written

Bounded languages are defined in Section 1, as is the asynchronous fellow traveller property.

We do not know of an example for which the above possibility (2) occurs. Certainly, one can exclude possibility (2) by placing restrictions on the class of groups considered. For example, it is shown in [1] that semihyperbolic groups, which are defined in terms of the type of combings which they admit, do not contain elements of the type described in possibility (2). (The class of semihyperbolic groups includes all biautomatic groups [8], and all groups which act properly and cocompactly by isometries on any 1-connected space of non-positive curvature, as all finitely generated virtually abelian groups do.)

Corollary B   *A semihyperbolic group* G *admits a combing by a bounded language with the asynchronous fellow traveller property if and only if* G *is virtually abelian.*

Theorem A is proved by reducing it to:

Theorem C   *If a finitely generated virtually nilpotent group* G *admits a combing by a bounded language, and if this combing satisfies the asynchronous fellow traveller property, then* G *is virtually abelian.*

Theorem C plays an important role in [3], where it is used to show that a virtually nilpotent group with a context free combing satisfying the asynchronous fellow traveller property is virtually abelian. In [3] we presented a purely algebraic proof of Theorem C, but this result is essentially a fact about the *geometry* of nilpotent groups. Here we try to present as accessible an account of this geometry as possible.

The results of this article were presented by one of the authors in April 1993 at the meeting on Geometric Methods in Group Theory hosted by the ICMS in Edinburgh. We would like to thank Andrew Duncan, Nick Gilbert and Jim Howie, not only for arranging such an enjoyable conference, but also for the courteous and efficient way in which they have behaved as editors of these proceedings.

1. Definitions and Preliminary Results

Throughout this paper A stands for a finite set and A^* for the free monoid on A. The length of a word $w \in A^*$ shall be denoted $|w|$. The empty word is denoted ϵ. A *formal language* is just a subset $L \subset A^*$. A language L is said to be *bounded* if there are words $w_1, \ldots, w_n$ in A^* such that every $w \in L$ can be written $w = w_1^{m_1} \ldots w_n^{m_n}$ for some choice of non-negative integers $m_i \in \mathbb{N}$. Bounded languages were introduced by Ginsburg and Spanier [7].

Let G be a finitely generated group. A *choice of generators* for G is a map $\mu : A \to G$ from a finite set which extends to a surjective monoid homomorphism $\mu : A^* \twoheadrightarrow G$. We assume that A is closed under formal inverse and that formal inverses are extended to A^* in the usual way. We shall usually write $\overline{w}$ rather than $\mu(w)$ for the image in G of a word w, and more generally $\overline{X}$ for the image of a set of words X.

A *combing* of G is a language $L \subset A^*$ which projects bijectively to G. We shall often use the letter $\mathcal{C}$ to denote a language which is a combing. If, in addition, $\mathcal{C}$ is a bounded language, then we shall refer to it as a combing of G by a bounded language. It is an easy exercise to check that if a finitely generated group admits a combing by a bounded language with respect to one choice of generators, then it admits such a combing with respect to every choice of generators.

Combings by bounded languages arise naturally when considering normal forms for polycyclic groups. For example, if $\{1\} = G_0 \subset \cdots \subset G_n = G$ is a normal tower for G, with each G_i/G_{i-1} cyclic, then arguing by induction on n one may assume that G_{n-1} admits a combing by a bounded language, and if we fix an element $a \in G - G_{n-1}$ whose image generates G_n/G_{n-1}, then by appending suitable powers of a to the words in the combing for G_{n-1} we obtain a combing of G by a bounded language.

Any choice of generators determines a word metric on G by $d(g,h) = \min\{|w| \mid w \in A^*, \mu(w) = g^{-1}h\}$. It is straightforward to check that d is indeed a metric and that it is left-invariant in the sense that $d(gh_1, gh_2) = d(h_1, h_2)$ for all $g, h_1, h_2 \in G$. The metrics d_1, d_2 determined by different choices of generators are Lipschitz equivalent; that is, $(1/c)\, d_1(g,h) \leq d_2(g,h) \leq c\, d_1(g,h)$ for some constant c.

For a given choice of generators, each word $w = a_1 \ldots a_n$ determines a discrete path $p_w : \mathbb{N} \to G$ given by

$$p_w(t) = \begin{cases} 1 & \text{if } n = 0 \\ \overline{a_1 \ldots a_t} & \text{if } 1 \leq t < n \\ \overline{w} & \text{if } n \geq t. \end{cases}$$

Henceforth we shall identify w with p_w and often talk of words as discrete paths in G. The synchronous distance D_s between two words $w, v \in A$ is the maximum separation of points (fellow-travellers) traversing the two corresponding paths at unit speed.

$$D_s(w,v) = \max_t \{d(p_w(t), p_v(t))\}.$$

There is also a notion of asynchronous distance in which each point is allowed to stop for a while. The standard technical device to encode this idea is the set $\mathcal{R}$ of all unbounded maps $\rho : \mathbb{N} \to \mathbb{N}$ such that $\rho(0) = 0$, and $\rho(n+1) = \rho(n)$ or $\rho(n) + 1$. The asynchronous distance is defined to be:

$$D_a(w,v) = \min_{\rho_1, \rho_2 \in \mathcal{R}} \{ \max_{n \geq 0} \{ d(p_w(\rho_1(n)), p_v(\rho_2(n))) \} \}. \tag{1.1}$$

We shall be extensively concerned with constraints of the form $D_a(w, v) \leq K$. This inequality can be rephrased as follows: There exist sequences of prefixes

$$x_0 = \epsilon, \, x_1, \ldots, x_N = w \quad y_0 = \epsilon, \, y_1, \ldots, y_N = v$$

of w and v respectively, such that for all i,

$$|x_i| \leq |x_{i+1}| \leq |x_i| + 1 \quad |y_i| \leq |y_{i+1}| \leq |y_i| + 1 \quad \text{and} \quad d(\overline{x_i}, \overline{y_i}) \leq K. \quad (1.2)$$

A language $L \subset A$ is said to satisfy the *synchronous fellow traveller property* if there exists a constant $K > 0$ such that for all $w, v \in L$, if $d(\overline{w}, \overline{v}) \leq 1$ then $D_s(w, v) \leq K$. The *asynchronous fellow traveller property* is defined analogously. Throughout this paper *we shall retain the symbol K to denote the constant in the definition of the (synchronous or asynchronous) fellow traveller property.*

Remark. Notice that if $L \subset A^*$ satisfies the asynchronous fellow traveller property, than so does every sublanguage of it. Since a sublanguage of a bounded language is itself bounded, we see that the hypothesized existence of the combing in Theorem A could be replaced by the existence of any bounded sublanguage of A^* which maps onto G and satisfies the asynchronous fellow traveller property.

We claim that D_s and D_a are pseudometrics on A^*; that is, they satisfy all the requirements of a metric except the requirement that distinct points be a positive distance apart. This assertion is an immediate consequence of the following lemma, whose proof we leave to the reader.

Lemma 1.1

   1. $\mathcal{R}$ is closed under composition.

   2. If $D_a(w, v)$ is realized by ρ_1 and ρ_2 as in (1.1), then it is also realized by $\rho_1 \circ \rho$ and $\rho_2 \circ \rho$, where $\rho \in \mathcal{R}$ is arbitrary.

   3. Given $\rho_1, \rho_2 \in \mathcal{R}$, there exist $\rho', \rho'' \in \mathcal{R}$ such that $\rho_1 \circ \rho' = \rho_2 \circ \rho''$.

We note some elementary properties of D_a.

Lemma 1.2

   1. If $\overline{w} = \overline{v}$, then $D_a(wu, vu') \leq \max\{D_a(w, v), D_a(u, u')\}$ for all $u, u' \in A^*$.

   2. If v is obtained from w by deleting any number of disjoint subwords xx^{-1}, then $D_a(w, v) \leq |x|$.

3. If $\overline{w} = \overline{x}\,\overline{v}\,\overline{x}^{-1}$, then $D_a(w^n, xv^nx^{-1}) \le D_a(w, xvx^{-1}) + |x|$ for all $n \in \mathbb{N}$.

Proof. From a geometric viewpoint, assertion (1) is clear. An algebraic proof can be obtained using (1.2): Let $K = \max\{D_a(w, v), D_a(u, u')\}$; multiply the sequences of prefixes used to compute $D_a(u, u')$ on the left by w and v, then append them to the sequences for $D_a(w, v)$ to get sequences of prefixes for wu and vu'; if u_i, u'_i is a pair of prefixes for u, u' then $d(\overline{wu_i}, \overline{vu'_i}) = d(\overline{u_i}, \overline{u'_i}) \le K$, because $\overline{w} = \overline{v}$ and d is left invariant.

(2) is proved by estimating $D_a(w, v)$ using the unit speed parametrization $\rho(n) = n$ for p_w and the parametrization for p_v which causes a point traversing the image of p_v to move with unit speed except for remaining stationary during the period in which p_w traces out the subword xx^{-1}. Part (3) follows from (1) , (2) and the triangle inequality for D_a.

It is shown in [3], Proposition 1.3 that every context free language of polynomial growth is bounded. We will need the following additional result on bounded languages [3], Lemma 1.4.

Lemma 1.3   *Every bounded language $L \subset A^*$ can be expressed as the union of finitely many bounded sublanguages L_l such that for each l there exists an integer r and a choice of words $u_{l,1}, \ldots, u_{l,r}, v_{l,0}, \ldots, v_{l,r} \in A^*$ with*

$$L_l = \{v_{l,0}u_{l,1}^{n_1}v_{l,1} \ldots u_{l,r}^{n_r}v_{l,r} \mid (n_1, \ldots, n_r) \in S_l\},$$

where $S_l \subset \mathbb{N}^r$ is empty if $r = 0$ (in which case $L_l = \{v_{l,0}\}$) and otherwise there exist r−tuples in S_l whose smallest entry is arbitrarily large.

2. Reduction of Theorem A to Theorem C

Throughout this section $\mathcal{C}$ will denote the combing hypothesized in Theorem A. We begin by showing that we are free to assume that $\mathcal{C}$ enjoys certain extra properties. Since $\mathcal{C}$ is bounded, there are words $w_i \in A^*$ such that

$$\mathcal{C} = \{w = w_1^{n_1} \ldots w_r^{n_r} \mid (n_1, \ldots, n_r) \in S\} \tag{2.1}$$

for some subset S of $\mathbb{N}^r$. We work in terms of this fixed decomposition of $\mathcal{C}$. We say that w_i has *bounded exponent* if n_i is bounded as w ranges over $\mathcal{C}$. Otherwise w_i is said to have *unbounded exponent*.

Lemma 2.1   *Without loss of generality, the combing $\mathcal{C}$ may be assumed to have the following additional properties:*

  1. *the exponent to which each w_i appears in any word of $\mathcal{C}$ is less than the order of $\overline{w_i}$ in G;*

> 2. if w_i and w_j are both of unbounded exponent, and if some positive powers
> of $\overline{w_i}$ and $\overline{w_j}$ are conjugate in G, then $w_i = w_j$.

Proof. We suppose that $\mathcal{C}$ has been chosen so that the number of distinct words w_i of unbounded exponent is minimal amongst all combings of G which are bounded languages and satisfy the asynchronous fellow traveller property. We emphasize that we have chosen our meaning carefully here; we have minimized the cardinality of the subset of $\{w_i\}$ consisting of words with unbounded exponent without counting multiplicities to account for the case where $w_i = w_j$ for some $i \neq j$.

Suppose that $\overline{w_i}$ has order m. If $m = 1$, we delete occurrences of w_i so as to decrease the integer r in the definition of $\mathcal{C}$. Otherwise, for each $w \in \mathcal{C}$, we replace the subword $w_i^{n_i}$ by w_i^q where $n_i = mp + q$ with $0 \leq q < m$. The image of w in G is not changed by this procedure, so $\mathcal{C}$, thus modified, is still a combing; and the number of distinct words of unbounded exponent has not increased. Moreover, by repeated application of Lemma 1.2 (1) , we have that $D_a(xw_i^{n_i}y, xw_i^m y) \leq D_a(w_i^m, \epsilon)$, and hence, by the triangle inequality for D_a, we see that $\mathcal{C}$ is still asynchronously bounded after being modified as above.

For (2) suppose for $0 < m \leq n$ and for some $x \in A^*$, $xw_i^m x^{-1}$ and w_j^n have the same image in G. If $w_i \neq w_j$, then in every word $w \in \mathcal{C}$ we replace $w_j^{n_j}$ by $xw_i^{mp}x^{-1}w_j^q$, where $n_j = np + q$, with $0 \leq q < n$. With these changes $\mathcal{C}$ is transformed into a combing $\mathcal{C}'$ which is still a bounded language. It again follows from Lemma 1.2 that $\mathcal{C}'$ is asynchronously bounded. We deduce that $w_i = w_j$, for otherwise $\mathcal{C}'$ would have fewer subwords w_i of unbounded exponent than $\mathcal{C}$, contradicting the minimality of $\mathcal{C}$.

Suppose that $\mathcal{C}$ has been chosen so as to satisfy the conclusions of Lemma 2.1. For each integer i with $1 \leq i \leq r$ we add a new generator a_i and its formal inverse to A, and define $\overline{a_i} = \overline{w_i}$. We then redefine $\mathcal{C}$ by replacing each of the words w_i in its definition by the corresponding new generator a_i. (Notice that since the index i was encoded in the definition of a_i, we have that $a_i \neq a_j$ even if $w_i = w_j$.) It follows from Lemma 1.2 and the fact that any two word metrics on G are Lipschitz equivalent that $\mathcal{C}$, redefined in this way, still satisfies the asynchronous fellow traveller property. The introduction of the new generators a_i allows us to refine Lemma 2.1:

Lemma 2.2 *Without loss of generality, the combing $\mathcal{C}$ may be assumed to have the following properties:*

> 1. $\mathcal{C} = \{w = a_1^{n_1} \dots a_r^{n_r} \mid (n_1, \dots n_r) \in S\}$ *for some subset S of $\mathbb{N}^r$ and distinct generators $a_i \in A$.*
>
> 2. *If $w = a_1^{n_1} \dots a_r^{n_r} \in \mathcal{C}$, then n_i is less than the order of $\overline{a_i}$ in G; in particular if a_i is of unbounded exponent, then $\overline{a_i}$ has infinite order.*

3. *If a_i and a_j have unbounded exponent and some positive powers of their images are conjugate in G, then $\overline{a_i} = \overline{a_j}$.*

Next we consider the restrictions which the asynchronous fellow traveller property places on bounded languages. We consider the situation $D_a(w,v) \leq K$ described by conditions (1.2), and maintain the notation established there. In (1.2) we allowed the possibility that for some i we have both $x_i = x_{i+1}$ and $y_i = y_{i+1}$. But clearly this possibility can be avoided simply by deleting all such pairs and reindexing. Likewise the simultaneous inequalities $x_i \neq x_{i+1}$ and $y_i \neq y_{i+1}$ may be avoided by interpolating an extra copy of x_i before x_{i+1} and an extra copy of y_{i+1} after y_i. One must also increase the constant K by 1. With these changes we have (with the notations of (1.2)): for all i,

$$\text{Either } x_i = x_{i+1} \text{ and } y_i \neq y_{i+1}, \text{ or vice versa.} \tag{2.2}$$

From now on we assume that the choices of prefixes expressing the condition $D_a(w,v) \leq K$ for words $w, v \in \mathcal{C}$ with $D(w,v) = 1$, satisfy condition (2.2). It is also convenient to introduce the following notation: Given $0 \leq i \leq j \leq N$ define $x_{i,j} \in A^*$ by $x_i x_{i,j} = x_j$, and define $y_{i,j}$ likewise. Thus, given any partition $0 \leq i_1 \leq i_2 \leq \cdots \leq i_s = N$ we have $w = x_{0,i_1} x_{i_1,i_2} \ldots x_{i_{s-1},N}$. Also, from (2.2) we have

$$|x_{i,j}| + |y_{i,j}| = j - i. \tag{2.3}$$

In particular $N = |w| + |v|$.

Lemma 2.3 *Suppose that G satisfies the hypotheses of Theorem A, suppose that $\mathcal{C}$ has been chosen as in Lemma 2.2, and suppose that Theorem A (2) does not hold. Then, there is a constant M such that for all $w, v \in \mathcal{C}$ with $d(\overline{w}, \overline{v}) \leq 1$, the prefixes of w and v described above satisfy $| |x_i| - |y_i| | \leq M$.*

Proof. We assume that the prefixes x_i for w, and y_i for v, are as in the preceding discussion. Because w can be written as in Lemma 2.2(1), we may write $w = x_{0,i_1} x_{i_1,i_2} \ldots x_{i_{r-1},i_r}$, where each x_{i_{p-1},i_p} is either a positive power of a_p or the empty word. We refine this decomposition a little: Each y_{i_{p-1},i_p} is a product of powers of a_1 to a_r in order, and we decompose it as such, then refine the above decomposition of w so as to make the comparison of prefixes notationally simple. Thus we factor w and v into products of powers of the generators a_λ.

$$w = x_{0,k_1} x_{k_1,k_2} \ldots x_{k_{r^2-1},k_{r^2}} \qquad v = y_{0,k_1} y_{k_1,k_2} \ldots y_{k_{r^2-1},k_{r^2}} \tag{2.4}$$

Define $\Delta(i) := | |x_i| - |y_i| |$. Clearly $\Delta(0) = 0$, and by (2.2) $\Delta(i+1) \leq \Delta(i) + 1$. To complete the proof of the lemma it suffices to show that there is a constant M' such that if $x_{i,j}$ and $y_{i,j}$ are both powers of generators then

$\Delta(j) \leq \Delta(i) + M'$. Indeed, if we exhibit such a constant, then we can set $M = r^2 M'$.

Assume $x_{i,j}$ and $y_{i,j}$ are powers of the generators a and b respectively. Let M_1 be an upper bound on the exponents of all generators a_λ of finite exponent. If a and b are both of bounded exponent, then it follows from (2.3) and Lemma 2.2(2) that $j - i < 2M_1$, whence $\Delta(j) < \Delta(i) + 2M_1$. Thus we may assume that a has unbounded exponent. Notice also that there is no loss of generality in assuming that $j - i$ is greater than a convenient constant. In the next stage of the proof we show that when $j - i$ is large neither $\bar{a}$ nor $\bar{b}$ can be of finite order.

For $i \leq k \leq j$ define $g_k := \overline{y_k}^{-1}\overline{x_k}$. As $d(1, g_j) = d(\overline{y_k}, \overline{x_k}) \leq K$, there are only a finite number of possibilities, say M_2, for g_k. Suppose b is of finite exponent; then $|y_{i,j}| < M_1$. If $j - i$ is large enough ($j - i > M_1 M_2$ suffices), then for some k, k' with $i \leq k \leq k' \leq j$ and $k' - k > M_2$, $y_k = y_{k+1} = \ldots = y_{k'}$. By choice of M_2, $g_{k_1} = g_{k_2}$ for some k_1, k_2 with $k \leq k_1 < k_2 \leq k'$ whence some power of $\bar{a}$ is conjugate to the identity in G. But this is impossible, by Lemma 2.2 (2) , because we are assuming that a is of unbounded exponent.

It remains to consider the case where both $\bar{a}$ and $\bar{b}$ have infinite order. We may assume $j - i > M_2$. Consequently, for some k, k' with $i \leq k \leq k' \leq j$, we have $g_k = g_{k'}$. Consider any such k, k'. The images in G of $x_{k,k'}$ and $y_{k,k'}$ are conjugate. In other words $\bar{a}$ raised to the power $|x_{k,k'}|$ is conjugate to $\bar{b}$ raised to the power $|y_{k,k'}|$. Since $k < k'$, condition (2.2) implies that at least one of these powers is nontrivial, and since neither $\bar{a}$ nor $\bar{b}$ is trivial, this implies that the other power is also nontrivial, hence we may apply Lemma 2.2(3) to deduce that $\bar{a} = \bar{b}$. We are assuming that Theorem A (2) does not hold, so we conclude that $|x_{k,k'}| = |y_{k,k'}|$ whenever $g_k = g_{k'}$. Hence $\Delta(k) = \Delta(k')$. Thus the number of different values of $\Delta(k)$ for $i \leq k \leq j$ is at most M_2. Condition (2.2) ensures that $\Delta(k)$ assumes every value between $\Delta(i)$ and $\Delta(j)$ as k ranges from i to j, consequently $\Delta(j) \leq \Delta(i) + M_2$.

A well known and very beautiful result of Gromov [6] states that a group with polynomial growth is virtually nilpotent. In light of Gromov's theorem, the following lemma completes the reduction of Theorem A to Theorem C.

Lemma 2.4   *Suppose that Theorem A (2) does not hold. Then:*

1. *C satisfies the synchronous fellow traveller property.*

2. *There exist constants M and Q such that for all $w \in C$,*

$$|w| \leq M d(1, \overline{w}) + Q.$$

3. *G has polynomial growth.*

Proof. Consider $w, v \in C$ with $d(\overline{w}, \overline{v}) = 1$ and let the sequences of prefixes $x_0, x_1 \ldots$ and $y_0, y_1, \ldots$ be as in the discussion prior to Lemma 2.3. If w_1 is any prefix of w, then $w_1 = x_i$ for at least one value of i. Thus $d(\overline{w_1}, \overline{y_i}) \leq K$. By Lemma 2.3, the difference between $|w_1| = |x_i|$ and $|y_i|$ is at most M. It follows that if v_1 is the prefix of v with length $\min\{|w_1|, |v|\}$, then $d(\overline{w_1}, \overline{v_1}) \leq d(\overline{w_1}, \overline{y_i}) + d(\overline{y_i}, \overline{v_1}) \leq K + M$. Thus (1) holds.

By Lemma 2.3 we have that $||w| - |v|| = ||x_N| - |y_N|| \leq M$. If we let Q denote the length of the word in $\mathcal{C}$ representing the identity, then a simple induction on $d(1, \overline{w})$ establishes (2). Finally, (2) implies that every $g \in G$ with $d(1, g) \leq m$ is represented in $\mathcal{C}$ by a word of length at most $Mm + Q$. By Lemma 2.2 (1) each such word is determined by an r-tuple $(n_1 \dots n_r)$ of integers each between 0 and $Mm + Q$. Thus there are at most $(Mm + Q + 1)^r$ such words.

3. The proof of Theorem C

Let G be a finitely generated virtually nilpotent group and let $\mathcal{C}$ be a combing of G by a bounded language. Suppose that $\mathcal{C}$ satisfies the asynchronous fellow traveller property and the conditions of Lemma 2.3. We claim that Theorem A (2) does not hold in G, so in particular we may apply Lemma 2.4. In order to see that this is the case, we consider a nilpotent subgroup H of finite index in G and suppose $g \in G$ is of infinite order with g^m conjugate to g^n for some m, n with $0 < m < n$. Since the finitely many conjugates of H intersect in a normal subgroup of finite index, there is an integer p such that all p-th powers lie in H. If g_1 conjugates g^m to g^n, then g_1^p conjugates g^{pm^p} to g^{pn^p}. Consequently, it suffices to show that Theorem A (2) cannot hold if G is nilpotent. Clearly G cannot be abelian. More generally, the subgroup generated by g^m must intersect $Z(G)$, the center of G, trivially whence the image of g in $G/Z(G)$ is of infinite order and we are done by induction on the nilpotency class of G.

At its core, our proof of Theorem C depends upon the geometry of conjugation in nilpotent groups. However, this geometry is somewhat obscured by the surrounding technicalities, so to clarify our exposition we concentrate on a case of particular interest, the 3-dimensional integral Heisenberg group:

$$\mathcal{H}_3 = \langle x, y, z \mid [x, y] = z, [x, z] = [y, z] = 1 \rangle.$$

After completing the proof in this case we shall see that the extension to arbitrary virtually nilpotent groups requires only a few observations about the structure of the proof in the 3-dimensional case, together with some elementary facts about nilpotent groups, in particular the structure of centralizers of elements in the penultimate term of the upper central series.

Remark. The import of Theorem C to 3-manifold topology is essentially contained in the case $G = \mathcal{H}_3$, because any group G which acts properly and cocompactly by isometries on the 3-dimensional geometry Nil contains $\mathcal{H}_3$ as a subgroup of finite index (see [9]).

The Case $G = \mathcal{H}_3$. Let $A \to G$ be a choice of generators for G and let $\mathcal{C} \subset A^*$ be a combining which satisfies Lemma 2.2 (1) - (3) and Lemma 2.4 (1) - (2) . In particular there is a constant K such that for all $w, v \in \mathcal{C}$ with $d(\overline{w}, \overline{v}) = 1$, the synchronous distance between the paths w and v is at most K. In other words, if w_1 is any prefix of w and v_1 is the prefix of v of length $\min\{|w_1|, |v|\}$, then $d(\overline{w_1}, \overline{v_1}) \leq K$.

Because $\mathcal{C}$ is a bounded language, Lemma 1.3 allows us to write it as the union of finitely many sublanguages $\mathcal{C}_i$ of the form:

$$\mathcal{C}_i = \{w = u_{i,0} b_{i,1}^{n_1} u_{i,1} \cdots b_{i,r}^{n_r} u_{i,r} \mid (n_1, \cdots, n_r) \in S_i\}$$

where $r = r(i) \in \mathbb{N}$, $b_{i,j} \in A$, and $u_{i,j} \in A^*$. (Each $b_{i,j}$ is one of the a_i's of Lemma 2.2 (1) .) The set S_i is empty if $r(i) = 0$ and otherwise $S_i \subset \mathbb{N}^r$ has the property that there exist r-tuples in S_i whose smallest entry is arbitrarily large. Given $w \in \mathcal{C}_i$ we define

$$\ell(w) = \min\{n_j\}.$$

In what follows, when referring to $w \in \mathcal{C}_i$ we shall assume that it is decomposed as in the above definition of $\mathcal{C}_i$.

We wish to use Euclidean geometry as a tool to analyze the geometry of the language $\mathcal{C}$. In order to do so, we consider $\widehat{G} = G/Z(G)$, the quotient of $G = \mathcal{H}_3$ by its center. $\widehat{G}$ is isomorphic to $\mathbb{Z}^2$ and so may be identified with the integer lattice of the Euclidean plane $\mathbb{E}^2$. Let $\pi : G \to \widehat{G}$ be the projection. Define $\widehat{g} = \pi(g)$ and $\widehat{w} = \pi(\overline{w})$; $\widehat{g}$ and $\widehat{w}$ are vectors in $\mathbb{E}^2$. In particular $\widehat{A} = \{\widehat{a} \mid a \in A\}$ spans $\mathbb{E}^2$. The integral translates of vectors in $\widehat{A}$ form the edges of a realization of the Cayley graph $\widehat{\Gamma}$ of $\widehat{G}$ corresponding to the choice of generators $\widehat{A} \to \widehat{G}$ induced by the choice of generators $A \to G$.

Let $\widehat{d}$ be the word metric in $\widehat{G}$ corresponding to the choice of generators $\widehat{A}$. Clearly $d(g_1, g_2) \geq \widehat{d}(\widehat{g_1}, \widehat{g_2})$, and it is straightforward to see that the ball of radius n around 1 in G projects onto the ball of radius n around $\widehat{1} = 0$ in $\widehat{G}$. Notice that if $\| \ \|$ denotes the usual Euclidean norm on $\mathbb{E}^2$, then there is a constant $\lambda > 1$ such that

$$\frac{1}{\lambda}\|\widehat{g_1} - \widehat{g_2}\| \leq \widehat{d}(\widehat{g_1}, \widehat{g_2}) \leq \lambda\|\widehat{g_1} - \widehat{g_2}\|.$$

The path in the Cayley graph Γ of G determined by $w \in A^*$ projects to a polygonal path $P(w)$ in $\widehat{\Gamma}$. If $\ell(w)$ is large, then qualitatively (to the distant observer) $P(w)$ looks like a concatenation of at most $r(i)$ long line segments corresponding to the $b_{i,j}$'s with $\widehat{b_{i,j}}$ nontrivial. Of course, upon closer examination one would see that these long segments of $P(w)$ were in fact interspersed with short line segments, translates of the $\widehat{u_{i,j}}$.

For each C_i we consider the sequence of vectors $\widehat{b_{i,1}}, \cdots, \widehat{b_{i,r(i)}}$, and pick C_i so that the number of subsequences $\widehat{b_{i,j}}, \cdots, \widehat{b_{i,j'}}$ with $0 \neq \widehat{b_{i,j}} \neq \widehat{b_{i,j'}} \neq 0$ and all intervening vectors 0 is maximal. For convenience we assume that this sublanguage is C_0. *For the remainder of this section s will denote the largest integer such that $\widehat{b_{0,s}} \neq 0$.*

Given $v = u_{i,0} b_{i,1}^{n_1} u_{i,1} \cdots b_{i,r}^{n_r} u_{i,r} \in C_i$, for each j with $\widehat{b_{i,j}} \neq 0$ we let

$$\alpha_{i,j} = u_{i,0} b_{i,1}^{n_1} u_{i,1} \cdots u_{i,j-1} \text{and} \beta_{i,j} = u_{i,0} b_{i,1}^{n_1} u_{i,1} \cdots u_{i,j-1} b_{i,j}^{n_j} \qquad (3.1)$$

respectively, and denote by $\mathcal{L}_{i,j}(v)$ the infinite line in $\mathbb{E}^2$ through the points $\widehat{\alpha_{i,j}}$ and $\widehat{\beta_{i,j}}$. The $\mathcal{L}_{i,j}(v)$'s are the lines determined by the long line segments mentioned above. The following proposition shows that for any given distance d_0, and for each $w \in C_0$ with $\ell(w)$ sufficiently large, if $v \in C_i$ satisfies $d(\overline{w}, \overline{v}) \leq d_0$, then $\mathcal{L}_{0,s}(w) = \mathcal{L}_{i,j'}(v)$ for some j'.

Proposition 3.1 *Let C and C_0 be as above. There exists a constant $B \geq 2$ and an unbounded function $\varphi : \mathbb{N} \to \mathbb{N}$ such that for all $w \in C_0$ and $v \in C_i$, if $d(\overline{w}, \overline{v}) \leq \varphi(\ell(w))$, then for some j'*

1. *$\mathcal{L}_{0,s}(w) = \mathcal{L}_{i,j'}(v)$;*

2. *the distance in $\mathbb{E}^2$ from $\widehat{w}$ to $\mathcal{L}_{0,s}(w)$ is at most B;*

3. *the distance in $\mathbb{E}^2$ from $\widehat{v}$ to $\mathcal{L}_{i,j'}(v)$ is at most B.*

The following consequence of Proposition 3.1 completes the proof of Theorem C in the case $G = \mathcal{H}_3$.

Corollary 3.2 *$\mathcal{H}_3$ does not admit a combing C as in Proposition 3.1.*

Proof. We shall argue by contradiction. Pick $w \in C_0$ with $\varphi(\ell(w)) \geq 2\lambda B$ and let $\mathcal{L} = \mathcal{L}_{0,s}(w)$. Clearly there is $v \in C$ with $\widehat{v}$ a distance greater than B from $\mathcal{L}$ and $\|\widehat{w} - \widehat{v}\| \leq 2B$. We choose v so that $d(\overline{v}, \overline{w}) = \widehat{d}(\widehat{w}, \widehat{v}) \leq \lambda \|\widehat{w} - \widehat{v}\|$. By Proposition 3.1, the distance from $\widehat{v}$ to $\mathcal{L}$ is at most B.

Proof of Proposition 3.1. The following proof is an elaboration of an argument of Epstein *et al.* ([5], 8.2.10). Recall that the combing $C \subset A^*$ is assumed to satisfy Lemma 2.2 (1) - (3) and Lemma 2.4 (1) - (2) . In particular there is a constant K such that for all $w, v \in C$ with $d(\overline{w}, \overline{v}) \leq 1$, the synchronous distance between the paths w and v is at most K.

Fix $w \in C_0$ and $v \in C_i$ with $d(\overline{w}, \overline{v}) \leq M$. We will show that if $\ell(w)$ is large enough, then all the conclusions of Proposition 3.1 hold. Let $\mathcal{L} = \mathcal{L}_{0,s}(w)$. Since $\widehat{b_{0,j}} = 0$ for all $j > s$, decomposing w as in (3.1) yields $\|\widehat{\beta_{0,s}} - \widehat{w}\| \leq$

$\lambda \hat{d}(\hat{\beta}_{0,s}, \hat{w}) \leq \lambda d(\hat{\beta}_{0,s}, \overline{w}) = d(1, \overline{u_{0,s}}) + \cdots + d(1, \overline{u_{0,r(0)}})$. Thus the definition $B = 2 + \sum_{i,j} |u_{i,j}|$ will suffice for Proposition 3.1 (2) .

The condition $d(\overline{w}, \overline{v}) \leq M$ implies (via the triangle inequality for D_s) that the synchronous distance between w and v is at most KM. In other words, if w_1 is any prefix of w and v_1 is the prefix of v of length $\min\{|w_1|, |v|\}$, then $d(\overline{w_1}, \overline{v_1}) \leq KM$. Let $\gamma(n)$ be the number of elements of G which can be expressed as words of length at most n in the generators, and assume

$$\ell(w) \geq 4\gamma(KM) + B \tag{3.2}$$

where B is as above. For each j with $0 \leq j \leq r(0)$ and $\widehat{b_{0,j}} \neq 0$ we let $x = \alpha_{0,j}$, where $\alpha_{0,j}$ is a prefix of w described in the notation of (3.1). Let $b = b_{0,j}$ (again, in the notation of (3.1)). Consider the prefixes $x, xb, \cdots, xb^{n_j}$ of w. By (3.2), $n_j \geq 2\gamma(KM) + B$. Let $z_0, z_1, \cdots, z_{n_j}$ be the corresponding prefixes of v with z_k of length $\min\{|xb^k|, |v|\}$. Of course, if one z_k equals v, then all of the subsequent ones do too.

Suppose that the last $\gamma(KM)$ prefixes z_k are all equal to z_{n_j}. Then, the distance in G from z_{n_j} to xb^k is at most KM for $n_j - \gamma(KM) \leq k \leq n_j$. It follows that for two distinct values of k the word differences $z_k^{-1} xb^k$ represent the same element of G; whence some positive power of $\overline{b}$ is conjugate in G to the identity, contrary to Lemma 2.2 (2) . Hence we may assume that $|xb^k| = |z_k|$ for $0 \leq k \leq 3\gamma(KM) + B$.

From the structure of $\mathcal{C}_i$ and the conclusion of the preceding paragraph it follows that for all k in some subinterval of $[0, \gamma(KM) + B]$ with length at least $\gamma(KM)$, the corresponding sequence of prefixes of v is

$$\alpha_{i,j'}, \ \alpha_{i,j'} b_{i,j'}, \ \alpha_{i,j'} b_{i,j'}^2, \ \cdots \ \alpha_{i,j'} b_{i,j'}^{\gamma(KM)} \tag{3.3}$$

for some j'. By the argument of the preceding paragraph, two distinct word differences in this subinterval must represent the same group element, and we conclude that some positive power of b is conjugate in G to a positive power of $b_{i,j'}$. In geometric terms, this means that there are segments of the lines $\mathcal{L}_{0,j}(w)$ and $\mathcal{L}_{i,j'}(v)$ which form opposite sides of a parallelogram in $\mathbb{E}^2$. The other sides of the parallelogram are translates of the image in $\mathbb{E}^2$ of the conjugating element. By Lemma 2.2, $\overline{b} = \overline{b_{i,j'}}$; and consequently the conjugating element centralizes $\overline{b}$. As $\overline{b} \in G - Z(G)$, the centralizer of $\overline{b}$ is an abelian subgroup of rank 2 containing $\overline{b}$ and $Z(G)$. It follows that the image in $\mathbb{E}^2$ of the conjugating element is parallel to $\overline{b}$. Thus the parallelogram described above is degenerate, and we conclude that $\mathcal{L}_{0,j}(w) = \mathcal{L}_{i,j'}(v)$.

For each subword $b_{0,j}^{n_j}$ in w with $\widehat{b_{0,j}} \neq 0$ we have found a corresponding subword $b_{i,j'}^{n_{j'}}$ in v with $b_{0,j} = b_{i,j'}$ and $\mathcal{L}_{0,j}(w) = \mathcal{L}_{i,j'}(v)$. Furthermore, the order in which the corresponding subwords occur in their respective ambient words is the same. It follows that $\mathcal{C}_i$ also satisfies the maximality condition on subsequences governing the choice of $\mathcal{C}_0$. In particular, if j' is now fixed so that $b_{0,s}^{n_s}$ corresponds to $b_{i,j'}^{n_{j'}}$, then for all $j \geq j'$, $\widehat{b_{i,j}}$ equals 0 or $\widehat{b_{i,j'}}$.

It follows in a straightforward way that Proposition 3.1 (3) holds. As we know $\mathcal{L}_{0,s}(w) = \mathcal{L}_{i,j'}(v)$, all the conclusions of Proposition 3.1 hold under the assumption (3.2). But γ is strictly increasing, so $\ell(w) = 4\gamma(KM) + B$ admits a strictly increasing solution $M = \varphi(\ell(w))$.

We shall now point out the adjustments which one must make to the preceding arguments in order to prove Theorem C in the general case. (For the necessary properties of nilpotent groups we refer the reader to [2].) First of all we must reduce from the case of virtually nilpotent groups to the case of nilpotent groups. This is a straightforward matter. Indeed there are standard techniques (see [5] or [3], Thm. 2.16) for showing that subgroups of finite index (or, more generally, quasiconvex subgroups) inherit combings from the ambient group, and these induced combings inherit the geometric and linguisitic characteristics of the combing of the ambient group. Similar but easier arguments apply to quotients of a given group by a finite normal subgroup. Such elementary arguments yield:

Lemma 3.3 *If a finitely generated group G admits a combing by a bounded language which satisfies the asynchronous fellow traveller property, then so too does every subgroup of finite index in G and every quotient of G by a finite normal subgroup.*

Consequently, it suffices to prove Theorem C in the case where G is torsion-free and nilpotent. In this setting we wish to imitate the preceding proof of the case $G = \mathcal{H}_3$. In this scheme the following lemma yields a projection which assumes the role previously played by the quotient map of $\mathcal{H}_3$ by its centre.

Lemma 3.4 *If G is a non-abelian, finitely generated, torsion-free, nilpotent group, then there exists an integer $n \geq 2$ and a projection $\pi : G \twoheadrightarrow \mathbb{Z}^n$ such that the image under π of the centralizer of every $g \in G - \ker\pi$ has rank at most $n - 1$.*

Proof. Let $1, Z_1, \ldots, Z_c = G$ be the ascending central series of G (recall that g is in Z_i if and only if it is central modulo Z_{i-1}). Each quotient G/Z_i is torsion free; and for each $g \in G$ the centralizer $C(g)$ is closed under taking roots. Thus G/Z_{c-1} is abelian; if it were cyclic, then G/Z_{c-2} would be abelian too, and hence $Z_{c-1} = Z_{c-2}$ would follow. But then G would be abelian, which is not the case. Thus G/Z_{c-1} is free abelian of rank $n \geq 2$.

Take π to be the projection $\pi : G \twoheadrightarrow G/Z_{c-1}$. It remains to check that the image under π of $C(g)$ has rank at most $n - 1$ if $g \in G - Z_{c-1}$. Without loss of generality we may pass to the quotient G/Z_{c-2}. In other words we may assume $Z_{c-2} = 1$ and $Z_{c-1} = Z(G)$, the center of G. It follows that $C(g)$ is a normal subgroup of G containing $Z(G)$. Because centralizers are closed under taking roots, $G/C(g)$ is free abelian. Since $g \notin Z(g)$, we deduce that $G/C(g) \cong (G/Z(G))/(C(g)/Z(G))$ has rank at least 1. $\square$

Henceforth we assume that G is a non-abelian torsion-free nilpotent group. Let $C_G(g)$ denote the centralizer of g in G. We continue our previous convention of viewing $\mathbb{Z}^n$ as the rectangular lattice in $\mathbb{E}^n$ and π as a map into $\mathbb{E}^n$. As before we write $\widehat{w}$ in place of $\pi(\overline{w})$ if $w \in A^*$; and also as before the word metric on $\widehat{G}$ is related to the Euclidean norm by $(1/\lambda)\|\widehat{g}-\widehat{h}\| \leq \widehat{d}(\widehat{g},\widehat{h}) \leq \lambda\|\widehat{g}-\widehat{h}\|$. We assume that the combing $\mathcal{C}$ and the sublanguages $\mathcal{C}_i$ are as described prior to Proposition 3.1. In particular we assume that $\mathcal{C}_0$ is as chosen there, and we keep the notations of (3.2).

If one attempts to generalize Proposition 3.1 to the present situation in the most naive way then one immediately encounters difficulties stemming from the fact that the image under π of $C_G(g)$, where $g \in \Gamma - \ker \pi$, is not necessarily cyclic. Isolating this difficulty leads one to focus on the correct generalization of Proposition 3.1; instead of working with the lines $\mathcal{L}_{i,j}(v)$ we consider their following higher dimensional analogues:

Given $b_{i,j}$ with $\widehat{b_{i,j}} \neq 0$, we let $\Lambda_{i,j}(v)$ denote the affine subspace of $\mathbb{E}^n$ obtained by translating to $\widehat{\alpha}_{i,j}$ the subspace spanned by $\pi(C_G(\overline{b}_{i,j}))$.

The argument of Corollary 3.2 reduces Theorem C to the following analogue of Proposition 3.1.

Proposition 3.5 *Let G be a finitely generated, non-abelian, torsion-free, nilpotent group, and let $\mathcal{C}$ and $\mathcal{C}_0$ be as above. Then, there exists a constant $B \geq 2$ and an unbounded function $\varphi : \mathbb{N} \to \mathbb{N}$ such that for all $w \in \mathcal{C}_0$ and $v \in \mathcal{C}_i$, if $d(\overline{w},\overline{v}) \leq \varphi(\ell(w))$, then for some j'*

1. $\Lambda_{0,s}(w) = \Lambda_{i,j'}(v)$;

2. *the distance in $\mathbb{E}^n$ from $\widehat{w}$ to $\Lambda_{0,s}(w)$ is at most B;*

3. *the distance in $\mathbb{E}^n$ from $\widehat{v}$ to $\Lambda_{i,j'}(v)$ is at most B.*

Proof. Modulo replacing each occurence of the symbol $\mathcal{L}$ by Λ, we can follow the proof of Proposition 3.1 *verbatim*, except for the fact that one does not deduce that the parallelogram discussed following expression (3.3) is degenerate, but rather one concludes that it is contained in the affine subspace $\Lambda_{0,j}(w)$. Thus $\Lambda_{0,j}(w)$ and $\Lambda_{i,j'}(v)$ intersect non-trivially. But these are both translates of the same subspace, because b and $b_{i,j'}$ are conjugate in G, and hence $\pi(C_G(b)) = \pi(C_G(b_{i,j'}))$. Thus, since they are parallel and intersect, $\Lambda_{0,j}(w)$ and $\Lambda_{i,j'}(v)$ must coincide. The proof then proceeds as in Proposition 3.2.

References

[1] J.M. Alonso and M.R. Bridson, Semihyperbolic groups, *Proc. London Math. Soc.*, to appear.

[2] G. Baumslag, *Lecture Notes on Nilpotent Groups*, Conf. Board of the Math. Sci. Regional Conf. Series in Math. **2** Amer. Math. Soc., Providence, RI, 1971

[3] M.R. Bridson and R.H. Gilman, Formal language theory and the geometry of 3-manifolds, Preprint, June 1993

[4] M.R. Bridson and R.H. Gilman, A remark about combings of groups, *Intl. J. Alg. Comp.* **3** (1993), 573-581.

[5] D.B.A. Epstein, J.W. Cannon, D.F. Holt, S.V.F. Levy, M.S. Paterson, W.P. Thurston, *Word processing in groups,* Jones and Bartlett, Boston MA., 1992

[6] M. Gromov, Groups of polynomial growth and expanding maps, *IHES Publ. Math. Sci.* **53**(1981) 53–78

[7] S. Ginsburg and E. H. Spanier, Bounded Algol-like languages, *Trans. A.M.S.* **113** (1964) 333-368

[8] S. M. Gersten and H. B. Short, Rational subgroups of biautomatic groups, *Annals of Math.* **134** (1990) 125–158

[9] G.P. Scott, The geometry of 3-manifolds, *Bull. LMS* **15** (1983) 401–487

Department of Mathematics
Princeton University
Princeton
NJ 08544
bridson@math.princeton.edu

Department of Mathematics
Stevens Institute of Technology
Hoboken
NJ 07030
rgilman@vaxc.stevens-tech.edu

Finitely presented groups and the finite generation of exterior powers

C.J.B. BROOKES[1]

1 Introduction

1.1 A group Γ is said to be *of type* $(FP)_m$, where m is a non-negative integer, if the trivial $\mathbb{Z}\Gamma$-module $\mathbb{Z}$ has a $\mathbb{Z}\Gamma$-projective resolution

$$\cdots \longrightarrow P_1 \longrightarrow P_0 \longrightarrow \mathbb{Z} \longrightarrow 0 \qquad (\dagger)$$

with P_i finitely generated for all $i \leq m$. The augmented chain complex of a $K(\Gamma, 1)$-Eilenberg-MacLane space with finite m-skeleton provides such a resolution and so, for example, finitely presented groups are of type $(FP)_2$; it is however unknown whether all groups of type $(FP)_2$ are finitely presented. If there is such a resolution ($\dagger$) with all P_i finitely generated then Γ is *of type* $(FP)_\infty$.

In [2] Bieri and Groves showed that for a metabelian group the condition that it is of type $(FP)_m$ implies a constraint on the homology of its commutator subgroup. Their argument [2, Proposition 5.3] applies for groups Γ with abelian normal subgroup A and polycyclic quotient $G = \Gamma/A$. If Γ is of type $(FP)_m$ we may use a resolution of the form ($\dagger$) to calculate the homology groups $H_i(A; \mathbb{Z})$. The resolution may be regarded as a $\mathbb{Z}A$-projective resolution of the trivial $\mathbb{Z}A$-module and so we wish to study the homology of

$$\cdots \longrightarrow P_1 \otimes_{\mathbb{Z}A} \mathbb{Z} \longrightarrow P_0 \otimes_{\mathbb{Z}A} \mathbb{Z} \longrightarrow \mathbb{Z} \longrightarrow 0.$$

By assumption P_i is a finitely generated $\mathbb{Z}\Gamma$-module for $i \leq m$ and so we know the $P_i \otimes_{\mathbb{Z}A} \mathbb{Z}$ to be finitely generated $\mathbb{Z}G$-modules. The integral group ring of a polycyclic group is Noetherian (Hall [16]) and so the homology group $H_i(A; \mathbb{Z})$, being a section of $P_i \otimes_{\mathbb{Z}A} \mathbb{Z}$, is also finitely generated as a $\mathbb{Z}G$-module for $i \leq m$. The abelian normal subgroup A may be regarded as a $\mathbb{Z}G$-module via conjugation and its i-th exterior power $\bigwedge^i A$ over $\mathbb{Z}$ inherits a $\mathbb{Z}G$-module structure as follows. By definition $\bigwedge^i A$ is the i-th homogeneous component of the exterior algebra obtained by taking the quotient of the tensor algebra $T(A)$ over $\mathbb{Z}$ by the 2-sided ideal I generated by the elements $a \otimes a$ with a in A. The diagonal G-action on $T(A)$ leaves I invariant and induces an action on each $\bigwedge^i A$. Proposition 5.1 of [2] says that for each integer $i \geq 0$ there is a natural embedding of $\bigwedge^i A$ in $H_i(A; \mathbb{Z})$. Thus we deduce that $\bigwedge^i A$ is a finitely generated $\mathbb{Z}G$-module for $i \leq m$ when Γ is of type $(FP)_m$.

The work of Bieri and Groves was concerned with metabelian groups and so with modules over the integral group ring of finitely generated abelian groups Q. For $\chi \in \mathrm{Hom}(Q, \mathbb{R})$ define Q_χ to be the monoid $\{q \in Q : \chi(q) \geq 0\}$. Then they define, for a $\mathbb{Z}Q$-module M, the subset Σ_M of $\mathrm{Hom}(Q, \mathbb{R})$ given by

---

[1]The author is an Alexander von Humboldt Fellow

$$\Sigma_M = \{ \chi : M \text{ is finitely generated over } \mathbb{Z}Q_\chi \}$$

and say that M is m-*tame* if whenever $\chi_1, \ldots, \chi_m \in \text{Hom}(Q, \mathbb{R})$ with $\chi_1 + \chi_2 \ldots + \chi_m = 0$, then at least one of the homomorphisms χ_i lies in Σ_M. It was conjectured in [2] that an extension Γ of A by G with both A and G abelian, is of type $(FP)_m$ if and only if A is m-tame as a $\mathbb{Z}G$-module. This was established for $m = 2$ by Bieri and Strebel [6] and for general m by Åberg in the case where A is of finite Prüfer rank [1]. Recently Noskov [19] has shown in the case where A is torsion-free and the extension splits that A is m-tame if Γ is of type $(FP)_m$.

A special case of Theorem C of [2] is that a $\mathbb{Z}Q$-module M of prime exponent p is m-tame if and only if all its exterior powers $\bigwedge^i M$ for $i \leq m$ are finitely generated. However in general, when M is not of finite exponent, the exterior powers and the homological approach described above do not tell the full story. For example if Q is infinite cyclic generated by q and $M \cong \mathbb{Z}Q/I$ where I is the ideal generated by $2q - 3$, the module is $\mathbb{Z}$-torsion-free of rank 1 and so its exterior i-th powers are zero for $i \geq 2$. However M is not 2-tame and all extensions Γ are known to be not of type $(FP)_2$. The main difficulties seem to arise in such finite rank examples, an area studied by Brewster in his thesis [8] when looking at the maximum condition on 2-sided ideals in integral group rings. The consideration of exterior powers is however sufficient to say that metabelian groups of type $(FP)_\infty$ have a subgroup of finite index of finite cohomological dimension [2, Theorem A]. Recently Kropholler has shown this to be true for all soluble groups of type $(FP)_\infty$ [17] and that they are constructible [18].

1.2 The point of this paper is to describe how the consideration of exterior powers is also good enough to see that even the $(FP)_2$ and $(FP)_3$ conditions on Γ impose very strong constraints when G is non-abelian.

Recent work of Brookes and Groves [13] and Brookes, Roseblade and Wilson [14] shows that it is impossible for many polycyclic groups to act faithfully on a module of prime exponent with finitely generated exterior square. The ensuing result is the following.

Theorem A [13,14] *Let Γ be an abelian-by-polycyclic group of type $(FP)_2$ with Fitting subgroup F. If Γ/F is virtually a d-generator group then it is virtually nilpotent of virtual class at most d.*

The main original content of the paper concerns the $(FP)_3$ condition; its imposition ensures that the virtual class of Γ/F is bounded, independent of the number of generators.

Theorem B *Let Γ be an abelian-by-polycyclic group of type $(FP)_3$ with Fitting subgroup F. Then Γ/F is virtually nilpotent of virtual class 2.*

It is unknown whether such a result also holds in the $(FP)_2$ case. As yet all examples of extensions Γ of type $(FP)_2$ have Γ/F virtually of class at most 2;

there are two examples of Robinson and Strebel [21] with Γ/F of class 2, and it is not clear whether Strebel's more general (unpublished) recipe [26] for finitely generated abelian-by-nilpotent groups includes examples with Γ/F of greater virtual class. In §3.2 it is shown that neither example of Robinson and Strebel is of type $(FP)_3$.

2 Krull Dimension

2.1 There are several notions of dimension from Noetherian ring theory available for use in the study of modules over polycyclic groups. They all have disadvantages as well as advantages. The best choice when the group is virtually nilpotent is that of Krull dimension as it behaves well under taking exterior powers, taking submodules over subgroups, and under induction. However the latter two virtues disappear for a general polycyclic group and we have to rely in §4 on an *ad hoc* approach estimating rates of growth.

Let R be a ring. The *Krull dimension*, $\dim M$, of an R-module M is defined inductively as follows: $\dim M = -1$ if and only if $M = 0$, and for $q \geq 0$, $\dim M \leq q$ if and only if in every descending series of submodules in M, all but finitely many of the factors have dimension at most $q - 1$. Thus $\dim M = q$ if and only if $\dim M \leq q$ and $\dim M \not\leq q - 1$. If there is no finite q with $\dim M = q$ we set $\dim M = \infty$ but in fact in our context all finitely generated modules have finite dimension.

The concept was introduced by Rentschler and Gabrièl in [20]. The general properties we shall need are the following.

(1) for $M_1 \leq M$, $\dim M = \max\{\dim M, \dim M/M_1\}$
(2) for a finitely generated R-module M, $\dim M \leq \dim R$
(3) any Noetherian R-module of dimension q has an image likewise of dimension q which is *critical*, that is, the image is a module all of whose proper quotients are of smaller dimension
(4) non-zero endomorphisms of critical modules are monomorphisms.

For group rings of abelian groups this definition coincides with the usual one for commutative rings. For example if Q is free abelian of finite rank $h(Q)$ and P is a prime ideal of $\mathbb{Z}Q$ then $\dim \mathbb{Z}Q/P$ is the maximal length r of a chain of prime ideals $P = P_0 < P_1 \ldots < P_r$ of $\mathbb{Z}Q$. In particular $\dim \mathbb{F}_p Q = h(Q)$, where $\mathbb{F}_p$ denotes the field of p elements, and $\dim \mathbb{Z}Q = 1 + h(Q)$. A theorem of Smith [25] says that

(5) for any polycyclic group H, $\dim \mathbb{F}_p H = h(H)$ and $\dim \mathbb{Z}H = 1 + h(H)$, where $h(H)$ denotes the Hirsch length of H.

Segal investigated the additional properties of integral group rings of finitely generated nilpotent groups H, the following being the most important.

(6) Let $L \leq H$ and V be a $\mathbb{Z}L$-submodule of a $\mathbb{Z}H$-module M.
Then $\dim_L V \leq \dim_H M$.
(7) Suppose $L \lhd H$ with H/L infinite cyclic and V is a critical
$\mathbb{Z}L$-module. Then $V \otimes_{\mathbb{Z}L} \mathbb{Z}H$ is a critical $\mathbb{Z}H$-module of dimension
$1 + \dim_L V$.

Lemma 1 *Let H be any group and M_1 and M_2 be non-zero $\mathbb{F}_p H$-modules
of finite Krull dimension. Then $\dim M_1 \otimes M_2 \geq \dim M_1 + \dim M_2$.*

Proof. This is a simple induction on $\dim M_1 + \dim M_2$. If M_1 and M_2 are both
Artinian, or in other words $\dim M_1 = 0 = \dim M_2$ then $M_1 \otimes M_2$ is non-zero
and so $\dim M_1 \otimes M_2 \geq 0$. Assume without loss of generality that $\dim M_1 > 0$.
Then there is a descending chain $M_1 = M_{10} > M_{11} > \ldots$ with infinitely
many factors of dimension $\dim M_1 - 1$. Using the inductive hypothesis and
the fact that $- \otimes M_2$ is exact for $\mathbb{F}_p$-vector spaces, there are infinitely many
factors of the chain $M_{10} \otimes M_2 \geq M_{11} \otimes M_2 \geq \ldots$ of dimension at least
$(\dim M_1 - 1) + \dim M_2$. Thus $\dim M_1 \otimes M_2 \geq (\dim M_1 - 1) + \dim M_2 + 1$.

Lemma 2 *Let H be a group and M be a $\mathbb{F}_p H$-module of finite Krull dimension. Suppose $\bigwedge^i M$ is non-zero. Then $\dim \bigwedge^i M \geq i \dim M$*

Proof. This is an inductive argument on i; the case $i = 1$ is trivial. Suppose
M has a submodule M_1 with $\dim M_1 > 0$. Then the image X of $M \otimes M_1 \otimes
M_1 \otimes \ldots \otimes M_1$ under the natural map $\bigotimes^i M \longrightarrow \bigwedge^i M$ contains the image Y
of $M_1 \otimes M_1 \otimes M_1 \otimes \ldots \otimes M_1$. Here Y is isomorphic to $\bigwedge^i M_1$ and X/Y to
$(M/M_1) \otimes \bigwedge^{i-1} M_1$. Thus

$$\dim(\bigwedge^i M / \bigwedge^i M_1) \geq \dim(X/Y) \geq \dim(M/M_1) + (i-1)\dim M_1$$

using the inductive hypothesis and Lemma 1.

Suppose $\dim M > 0$ and we have a descending chain of submodules
$M > M_1 > \ldots$ with infinitely many factors of dimension $\dim M - 1$. It
follows that $\dim M_j = \dim M > 0$ for all j. By the preceding paragraph the
descending chain $\bigwedge^i M > \bigwedge^i M_1 > \ldots$ has infinitely many factors of dimension
at least $(\dim M - 1) + (i-1)\dim M$, and so $\bigwedge^i M$ has dimension at least
$i \cdot \dim M$.

2.2 These estimates are also true for tensor products and exterior powers
of kH-modules over any coefficient field k. However the story is more complicated for modules over integral group rings, but Segal's results allow us to
deduce something for general modules with finitely generated exterior powers
if H is virtually nilpotent.

Lemma 3 *Let H be a polycyclic group and M be a finitely generated $\mathbb{Z}H$-
module with $\bigwedge^i M$ finitely generated for some i.*

(i) If M is of finite exponent then $i \cdot \dim M \leq h(H)$
(ii) If H is virtually nilpotent then $i(\dim M - 1) \leq h(H)$

Proof. In either case since M is Noetherian it has a critical image N of dimension $\dim M$. By (4), since multiplication by an integer induces an endomorphism of N, we know N to be either $\mathbb{Z}$-torsion-free or of prime exponent p. In case (i) N must be an $\mathbb{F}_p H$-module and, since $\bigwedge^i N$ is an image of $\bigwedge^i M$, we deduce from (1), (2) and Lemma 2 that $i \cdot \dim M \leq h(H)$. In case (ii) either N is again an $\mathbb{F}_p H$-module and we can deduce the same inequality, or we invoke Lemma 8 and Theorem F of Segal [24] to see that $\dim(N/Np) = \dim M - 1$ for some prime p. Case (i) now applies to deduce that $i \cdot \dim(N/Np) \leq h(H)$ and we are done.

Combining with the homological arguments and using the notation of §1 we get

Proposition 4   *(i) Suppose Γ is of type $(FP)_m$ and A is of finite exponent. Then $m \cdot \dim A \leq h(G)$.*
(ii) Suppose Γ is of type $(FP)_m$ and G is virtually nilpotent. Then

$$m(\dim A - 1) \leq h(G).$$

Finitely presented groups are necessarily of type $(FP)_2$ and so the case $m = 2$ applies. More generally using Wilson's Golod-Shafarevic-type theorem [27], Groves and Wilson [15] established these $(m = 2)$ inequalities for all abelian-by-nilpotent Γ which are quotients of finitely presented groups Γ_1 with Γ_1 having no infinite torsion images. For example these inequalities hold for all abelian-by-nilpotent images of finitely presented soluble groups.

3   Impervious modules over nilpotent groups

3.1 The next stage is to study for nilpotent groups the structure of modules with finitely generated exterior powers. This will, in combination with Segal's results, then provide lower bounds for the dimension of such modules. The starting point is to consider *impervious* modules, that is those modules with no non-zero submodules induced from a module over a subgroup of infinite index.

Lemma 5   *Let H be a finitely generated nilpotent group and M be a non-zero $\mathbb{Z}H$-module, which is either $\mathbb{Z}$-torsion-free or of prime exponent, and with $\bigwedge^i M$ finitely generated for some $i \geq 2$. Then M is impervious.*

Proof. Suppose M contains a submodule M_1 of the form $V \otimes_{\mathbb{Z}L} \mathbb{Z}H$ for some L of infinite index in H. We may assume that L is normal with H/L infinite cyclic, since any subgroup of infinite index is contained in such an L. Choose

h so that $H = L\langle h \rangle$. Since M is assumed to be either $\mathbb{Z}$-torsion-free or of prime exponent, the natural map $\otimes^i M_1 \longrightarrow \otimes^i M$ induces an embedding of $\wedge^i M_1$ in $\wedge^i M$ and so $\wedge^i M_1$ inherits finite generation. But for example $\wedge^2 M_1$ is clearly not finitely generated; it contains an infinite direct sum $\bigoplus\limits_{l \geq 1} W_l$ where W_l is the $\mathbb{Z}H$-module image of $\bigoplus\limits_{j-k=l} (V \otimes h^j) \otimes (V \otimes h^k)$ under the natural map $\otimes^2 M_1 \longrightarrow \wedge^2 M_1$. A similar argument for larger i also gives a contradiction.

In fact if an abelian-by-nilpotent group Γ is the image of a group Γ_1 with Γ_1 of type $(FP)_2$ containing no non-abelian free subgroups then the underlying module A is necessarily impervious; as observed in [15] the argument of [10, Proposition 2] shows how to deduce this from looking at the metabelian images of Γ and using Theorem C of Bieri, Neumann and Strebel [5] (see in addition 1.5.7 of the forthcoming book of Bieri and Strebel [7]).

3.2 In this section we consider finitely generated nilpotent groups H with infinite cyclic commutator subgroup H' such that H/H' is free abelian. The commutator subgroup is thus central. If centre and commutator subgroup coincide we have the extraspecial groups H_n,

$$\langle x_1, y_1, x_2, y_2, \ldots, x_n, y_n, z \ : \ [x_i, y_i] = z, \ [x_i, z] = 1 = [y_i, z], \ (1 \leq i \leq n) \rangle$$

for example, the discrete Heisenberg group H_1.

Groves and Wilson showed in [15] that an impervious $\mathbb{Z}H_n$-module on which no non-trivial central group elements of H_n act unipotently is of dimension at least $n+1$. But $h(H_n) = 2n+1$ and so by Lemma 3(i) such a module, if it has finite exponent, cannot have finitely generated exterior square. However the underlying module of the first example of Robinson and Strebel [21] shows that the finite exponent condition cannot be dropped; their second example shows that the conditions imposed on H here are not sufficient to rule out the existence of $\mathbb{F}_p H$-modules with finitely generated exterior square and with no non-trivial central group element acting unipotently.

For $n \geq 2$ we can deduce immediately from Lemma 3(ii) that a $\mathbb{Z}H_n$-module, on which no non-trivial central group element acts unipotently, does not have a finitely generated exterior cube. In fact this is also true for $n = 1$; Lemma 3(i) deals with the finite exponent case as before but we have to try a little harder at the end of the proof of Lemma 3(ii) with the $\mathbb{Z}$-torsion-free critical N. We know, since N/Np has finite exponent and finitely generated exterior cube, that some central group element acts unipotently and so H'_n acts finitely on N/Np. Thus H_n acts as a group of Hirsch length at most $2n$ on N/Np and we see that $3 \cdot \dim(N/Np) \leq 2n$ and hence $\dim M \leq 2n/3 + 1$ in this case, in contradiction of the lower bound, $\dim M \geq n + 1$.

It is possible, though not as easy, to rule out the finite generation of exterior cubes under the more general conditions of this section. In so doing one sees that neither example of Robinson and Strebel is of type $(FP)_3$. The ingredients of the proof are similar to those used for $\mathbb{Z}H_n$-modules where the lower bound stemmed from consideration of a torsion-free abelian normal subgroup B containing the centre Z of H with $h(B/Z) = n$ and $h(H/Z) = 2n$. This exists since the map $H \times H \longrightarrow H'$ given by $(g,h) \longmapsto [g,h]$ induces an integral-valued alternating form on H/H'.

Lemma 6   *With H satisfying the hypotheses of this section, let M be a finitely generated $\mathbb{Z}H$-module with finitely generated exterior cube. Then some positive power of the augmentation ideal of some characteristic subgroup of H' of finite index annihilates M. In particular if M is of finite exponent then H' acts finitely.*

Proof. Suppose not. Since we are looking at Noetherian modules, by passing to an image we may assume we have a counterexample M all of whose quotients satisfy the conclusions of the lemma. Clearly we can take the action of H on M to be faithful.

Every non-zero endomorphism of M must be a monomorphism; otherwise both its cokernel and its image would be isomorphic to proper quotients of M and some positive power of the augmentation ideal of some characteristic subgroup L of finite index in H' would kill both, and hence M. Every element of $\mathbb{Z}Z$, where Z is the centre of H, induces by multiplication an endomorphism of M. It follows that the annihilator P of M in $\mathbb{Z}Z$ is prime and that M is $\mathbb{Z}Z/P$-torsion-free. The faithfulness of the action of H implies that $(1 + P) \cap Z = 1$. It follows from Lemma 5 that M is impervious.

Now pick a non-zero m in M with $X = \text{ann}_{\mathbb{Z}B}m$ maximal among such annihilators. Thus X is a prime ideal of $\mathbb{Z}B$ with $X \cap \mathbb{Z}Z = P$. Clifford theory (Roseblade [22, Lemma 3]) says that the submodule M_1 of M generated by m is of the form $m(\mathbb{Z}N \otimes_{\mathbb{Z}N} \mathbb{Z}H)$, where N is the normaliser of X under conjugation. Because M is impervious N must have finite index in H and the centraliser of N in H is just Z. But $(1+X) \cap B$ is N-invariant and if non-trivial would have non-trivial intersection with Z. So $(1 + X) \cap B = 1$. Roseblade's Theorem D of [23], applied directly when $Z \cap X = (p)$, or to the prime ideal $X \otimes_{\mathbb{Z}B} \mathbb{Q}B$ of $\mathbb{Q}B$ in the $Z \cap X = 0$ case, says that $X = P.\mathbb{Z}B$. Thus M contains a copy V of $\mathbb{Z}B/X \cong (\mathbb{Z}Z/P) \otimes_{\mathbb{Z}Z} \mathbb{Z}B$. A mild variation on Segal's Lemma 9 [24] (with an almost identical proof) states there is a maximal ideal I of $\mathbb{Z}Z$ containing P such that $V > VI = MI \cap V$ and thus V/VI is non-zero and embeds in M/MI. But $V/VI \cong (\mathbb{Z}Z/I) \otimes_{\mathbb{Z}Z} \mathbb{Z}B$. Since $\mathbb{Z}Z/I$ is critical, an induction using property (7) gives $\dim V/VI = h(B/Z) = n$. Also (6) says $\dim M/MI \geq \dim V/VI$. Since I is maximal it is cofinite, and so H acts on M/MI as a group of Hirsch length at most $h(H/Z) = 2n$, and M/MI is of exponent p for some prime p. Lemma 3(i) applied to M/MI, a module with finite exterior cube, gives $3.\dim M/MI \leq 2n$, a contradiction.

3.3   For general finitely generated nilpotent groups the theorem that yields lower bounds for the dimension of modules is the following.

Theorem C [13, Theorem 3.2] *Let H be a finitely generated torsion-free nilpotent group with centre Z and commutator H'. Let M be an impervious $\mathbb{Z}H$-module which is $\mathbb{Z}Z/P$-torsion-free for some prime ideal P of $\mathbb{Z}Z$ with $(1+P) \cap Z = 1$. Let H_0 be a normal subgroup of H containing $H'Z$ and with $H_0/H'Z$ finite-by-cyclic. Then M is $\mathbb{Z}H_0/P.\mathbb{Z}H_0$-torsion-free.*

The proof of this uses Clifford theory much as in the proof of Lemma 6. However due to the non-commutativity of the situation we have to consider the theory of indecomposable injective modules developed by Brookes and Brown [10, 11, 12]. This means that we require information about the stabilisers of isomorphism classes of indecomposable injectives rather that about normalisers of prime ideals. One way of proving the result of Roseblade [23, Theorem D] used in the lemma is to make use of the geometry of the subset $\Sigma_{\mathbb{Z}Q/P}$ of $\text{Hom}(Q, \mathbb{R})$ defined in §1.1 for a prime ideal P of the integral group ring of a free abelian group Q of finite rank. Bieri and Groves [3] showed that the union of the origin and the complement of $\Sigma_{\mathbb{Z}Q/P}$ in $\text{Hom}(Q, \mathbb{R})$ is a closed rationally defined polyhedron (see also [4]). The action on Q of the normaliser N of the prime ideal induces an action on $\text{Hom}(Q, \mathbb{R})$ under which $\Sigma_{\mathbb{Z}Q/P}$ is invariant. In the proof of Theorem C a crucial step is to define an analogous geometric set of $\text{Hom}(Q, \mathbb{R})$, but this time for a module over a crossed product of a division ring by a free abelian group Q. Unlike the group ring, in general such a crossed product is non-commutative and we have only been able to prove a weaker result about the geometric structure of the set. There remain several open questions.

For a module M as in Theorem C we know we have a copy of $\mathbb{Z}H_0/P.\mathbb{Z}H_0$ embedded as a $\mathbb{Z}H_0$-submodule and this submodule is induced from the critical $\mathbb{Z}Z$-module $\mathbb{Z}Z/P$. Property (6), of Section 2, gives

$$\dim M \geq \dim \mathbb{Z}H_0/P.\mathbb{Z}H_0$$

and property (7), applied $h(H_0/Z)$ times, gives

$$\dim(\mathbb{Z}H_0/P.\mathbb{Z}H_0) = \dim \mathbb{Z}Z/P + h(H_0/Z).$$

But the faithfulness of P ensures $\dim \mathbb{Z}Z/P \geq 1$ and using a choice of H_0 with $h(H_0/H'Z) = 1$ we get

Corollary C [13, Corollary 2] *With H and M as in the theorem, suppose H is non-abelian. Then $\dim M \geq h(H'Z) - h(Z) + 2$.*

Instead of using the corollary to give a lower bound on dimension one can apply Theorem C directly to produce the following.

Theorem D [13, Theorem 3.4] *Let Γ be an abelian-by-nilpotent group of type $(FP)_2$ with Fitting subgroup F. Suppose that Γ/F is virtually a d-generator group. Then Γ/F is virtually of class at most d.*

The corresponding module result has a similar proof.

Proposition 7   *Let H be a virtually d-generator nilpotent group and M be a finitely generated $\mathbb{Z}H$-module. Suppose that $\bigwedge^2 M$ is also finitely generated. Then there is a normal subgroup K of H with H/K virtually of class d so that some positive power of the augmentation ideal of K annihilates M.*

3.4   Theorem C can also be used to prove the following module result which has the abelian-by-nilpotent case of Theorem B, stated in the introduction, as an immediate corollary. The idea of the proof is to reduce a counterexample to one where Theorem C applies and then to observe that this counterexample would have an image ruled out by Lemma 6.

Proposition 8   *Let H be a finitely generated nilpotent group and M be a finitely generated $\mathbb{Z}H$-module. Suppose that $\bigwedge^3 M$ is also finitely generated. Then there is a normal subgroup K of H with H/K virtually of class 2 so that some positive power of the augmentation ideal of K annihilates M.*

Proof.   By passing to a subgroup of finite index we may assume that both H and H/H' are torsion-free.

Suppose the proposition is false. As in Lemma 6 we may consider a counterexample M with faithful action, all of whose proper quotients obey the conclusions of the proposition, and conclude that M is impervious and $\mathbb{Z}Z/P$-torsion-free for some prime ideal P of $\mathbb{Z}Z$ with $(1 + P) \cap Z = 1$.

We now apply Theorem C to find that M is in fact $\mathbb{Z}(H'Z)/P.\mathbb{Z}(H'Z)$-torsion-free. Because M is a counterexample, H itself cannot be virtually of class 2 and so H'/C is infinite where $C = H' \cap Z$. Indeed there is a normal subgroup L of H with $C \leq L < H'$ and H'/L infinite cyclic. Note that H/L is the sort of group considered in §3.2.

M contains a copy U of $\mathbb{Z}H'/X.\mathbb{Z}H'$ where $X = P \cap \mathbb{Z}C$. Thus U is of the form $V \otimes_{\mathbb{Z}L} \mathbb{Z}H'$ where $V = \mathbb{Z}L/X.\mathbb{Z}L$. Observe in fact that M is $\mathbb{Z}L/X.\mathbb{Z}L$-torsion-free. We can find an H-invariant ideal I of $\mathbb{Z}L$, with $\mathbb{Z}L/I$ finite, such that $U > U \cap MI = UI$ and hence U/UI is non-zero and embeds in M/MI. To do this we need a result similar to that used in Lemma 6; it is a halfway house between Lemma 9 of [24] and Lemma 14 of Brookes [9]. Using Theorem C* of Roseblade [22] (see also Lemma 13 of [9]) there is a finite class $\mathcal{S} = \{ \mathbb{Z}L/\Lambda_0,\ \mathbb{Z}L/\Lambda_1,\ \ldots\ ,\ \mathbb{Z}L/\Lambda_t \}$ of cyclic $\mathbb{Z}L$-modules, with $\Lambda_0 = X\,\mathbb{Z}L$ and $\Lambda_i > X.\mathbb{Z}L$ for $1 \leq i \leq t$, such that M/U is the union of an ascending series of $\mathbb{Z}L$-submodules with each factor H-conjugate to some member of $\mathcal{S}$. Set $\Lambda = \Lambda_1 \cap \ldots \cap \Lambda_t$. The argument of Lemma 14 of [9] shows that, in the terminology of [9], if an ideal I of $\mathbb{Z}L$, H-invariant under conjugation, culls $\Lambda/X.\mathbb{Z}L$ in $\mathbb{Z}L/X.\mathbb{Z}L$ then $U \cap MI = UI < U$. Since $\mathbb{Z}C/X$ is uniform and H-ideal critical, Proposition 1 of [9] (with $\Gamma = H$) and induction on a normal series

$$C = L_0 < L_1 < \ldots < L_s = L$$

with infinite cyclic factors show that $\mathbb{Z}L/X.\mathbb{Z}L = (\mathbb{Z}C/X) \otimes_{\mathbb{Z}C} \mathbb{Z}L$ is also uniform and H-ideal critical. But $\Lambda > X.\mathbb{Z}L$ and so there is such a culling ideal I. (In fact the minimality of M as a counterexample implies that M is of prime exponent and L is abelian. For the latter one shows that H/C is virtually of class 2; but it is also torsion-free and hence itself of class 2, and so H' is abelian.)

Now, because $\mathbb{Z}L/I$ is finite, some characteristic subgroup of finite index in L acts trivially on M/MI. So there is some subgroup of finite index in H which acts on M/MI as a group of the sort considered in §3.2. Moreover, because M/MI contains $U/UI \cong V/VI \otimes_{\mathbb{Z}L} \mathbb{Z}H'$, the action of H' on M/MI is not finite. This contradicts Lemma 6 and the proposition is proved.

4 Modules over polycyclic groups

4.1 The main module result of Brookes, Roseblade and Wilson [14] is the following.

Theorem E [14] *Let H be a polycyclic group and M be a finitely generated $\mathbb{Z}H$-module with $\bigwedge^i M$ finitely generated for some $i \geq 2$. Then there is a normal subgroup K of H with H/K virtually nilpotent such that some positive power of the augmentation ideal of K annihilates M.*

In the proof the key special case is when H contains a normal free abelian subgroup B with H/B also free abelian, and M is an $\mathbb{F}_p H$-module which is $\mathbb{F}_p B$-torsion-free.

Suppose $i = 2$ and instead of the exterior square consider for a moment the tensor square $N = \bigotimes^2 M$. It is a $\mathbb{F}_p(H \times H)$-module which is $\mathbb{F}_p(B \times B)$-torsion-free. Suppose it is finitely generated under the diagonal action of H. Thus

$$\dim_H N \leq \dim \mathbb{F}_p H = h(H) = h(B) + h(H/B).$$

From the definition it is clear that the dimension of N as a $H \times H$-module is at most that as a (diagonal) H-module. If H were nilpotent then (6) applied to an embedded copy of $\mathbb{F}_p(B \times B)$ in N would ensure that

$$\dim_{H \times H} N \geq \mathbb{F}_p(B \times B) = 2h(B)$$

and thus we could deduce that $h(B) \leq h(H/B)$. However we are interested in the case where H is not virtually nilpotent and there is no result like (6) available. However by *ad hoc* growth arguments one can still show that $h(B) \leq h(H/B)$. This inequality is also valid under the weaker assumption that $\bigwedge^2 M$ is finitely generated. The argument is similar but there is the added complication that $\bigwedge^2 M$, rather than being a module over $\mathbb{F}_p(H \times H)$, the tensor square of $\mathbb{F}_p H$, is only a module over the symmetric square, the subring of the tensor square consisting of elements fixed under the canonical involution.

In fact the basic case in the proof of Theorem E is when H/B acts faithfully and rationally irreducibly on B. In other words H/B is isomorphic to a group of units of the ring of integers of a number field with additive group isomorphic to $B \otimes_{\mathbb{Z}} \mathbb{Q}$. In our context we deduce from Dirichlet's unit theorem that $h(H/B) < h(B)$. This contradicts the inequality arising from the finite generation of $\bigwedge^2 M$.

4.2 Theorem B is an immediate corollary of the following module result, the nilpotent case of which was Proposition 8.

Proposition 9 *Let H be a polycyclic group and M be a finitely generated $\mathbb{Z}H$-module with $\bigwedge^3 M$ finitely generated. Then there is a normal subgroup K of H with H/K virtually nilpotent of class 2 so that some positive power of the augmentation ideal of K annihilates M.*

Proof. We suppose otherwise and consider a minimal counterexample M in the sense that all of its images satisfy the conclusions of the proposition. As usual M is either $\mathbb{Z}$-torsion-free or of prime exponent, and so for any non-zero submodule M_1 we have $\bigwedge^3 M_1$ embedded in $\bigwedge^3 M$. Thus M_1 inherits finite generation of its exterior cube from M and is also a minimal counterexample to the proposition.

From Theorem E there is such an M_1 upon which H acts as a virtually nilpotent group. By passing to a subgroup of finite index we see that M_1 is a counterexample for some finitely generated nilpotent group, contradicting Proposition 8.

References

[1] H. Åberg, Bieri-Strebel valuations (of finite rank), Proc. London Math Soc. (3) 52 (1986), 269-304.

[2] R. Bieri and J.R.J. Groves, Metabelian groups of type $(FP)_\infty$ are virtually of type (FP), Proc. London Math. Soc. (3) 45 (1982), 365-384.

[3] R. Bieri and J.R.J. Groves, On the geometry of the set of characters induced by valuations, J. Reine Angew. Math. 347 (1984), 168-195.

[4] R. Bieri and J.R.J. Groves, A rigidity property for the set of all characters induced by valuations, Trans. Amer. Math. Soc. 294 (1986), 425-434.

[5] R. Bieri, W.D. Neumann and R. Strebel, A geometric invariant of discrete groups, Invent. Math. 90 (1987), 451-477.

[6] R. Bieri and R. Strebel, Valuations and finitely presented metabelian groups, Proc. London Math. Soc (3) 41 (1980), 439-464.

[7] R. Bieri and R. Strebel, Geometric invariants for discrete groups, de Gruyter, Berlin, to appear.

[8] D.C. Brewster, The maximum condition on ideals of the group ring, Ph.D. thesis, University of Cambridge, 1976.

[9] C.J.B. Brookes, Modules over polycyclic groups, Proc. London Math. Soc. (3) 57 (1988), 88-108.

[10] C.J.B. Brookes, Stabilisers of injective modules over nilpotent groups, Group Theory (Singapore, 1987), de Gruyter, Berlin, 1989, pp. 275-291.

[11] C.J.B. Brookes and K.A. Brown, Primitive group rings and Noetherian rings of quotients, Trans. Amer. Math. Soc. 288 (1985), 605-623.

[12] C.J.B. Brookes and K.A. Brown, Injective modules, induction maps and endomorphism rings, Proc. London Math. Soc. (3) 67 (1993), 127-158.

[13] C.J.B. Brookes and J.R.J. Groves, Modules over nilpotent groups, J. London Math. Soc., to appear.

[14] C.J.B. Brookes, J.E. Roseblade and J.S. Wilson, Modules for group rings of polycyclic groups, submitted.

[15] J.R.J. Groves and J.S. Wilson, Finitely presented metanilpotent groups, J. London Math. Soc., to appear.

[16] P.Hall, Finiteness conditions for soluble groups, Proc. London Math. Soc. (3) 4 (1954), 419-436.

[17] P.H. Kropholler, Soluble groups of type $(FP)_\infty$ have finite torsion-free rank, Bull. London Math. Soc., 25 (1993), 558-566.

[18] P.H. Kropholler, On groups of type $(FP)_\infty$, J. Pure Appl. Algebra 90 (1993), 55-67.

[19] G.A. Noskov, The Bieri-Strebel invariant and homological finiteness properties of metabelian groups, preprint, Universität Bielefeld, 1993.

[20] R. Rentschler and P. Gabrièl, Sur la dimension des anneaux et ensembles ordonnés, C.R. Acad. Sci. Paris Sér. A 265 (1967), 712-715.

[21] D.J.S. Robinson and R. Strebel, Some finitely presented soluble groups which are not nilpotent by abelian by finite, J. London Math. Soc. (2) 26 (1982), 435-440.

[22] J.E. Roseblade, Group rings of polycyclic groups, J. Pure Appl. Algebra 3 (1973), 307-328.

[23] J.E. Roseblade, Prime ideals in group rings of polycyclic groups, Proc. London Math. Soc. (3) 36 (1978), 385-447.

[24] D. Segal, On the residual simplicity of certain modules, Proc. London Math. Soc. (3) 34 (1977), 327-353.

[25] P.F. Smith, On the dimension of group rings, Proc. London Math. Soc. (3) 25 (1972), 288-302.

[26] R. Strebel, On finitely presented abelian by nilpotent groups, preprint, McGill University, 1981.

[27] J.S. Wilson, Finite presentations of pro-p groups and discrete groups, Invent. Math. 105 (1991), 177-183.

Fachbereich Mathematik
Universität Frankfurt
Robert Mayer Str. 6-10
D-60054 Frankfurt am Main
Germany

Semigroup presentations and minimal ideals

C. M. CAMPBELL, E. F. ROBERTSON, N. RUŠKUC & R. M. THOMAS

1 Introduction

The purpose of this paper is first to give a survey of some recent results concerning semigroup presentations, and then to prove a new result which enables us to describe the structure of semigroups defined by certain presentations.

The main theme is to relate the semigroup S defined by a presentation Π to the group G defined by Π. After mentioning a result of Adjan's giving a sufficient condition for S to embed in G, we consider some cases where S maps surjectively (but not necessarily injectively) onto G. In these examples, we find that S has minimal left and right ideals, and it turns out that this is a sufficient condition for S to map onto G. In this case, the kernel of S (i.e. the unique minimal two-sided ideal of S) is a disjoint union of pairwise isomorphic groups, and we describe a necessary and sufficient condition for these groups to be isomorphic to G.

We then move on and expand on these results by proving a new result (Theorem 9), which is a sort of rewriting theorem, enabling us to determine the presentations of the groups in the kernel in certain cases. We finish off by applying this new result to certain semigroup presentations and by pointing out its limitations.

2 Semigroup and Group Presentations

We will be considering presentations Π of the form

$$< a_1, a_2, ..., a_n : \alpha_1 = \beta_1, \alpha_2 = \beta_2, ..., \alpha_m = \beta_m >,$$

where $n \geq 1$, $m \geq 0$ and each α_i and each β_i is a non-empty word in the letters $a_1, a_2, ..., a_n$. If S is the semigroup, and G is the group, defined by Π, then we have a natural semigroup homomorphism ϕ from S to G mapping each generator of S onto the corresponding generator of G.

It is clear that ϕ need not be surjective in general; for example, if $m = 0$, then we have a free semigroup, and the mapping ϕ is injective but not surjective. This is a particular case of a more general situation. If Π is a presentation as above, we can form a graph Γ_L by taking $\{a_1, a_2, ..., a_n\}$ to be the set of vertices of Γ_L, and then joining a_i to a_j in Γ_L if a_i and a_j occur as the initial letters of the words α_k and β_k for some relation $\alpha_k = \beta_k$ in Π. We call Γ_L the *left Adjan graph* of Π. The *right Adjan graph* Γ_R of Π is defined similarly, but we take a_i to be joined to a_j here if a_i and a_j occur as the final letters of α_k and β_k for some relation $\alpha_k = \beta_k$. The following result was proved in [1] :

Theorem 1 *Let S and G be the semigroup and group respectively defined by the presentation Π, ϕ be the natural homomorphism from S to G, and Γ_L and Γ_R be the left and right Adjan graphs of Π. If neither Γ_L nor Γ_R contains a cycle, then ϕ is injective.*

A geometric proof of this result was given in [10]. The case where $m = 0$, so that Γ_L and Γ_R are null graphs, is therefore a special case of Theorem 1.

However, there are situations where the mapping ϕ is surjective. One example where this occurs comes from the class of presentations for the *Fibonacci groups*. Recall that the Fibonacci group $F(r, n)$ is the group defined by the presentation

$$< a_1, a_2, ..., a_n : a_1 a_2 ... a_r = a_{r+1}, a_2 a_3 ... a_{r+1} = a_{r+2}, ...$$
$$a_{n-1} a_n ... a_{r-2} = a_{r-1}, a_n a_1 ... a_{r-1} = a_r > .$$

We let $S(r, n)$ denote the semigroup defined by this presentation. Before we explain the relationship between $F(r, n)$ and $S(r, n)$, we need some terminology.

If S is a semigroup, then a subset R of S is said to be a *right ideal* of S if $Rx \subseteq R$ for all x in S, and a subset L is said to be a *left ideal* if $xL \subseteq L$ for all x in S. Given this, we can state a result from [3] about the Fibonacci groups and semigroups :

Theorem 2 *If $r \geq 2$, $n \geq 1$ and $d = gcd(r, n)$, then $S(r, n)$ is the union of d pairwise disjoint right ideals, each of which is a subgroup isomorphic to $F(r, n)$.*

This answers, in the affirmative, a conjecture made in [6], and also, independently, by Leonard Soicher. The natural homomorphism ϕ from $S(r, n)$ to $F(r, n)$ induces a group isomorphism between each of the right ideals and $F(r, n)$. In particular, ϕ is surjective (though obviously not injective if $d > 1$).

In fact, Theorem 2 is a special case of a more general result. We define the *generalized Fibonacci group* $F(r, n, k)$ to be the group defined by the presentation

$$< a_1, a_2, ..., a_n : a_1 a_2 ... a_r = a_{r+k}, a_2 a_3 ... a_{r+1} = a_{r+k+1}, ...$$
$$a_{n-1} a_n ... a_{r-2} = a_{r+k-2}, a_n a_1 ... a_{r-1} = a_{r+k-1} >,$$

and let $S(r, n, k)$ denote the corresponding semigroup; see [11] for a survey of the Fibonacci and generalized Fibonacci groups. Then we have from [3] :

Theorem 3 *(i) If $r \geq 2$, $n \geq 1$, n and $r + k - 1$ are coprime, and $d = gcd(n, k)$, then $S(r, n, k)$ is the union of d pairwise disjoint left ideals, each of which is a subgroup isomorphic to $F(r, n, k)$.*

(ii) If $r \geq 2$, $n \geq 1$, n and k are coprime, and $d = gcd(n, r + k - 1)$, then $S(r, n, k)$ is the union of d pairwise disjoint right ideals, each of which is a subgroup isomorphic to $F(r, n, k)$.

Since $F(r, n, 1)$ is just $F(r, n)$, Theorem 2 is just a special case of Theorem 3. In fact, the proof of Theorem 3 yields the following more general result :

Theorem 4 *Let S be the semigroup and G be the group defined by the presentation*

$$\Pi = < a_1, a_2, ..., a_n : \alpha_1 = a_1, \alpha_2 = a_2, ..., \alpha_n = a_n > .$$

(i) Suppose that each α_j is a word of length two or more in the letters a_1, a_2, ... , a_n and each a_i occurs as the first, second and last letters of three of the α_j. Assume further that the right Adjan graph of Π is connected and let d denote the number of components in the left Adjan graph of Π. Then S is a union of d pairwise disjoint right ideals, each of which is a subgroup of S isomorphic to G.

(ii) Suppose that each α_j is a word of length two or more in the letters a_1, a_2, ... , a_n and each a_i occurs as the first, second-to-last and last letters of three of the α_j. Assume further that the left Adjan graph of Π is connected and let d denote the number of components in the right Adjan graph of Π. Then S is a union of d pairwise disjoint left ideals, each of which is a subgroup of S isomorphic to G.

The proofs of Theorems 2, 3 and 4 in [3] are couched in terms of the *Green's relations* in the semigroup. If S is a semigroup, we define a relation $\mathcal{R}$ on S by $a \mathcal{R} b$ if a and b generate the same right ideal of S. We also define a similar relation $\mathcal{L}$ in terms of left ideals, and set $\mathcal{H} = \mathcal{R} \cap \mathcal{L}$. Green's Theorem [9] states that, if H is an $\mathcal{H}$-class, then either $H^2 \cap H = \emptyset$ or else H is a subgroup of S. The proof in [3] essentially proceeds in two stages; first one identifies the $\mathcal{R}$ and $\mathcal{L}$-classes to get a decomposition of the semigroup into subgroups, and then one shows that the subgroups are isomorphic to the groups defined by the same presentation.

A similar sort of result to Theorem 4 was obtained in [5], where the semigroups S defined by the presentations

$$< a_1, a_2, ..., a_n : a_i^{m+1} = a_i \ (1 \leq i \leq n), a_i a_j^2 = a_j a_i^2 \ (1 \leq i < j \leq n) >$$

were considered, and a result was obtained linking the structure of the semigroups (when finite) to that of the corresponding groups G. If m is odd, then

S is finite if m and 3 are coprime, or if $n \leq 3$, and there is a unique minimal left ideal, which is also the unique minimal right ideal and is isomorphic to G. If m is even, then S is finite if and only if $m \leq 4$. In this case, we have several minimal left ideals, each of which is isomorphic to G, and their union is the unique minimal right ideal of S.

3 Minimal Ideals

We have, so far, described some specific cases where the semigroup maps onto the corresponding group. In many other cases we looked at, we found similar results, in that the semigroup defined by a presentation Π not only mapped onto the group G defined by Π, but also contained one or more copies of G. However, we discovered examples where the semigroup defined by Π was made up of copies of a group other than that defined by Π. We want now to discuss this situation and to put the results we have mentioned so far into a more general context. With this in mind, we introduce some more terminology.

A subset of a semigroup S which is both a right ideal and a left ideal is said to be a *two-sided ideal* of S. A semigroup S can have at most one minimal two-sided ideal; if such an ideal K exists, then K is said to be the *kernel* of S. In the cases we shall be considering, the semigroups will usually have both minimal left and right ideals, and, as we shall see, this is enough to force the map ϕ to be surjective. (This contrasts with the case of the free semigroups, for example, mentioned above, which do not have minimal left or minimal right ideals.)

We have the following standard result from [7] :

Theorem 5 *Let S be a semigroup with minimal right ideals $\{R_i : i \in I\}$. Then $R_i \cap R_j = \emptyset$ for $i \neq j$ and S has a kernel $K = \bigcup \{R_i : i \in I\}$.*

There is (hardly surprisingly) an analogous result for left ideals. We then have :

Theorem 6 *Let S be a semigroup with minimal right ideals $\{R_i : i \in I\}$ and minimal left ideals $\{L_j : j \in J\}$, and let $H_{i,j} = R_i \cap L_j$ for all $(i,j) \in I \times J$. Then :*

(i) *$H_{i,j}$ is a group for all i and j, and $H_{i,j}$ is isomorphic to $H_{k,l}$ for all i, j, k and l.*

(ii) *If $e_{i,j}$ is the identity of the group $H_{i,j}$, then $R_i = e_{i,j}S$, $L_j = Se_{i,j}$ and $H_{i,j} = R_i L_j = R_i \cap L_j = e_{i,j} S e_{i,j}$ for any i and j.*

We see from Theorems 5 and 6 that, provided our semigroup S contains minimal right and left ideals, the kernel K of S exists and consists of a collection of pairwise disjoint isomorphic subgroups. If S is defined by the presentation Π, G is the group defined by Π, and the groups in K are all isomorphic to H, then we want to investigate the relationship between G and H. This is described in the following result from [4] :

Theorem 7 *Let S be the semigroup, and G be the group, defined by the presentation Π, and suppose that S possesses both minimal left and minimal right ideals. Let ϕ be the natural homomorphism from S to G, L be a minimal left ideal of S, R be a minimal right ideal, and let H be the group $R \cap L$. Let E denote the set of idempotents in K. Then $\phi|_H : H \to G$ is a group epimorphism. Moreover, $\phi|_H : H \to G$ is a group isomorphism if and only if E is a subsemigroup of S.*

By an *idempotent*, we mean an element e such that $e^2 = e$. It was noted in [4] that, since the idempotents in a minimal left (or right) ideal are necessarily closed, Theorem 7 gives :

Corollary 8 *Let S be the semigroup, and G be the group, defined by the presentation Π, and suppose that S possesses minimal left ideals and a unique minimal right ideal (or minimal right ideals and a unique minimal left ideal). Then every minimal left ideal (respectively every minimal right ideal) is a subgroup isomorphic to G.*

This result puts Theorems 2, 3 and 4 into perspective. In those cases, we had a unique minimal left or minimal right ideal, and so the subgroups in the semigroup S defined by the presentation Π have to be isomorphic, by Corollary 8, to the group defined by Π. On the other hand, many presentations Π define semigroups where the idempotents in the kernel do not form a subsemigroup, so that, by Theorem 7, the kernel is composed of subgroups not isomorphic to the group defined by Π. In a case such as this, we want a method for determining a presentation for these subgroups.

4   A Rewriting Theorem

We now demonstrate a method for determining the presentation of the groups in the kernel of a semigroup. We show that, under the additional hypothesis that the semigroup contains an idempotent satisfying certain properties, there is a reasonably straightforward way of deriving the required presentation. To be more precise, we shall prove the following result :

Theorem 9 *Let S be the semigroup defined by the presentation*

$$\Pi = \; < a_1, a_2 : a_1^{n_1+1} = a_1, a_2^{n_2+1} = a_2, \alpha_1 = \beta_1, \ldots\ldots\ldots, \alpha_m = \beta_m >,$$

where $n_1 > 0$, $n_2 > 0$, $m > 0$, and each α_i and each β_i is a non-empty word in a_1 and a_2. Suppose further that L is a minimal left ideal of S, R is a minimal right ideal of S, H is the subgroup $L \cap R$, and that the idempotent e in H satisfies :

(E1) $ea_1^i ea_2 = ea_1^i a_2$ *for* $1 \leq i \leq n_1$;

(E2) $ea_2^i ea_1 = ea_2^i a_1$ *for* $1 \leq i \leq n_2$.

Let $e = a_t^p \gamma$, *where* $t \in \{1,2\}$, $p > 0$, *and* γ *does not start with* a_t. *Let* $A = \{a_1, a_2\}$ *and* B *be the alphabet* $\{x_{j,k} : 1 \leq j \leq 2, 1 \leq k \leq n_j\}$, *with the convention that* $x_{i,j} = x_{i,k}$ *if* $j \equiv k$ *(mod n_i). We define a map* $\phi : A^+ \to B^+$ *by :* $a_i^j \phi = x_{i,j}$ *and* $(\eta_1 a_i a_j \eta_2)\phi = (\eta_1 a_i)\phi(a_j \eta_2)\phi$ *if* $i \neq j$. *Then* H *has a presentation* $< B : \Re >$, *where* $\Re$ *is the set of relations :*

(i) $(a_{j_1}^{k_1} \alpha_i a_{j_2}^{k_2})\phi = (a_{j_1}^{k_1} \beta_i a_{j_2}^{k_2})\phi$, $j_p \in \{1,2\}$, $1 \leq k_p \leq n_{j_p}$, $1 \leq i \leq m$;

(ii) $(a_j^k e)\phi = a_j^k \phi$, $j \in \{1,2\}$, $1 \leq k \leq n_j$;

(iii) $(e a_j^k)\phi = a_j^k \phi$, $j \in \{1,2\}$, $1 \leq k \leq n_j$;

(iv) $x_{t,q+p} = x_{t,q} x_{t,p}$ *if e is of the form* $a_t^p \delta a_t^q$ *(else no relation here).*

To prove Theorem 9, we define a map $\psi : B^+ \to A^+$ by $x_{i,j}\psi = ea_i^j e$, $(\eta \zeta)\psi = (\eta \psi)(\zeta \psi)$. We proceed via three lemmas; to avoid undue repetition, we will not repeat the hypotheses of Theorem 9 in each case.

Lemma 10 *For any* $\eta \in A^+$, $\eta \phi \psi = e \eta e$ *in* S.

Proof. We proceed by induction on the length of η, where we are taking the word $a_{i_1}^{p_1} a_{i_2}^{p_2} .... a_{i_k}^{p_k}$ to have length k. If $\eta = a_i^j$, then $\eta \phi \psi = x_{i,j}\psi = ea_i^j e$. If $\eta = a_i^j \zeta$, where $\zeta \phi \psi = e \zeta e$ in S and ζ does not begin with a_i, then $\eta \phi \psi = (x_{i,j}\psi)(\zeta \phi \psi) = ea_i^j ee\zeta e = ea_i^j e\zeta e = ea_i^j \zeta e = e\eta e$ as required.

Lemma 11 *If G is the group defined by the presentation* $< B : \Re >$, *then* ψ *induces a group epimorphism* $\theta : G \to H$.

Proof. From the properties (E1) and (E2), it follows easily that the $x_{i,j}\psi$ generate H. We now need to show that, for every relation $\alpha = \beta$ in $\Re$, $\alpha \psi = \beta \psi$ in H. This follows for the relations (i), (ii) and (iii), using the relations in Π, $e^2 = e$ and $e\phi \psi = e\eta e$. To prove the result for (iv) (when e is of the form $a_t^p \gamma = a_t^p \delta a_t^q$), note that, as e is in the kernel of S, there exists ξ such that $e\gamma \xi = e$. Now $x_{t,q+p}\psi = ea_t^{q+p} e = ea_t^{q+p} e\gamma \xi = ea_t^{q+p} \gamma \xi = ea_t^q e\xi = ea_t^q ee\xi = ea_t^q ea_t^p \gamma \xi = ea_t^q ea_t^p e\gamma \xi = ea_t^q ea_t^p e = (x_{t,q}\psi)(x_{t,p}\psi)$ as required.

Note that, since Π is a presentation for S, it follows from the relations in (i) that, if $\alpha = \beta$ in S, then $\alpha \phi = \beta \phi$ in G, so that ϕ induces a map from S to G. Moreover, by relations (ii) and (iii), we have that $(ew)\phi = (we)\phi = w\phi$ for any w.

Lemma 12 *If G is the group defined by the presentation $< B : \Re >$ and $\eta \in B^+$, then $\eta\psi\phi = \eta$ in G.*

Proof. We proceed by induction on the length of η. If $\eta = x_{i,j}$, then $\eta\psi\phi = (ea_i^j e)\phi = a_i^j \phi = x_{i,j} = \eta$. So let $\eta = x_{i,j}\zeta$ with $\zeta\psi\phi = \zeta$ in G. If the last letter of e is not a_t, then $\eta\psi\phi = (ea_i^j e(\zeta\psi))\phi = (ea_i^j ee(\zeta\psi))\phi = (ea_i^j e)\phi(e(\zeta\psi))\phi = (a_i^j \phi)(\zeta\psi\phi) = x_{i,j}\zeta = \eta$. Note that we did not need relation (iv) here. On the other hand, if $e = a_t^p \delta a_t^q$, then

$$\eta\psi\phi = (ea_i^j e(\zeta\psi))\phi = (ea_i^j ee(\zeta\psi))\phi = (ea_i^j a_t^p \delta a_t^q a_t^p \delta a_t^q(\zeta\psi))\phi$$
$$= (ea_i^j a_t^p \delta)\phi x_{t,q+p}(\delta a_t^q(\zeta\psi))\phi = (ea_i^j a_t^p \delta)\phi x_{t,q} x_{t,p}(\delta a_t^q(\zeta\psi))\phi =$$
$$(ea_i^j a_t^p \delta a_t^q)\phi(a_t^p \delta a_t^q(\zeta\psi))\phi = (ea_i^j e)\phi(e(\zeta\psi))\phi = (a_i^j \phi)(\zeta\psi\phi) = x_{i,j}\zeta = \eta.$$

Hence we have the result.

Proof of Theorem 9. By Lemma 11, ψ induces an epimorphism θ from G to H, and, if $u\psi = v\psi$ in H, then $u = u\psi\phi = v\psi\phi = v$ in G by Lemma 12, so that θ is injective. Hence θ is a group isomorphism.

5   Application of Rewriting Theorem

We will now demonstrate an application of Theorem 9. We consider semigroups defined by presentations of the form

$$\Pi = < r, s : r^3 = r, s^{a+1} = s, rs^c rs^b = s^d r > .$$

In order to prove the existence of a suitable idempotent, we need the following (somewhat technical) result :

Proposition 13 *Let S be a semigroup containing elements r and s satisfying the relations $r^3 = r$, $s^{a+1} = s$, and $rs^c rs^b = s^d r$. Let $m = gcd(a,b)$ and $n = gcd(a,d)$. Then :*

  (i) $srs^a = sr$;

  (ii) $sr^2 s^a = sr^2$;

  (iii) $r^2 s^{in} r = s^{in} r$ for $i \geq 1$;

  (iv) $srs^{im} r^2 = srs^{im}$ for $i \geq 1$;

  (v) $sr^2 s^{im} r^2 = sr^2 s^{im}$ for $i \geq 1$.

Proof. (i) Using the relations $s^{a+1} = s$ and $rs^c rs^b = s^d r$, we have that
$srs^a = s^{a+1-d} s^d rs^a = s^{a+1-d} rs^c rs^b s^a = s^{a+1-d} rs^c rs^b = s^{a+1-d} s^d r = sr$.

(ii) Using (i), we get that $sr^2 s^a = srs^a rs^a = srs^a r = sr^2$.

(iii) First note that $r^2 s^d r = r^2 rs^c rs^b = rs^c rs^b = s^d r$. Now, if $r^2 s^{kd} r = s^{kd} r$,
then $r^2 s^{(k+1)d} r = r^2 s^{kd} s^d r = r^2 s^{kd} rs^c rs^b = s^{kd} rs^c rs^b = s^{kd} s^d r = s^{(k+1)d} r$,
so that $r^2 s^{id} r = s^{id} r$ for all i, and hence, since $s^{a+1} = s$, we have that
$r^2 s^{in} r = s^{in} r$ for all i.

(iv) If $f = a - b$, then $srs^f r^2 = s^{a-d+1} s^d rs^f r^2 = s^{a-d+1} rs^c rs^a r^2 = s^{a-d+1} rs^c r$ [by (i)] $= s^{a-d+1} rs^c rs^a$ [by (i)] $= s^{a-d+1} s^d rs^f = srs^f$. Then note
that, if $srs^{kf} r^2 = srs^{kf}$, we have $srs^{(k+1)f} r^2 = srs^{kf} s^f r^2 = srs^{kf} r^2 s^f r^2 = srs^{kf} r^2 s^f = srs^{kf} s^f = srs^{(k+1)f}$. Hence $srs^{if} r^2 = srs^{if}$ for all i, and so
$srs^{im} r^2 = srs^{im}$ for all i as required.

(v) $sr^2 s^{im} r^2 = srs^a rs^{im} r^2$ [by (i)] $= srs^a rs^{im}$ [by (iv)] $= sr^2 s^{im}$.

Having established these facts, we can now find a suitable idempotent in
our semigroup.

Proposition 14 *Let S be a semigroup containing elements r and s satisfying
the relations $r^3 = r$, $s^{a+1} = s$, and $rs^c rs^b = s^d r$. Let $e = s^a r^2$. Then e is an
idempotent, and*

(i) $er^i es = er^i s$ for $1 \le i \le 2$;

(ii) $es^i er = es^i r$ for $1 \le i \le a$.

Proof. First Proposition 13 (ii) gives $e^2 = s^a r^2 s^a r^2 = s^a r^2 r^2 = e$. Then
Proposition 13 (i) and (ii) give $er^i es = s^a r^{2+i} s^a r^2 s = s^a r^{2+i} r^2 s = er^i s$, so
that (i) holds. We also have $es^i er = s^a r^2 s^i s^a r^2 r = s^a r^2 s^i r = es^i r$, so that (ii)
holds.

The upshot of Proposition 14 is that, if S is any semigroup defined by a
presentation of the form

$$\Pi = < r, s : r^3 = r, s^{a+1} = s, rs^c rs^b = s^d r >,$$

then we may apply Theorem 9 to find a presentation for the groups in the
kernel of S provided that sr generates a minimal left and a minimal right
ideal. If we let m and n be defined as in Proposition 13, then we do, in fact,
have minimal left and right ideals if $m = 1$ or $n = 1$; however, we find that
there is a unique minimal left ideal if $m = 1$, and a unique minimal right ideal
if $n = 1$, and so Corollary 8 gives that the groups in the kernel are defined
by Π. We would like to consider a case where Corollary 8 need not apply,
and, to that end, we will consider the case $b = d = 2$ with a even (so that
$m = n = 2$). We also need that c is odd; we shall see later (Theorem 22)
that, if c is even, we do not have minimal left or right ideals. However, for
odd c, we have :

Proposition 15 *Let S be the semigroup defined by the presentation*

$$\Pi = \; < r, s : r^3 = r, s^{a+1} = s, rs^c rs^2 = s^2 r >,$$

and suppose that a is even and c is odd. Then S has two minimal left ideals and three minimal right ideals.

The proof of this result is a little involved, and so we break it up into a sequence of lemmas; again, we will not repeat the full hypotheses in each case. Note that the set I of elements represented by a word containing sr is an ideal of S, and so contains any minimal left or right ideal.

Lemma 16 *s^2 commutes with sr^2 and $(rs^c r)^2 = s^a r^2$.*

Proof. By Proposition 13 (iii) and (v), we have $r^2 s^2 r = s^2 r$ and $sr^2 s^2 r^2 = sr^2 s^2$, so that $s^2 sr^2 = sr^2 s^2$. Then Proposition 13 (i) gives

$$(rs^c r)^2 = (rs^c rs^2 s^{a-2})^2 = (s^2 rs^{a-2})^2 = s^2 rs^a rs^{a-2} = s^2 r^2 s^{a-2} = s^a r^2.$$

Lemma 17 *Let L_1 and L_2 be the left ideals Ssr and $Ssrs$ respectively. Then $I = L_1 \cup L_2$.*

Proof. Since any word involving sr ends in $sr^i s^{2j} = sr^i s^{2j} r^2$ [Proposition 13 (iv) or (v)] $= sr^i s^{a-2+2j} s^2 rr = sr^i s^{a-2+2j} rs^c rs^2 r$ or $sr^i s^{2j+1} = sr^i s^{2j} r^2 s$ [Proposition 13 (iv) or (v)] $= sr^i s^{a-2+2j} rs^c rs^2 rs$, any element of I lies in L_1 or L_2.

Lemma 18 *Let R_1, R_2 and R_3 be the right ideals srS, $rsrS$ and $r^2 srS$ respectively. Then $I = R_1 \cup R_2 \cup R_3$.*

Proof. Any word in I starts with $r^i s^{2j} r$ or $r^i s^{2j+1} r$ for some $i \geq 0$ and $j \geq 1$, where we interpret r^0 as the empty word. Now $r^i s^{2j+1} r = r^i s^{2j-1} s^2 r = r^i s^{2j-1} rs^c rs^2 = ... = r^i sr\zeta$ (for some ζ) and, given this, we have that $r^i s^{2j} r = r^i s^{2j-2} s^2 r = r^i s^{2j-2} rs^c rs^2 = ... = r^i s^2 r\zeta = r^{i+1} s^c r\eta = r^{i+1} sr\vartheta$ (for some ζ, η and ϑ).

Lemma 19 *L_1 and L_2 are distinct minimal left ideals.*

Proof. We prove that, given any word ζ, there is a word η such that $\eta\zeta sr = sr$ in S; since L_1 and L_2 are Ssr and $Ssrs$, this gives minimality. We proceed by induction on the length of ζ (where we are taking length as in Lemma 10). Premultiplying by r if necessary, we may assume that ζsr is of the form $s^i rs^k r\vartheta$, $s^i r^2 s^k r\vartheta$ or $r^2 s^k r\vartheta$. We can premultiply $s^i rs^k r\vartheta$ by srs^{a-i} to get $srs^a rs^k r\vartheta = sr^2 s^k r\vartheta$ [using Proposition 13 (i)]. So we are left with the cases

$s^i r^2 s^k r\vartheta$ and $r^2 s^k r\vartheta$. There is no problem if k is even by Proposition 13 (iii); so we assume that k is odd.

If we premultiply $s^i r^2 s^k r\vartheta$ by $s^{2c-1} r^2 s^{a+1-i}$, then we get (using Lemma 16 and the fact that $c - 1$ and $a + k - c$ are both even) that $s^{2c-1} r^2 s r^2 s^k r\vartheta = s^c r^2 s^c r^2 s^k r\vartheta = s^k r^2 s^c r^2 s^c r\vartheta = s^k r (r s^c r)^2 \vartheta = s^k r s^a r^2 \vartheta = s^k r\vartheta$ [by Proposition 13 (i)]. Similarly, we may premultiply $r^2 s^k r\vartheta$ by $s^{2c-1} r^2 s$ to get $s^k r$ as required. For distinctness, note that words of the form $\zeta s r^i$ or $\zeta s r^i s^{2j}$ represent elements of L_1 and words of the form $\zeta s r^i s^{2j+1}$ represent elements of L_2. It is now clear that any relation in Π transforms a word representing an element of L_i ($i = 1$ or 2) to another word of the same form, so that the L_i are distinct.

We have a similar result for right ideals :

Lemma 20 R_1, R_2 and R_3 are distinct minimal right ideals.

Proof. As in the proof of Lemma 19, we show that, given any word ζ, there is a word η such that $sr\zeta\eta = sr$ in S. We again proceed by induction on the length of ζ.

Postmultiplying by r if necessary, we may assume that $sr\zeta$ is of the form $\vartheta s r^i s^j$ or $\vartheta s r^i s^j r^2$. Postmultiplying $\vartheta s r^i s^j$ by s^{a-j} yields $\vartheta s r^i s^a = \vartheta s r^i$ [by Proposition 13 (i) or (ii)]. If j is even, then $\vartheta s r^i s^j r^2 = \vartheta s r^i s^j$ [by Proposition 13 (iv) or (v)]; so we assume j is odd. By Lemma 16, $\vartheta s r^i s^j r^2 = \vartheta s r^i s^c r^2 s^{a+j-c}$; postmultiplying by $s^{a-j+c} r s^2 r$ yields that $\vartheta s r^i s^c r^2 s^a r s^2 r = \vartheta s r^i s^c r s^2 r$ [Proposition 13 (ii)] $= \vartheta s r^{i+1} (r s^c r s^2) r = \vartheta s r^{i+1} s^2 r^2 = \vartheta s r^{i+1} s^2$ [Proposition 13 (iv) or (v)].

For distinctness, we argue as in the proof of Lemma 19, noting that words of the form $s^{2i+1} r\zeta$ represent elements of R_1, words of the form $r s^{2i+1} r\zeta$ or $s^{2i} r\zeta = r^2 s^{2i} r\zeta$ represent elements of R_2, and words of the form $r^2 s^{2i+1} r\zeta$ or $r s^{2i} r\zeta$ represent elements of R_3.

Lemmas 19 and 20 complete the proof of Proposition 15. We may now use Theorem 9 to derive a presentation for the subgroups in the kernel of our semigroup.

Suppose that we have a semigroup S defined by the presentation

$$\Pi = \;< r, s : r^3 = r, s^{a+1} = s, r s^c r s^2 = s^2 r >,$$

where a is even and c is odd. To avoid confusion in what follows, we take $1 \leq c < a$. As in Theorem 9, using the idempotent $e = s^a r^2$ (which satisfies the appropriate hypotheses by Proposition 14), we introduce a new alphabet $B = \{x_1, x_2, y_1, y_2, ..., y_a\}$, and define $\phi : \{r, s\}^+ \to B^+$ by $r^i \phi = x_i$, $s^i \phi = y_i$, and $(\eta_1 u v \eta_2)\phi = (\eta_1 u)\phi(v\eta_2)\phi$ if $\{u, v\} = \{r, s\}$. Then the kernel K of S is a disjoint union of subgroups with presentation $< B : \Re >$, where $\Re$ is the set of relations

$$(t^i r s^c r s^2 u^j)\phi = (t^i s^2 r u^j)\phi, (t^i e)\phi = t^i \phi, (e t^i)\phi = t^i \phi \quad \text{if} \; t, u \in \{r, s\}.$$

The relations of the form $(t^i e)\phi = t^i \phi$ and $(et^i)\phi = t^i \phi$ are easily seen to be equivalent to $x_2 = y_a = 1$. The relations of the form $(t^i r s^c r s^2 u^j)\phi = (t^i s^2 r u^j)\phi$ are then

$$y_c x_1 y_2{}_{+j} = x_1 y_2 x_1 y_j, x_1 y_c x_1 y_2{}_{+j} = y_2 x_1 y_j, y_c x_1 y_2 = x_1 y_2 x_1,$$

$$x_1 y_c x_1 y_2 = y_2 x_1, y_c x_1 y_2 x_1 = x_1 y_2, x_1 y_c x_1 y_2 x_1 = y_2,$$

$$y_i x_1 y_c x_1 y_2 x_1 = y_2{}_{+i}, y_i x_1 y_c x_1 y_2 = y_2{}_{+i} x_1,$$

$$y_i x_1 y_c x_1 y_2{}_{+j} = y_2{}_{+i} x_1 y_j, x_1 y_c x_1 y_2 = y_2 x_1, y_c x_1 y_2 = x_1 y_2 x_1,$$

where we have deleted all occurrences of the trivial generator x_2 and where i and j range over $1, 2, ..., a$. The first two relations here imply that $x_1^2 = 1$. If we add this relation (and write x_1 as x to cut down on the subscripts), we find that our presentation is

$$< x, y_1, y_2, ..., y_a : x^2 = 1, y_c x y_2{}_{+j} = x y_2 x y_j (1 \le j \le a), y_c x y_2 = x y_2 x,$$
$$y_i x y_c x y_2 x = y_2{}_{+i} (1 \le i \le a), y_i x y_c x y_2{}_{+j} = y_2{}_{+i} x y_j (1 \le i, j \le a), y_a = 1 > .$$

The second and fifth relations give that $y_i y_2 = y_2{}_{+i}$, and adding this new relation makes the fifth relation redundant. The second and third relations give that $y_2{}_{+j} = y_2 y_j$, and adding this new relation makes the second relation redundant. So we now have

$$< x, y_1, y_2, ..., y_a : x^2 = 1, y_2{}_{+j} = y_2 y_j (1 \le j \le a), y_c x y_2 = x y_2 x,$$
$$y_i x y_c x y_2 x = y_2{}_{+i} (1 \le i \le a), y_i y_2 = y_2{}_{+i} (1 \le i \le a), y_a = 1 > .$$

The relations $y_i x y_c x y_2 x = y_2{}_{+i}$ are now easily seen to be redundant via the remaining relations. We may eliminate $y_3 = y_1 y_2$, $y_5 = y_1 y_2^2$, ... , $y_{a-1} = y_1 y_2^{a/2-1}$, $y_4 = y_2^2$, $y_6 = y_2^3$, ... , $y_a = y_2^{a/2}$ to get

$$< x, y_1, y_2 : x^2 = y_2^{a/2} = 1, y_1 y_2 = y_2 y_1, y_1 y_2^{(c-1)/2} x y_2 = x y_2 x > .$$

Write y_2 as z, add $t = y_1 z^{(c-1)/2}$, and delete $y_1 = t z^{(1-c)/2}$ to get :

$$< x, t, z : x^2 = 1, tz = zt, xtxz = zx, z^{a/2} = 1 >,$$

which, on eliminating $t = xzxz^{-1}x$, becomes

$$< x, z : x^2 = z^{a/2} = xzxz^{-1}(xz^{-1}xz)^2 = 1 > .$$

Let H be the group defined by this presentation. If we introduce $p = xz^{-1}xz$, we get

$$< x, z, p : x^2 = z^{a/2} = 1, p = xz^{-1}xz, xpx^{-1} = p^{-1}, zpz^{-1} = p^2 >,$$

from which it is easy to deduce that H is metabelian of order $a(2^{a/2} - 1)$. On the other hand, if G is the group defined by II, i.e. G is defined by

$$< r, s : r^2 = s^a = 1, rs^c rs^2 = s^2 r >,$$

then the last relation implies that s^c is conjugate to r in G, and so $s^{2c} = 1$. If we let $f = \gcd(\frac{a}{2}, c)$, we see that $s^{2f} = 1$. Now c is odd, so that $c \equiv f \pmod{2f}$, and so we have

$$< r, s : r^2 = s^{2f} = 1, rs^f rs^2 = s^2 r >,$$

which defines the group $H = H^{f,2,-2}$ (in the notation of [2]), which is metabelian of order $2f(2^f - 1)$ by Theorem 8.2 of [2]. So we have proved

Theorem 21 *Let S be the semigroup and G be the group defined by the presentation*

$$< r, s : r^3 = r, s^{a+1} = s, rs^c rs^2 = s^2 r >,$$

where a is even, c is odd and $1 \le c < a$, and let $f = \gcd(\frac{a}{2}, c)$. Then G is metabelian of order $2f(2^f - 1)$, while the kernel of S is a disjoint union of six copies of a metabelian group H of order $a(2^{a/2} - 1)$. In particular, H is isomorphic to G if and only if $a = 2c$.

6   Limitations of Rewriting Theorem

It should be pointed out, however, that Theorem 9 does not apply to all semigroups defined by presentations of the form

$$< r, s : r^3 = r, s^{a+1} = s, rs^c rs^b = s^d r >$$

as the next result shows.

Theorem 22 *Let S be the semigroup defined by the presentation*

$$\Pi = < r, s : r^3 = r, s^{a+1} = s, rs^c rs^b = s^d r >,$$

and suppose that $\gcd(a, b, c, d) > 1$. Then S does not have minimal left or right ideals.

Proof. Let $e = \gcd(a, b, c, d)$, and add the relations $s^{e+1} = s$, $r^2 = r$ and $rs^e = s^e r = r$ to Π to get

$$< r, s : r^2 = r, s^{e+1} = s, rs^e = s^e r = r > .$$

This is a monoid M with identity s^e, and, as a monoid, is presented by $< r, s : r^2 = r, s^e = 1 >$. M does not have minimal left or right ideals, and, since M is a homomorphic image of S, the result follows.

Theorem 22 explains why the assumption that c is odd is rather important in Proposition 15. We should also point out that a semigroup may not possess a suitable idempotent even when it does have minimal left and right ideals. For example, if we consider the semigroup S defined by the presentation

$$< r, s : r^3 = r, s^3 = s, (rs)^3 = (rs)^2, rs^2rsr^2s = rs^2r^2s,$$
$$srs^2rsr = (sr)^2, sr^2s^2r^2sr = sr^2sr, rs^2rsrs = rs^2rs, rsrs^2rs = rsrs,$$
$$(s^2r^2)^3 = s^2r^2, (rs^2)^3 = (rs^2)^2, (sr^2)^2(sr)^2(rs)^2r^2s^2 = (sr^2)^2s^2 >,$$

we notice that a word of the form $u^iv^j\alpha w^k x^l$, where $\{u, v\} = \{w, x\} = \{r, s\}$, can only represent the same element of S as another word if that word is also of the form $u^iv^j\beta w^k x^l$ for some β. So there cannot be an idempotent e such that $er^ies = er^is$ and $es^ier = es^ir$. However, we used the programs described in [12] to show that S is a finite semigroup of order 224, and so does possess minimal left and right ideals. In fact, the kernel of S has order 144, there being 6 minimal right and 12 minimal left ideals, so that the kernel is the union of 72 subgroups of order 2.

Acknowledgement. The fourth author would like to thank Hilary Craig for all her help and encouragement.

References

[1] S. I. Adjan, Defining relations and algorithmic problems for groups and semigroups, *Proceedings of the Steklov Institute of Mathematics* **85** (1966) (American Mathematical Society, Providence, Rhode Island, 1967); translated from *Trudy. Mat. Inst. Steklov.* **85** (1966).

[2] C. M. Campbell, H. S. M. Coxeter and E. F. Robertson, Some families of finite groups having two generators and two relations, *Proc. Royal Soc. London* **357A** (1977), 423–438.

[3] C. M. Campbell, E. F. Robertson, N. Ruškuc and R. M. Thomas, Fibonacci Semigroups, *J. Pure Appl. Algebra*, to appear.

[4] C. M. Campbell, E. F. Robertson, N. Ruškuc and R. M. Thomas, Semigroup and group presentations, *Bull. London Math. Soc.*, to appear.

[5] C. M. Campbell, E. F. Robertson, and R. M. Thomas, On a class of semigroups with symmetric presentations, *Semigroup Forum* **46** (1993), 286–306.

[6] C. M. Campbell, E. F. Robertson and R. M. Thomas, Semigroup presentations and number sequences, *in* G. E. Bergum, A. N. Phillipou and A. F. Horadam (eds.), *Applications of Fibonacci Numbers* (Kluwer Academic Publishers, 1993), 77–83.

[7] A. H. Clifford, Semigroups containing minimal ideals, *Amer. J. Math.* **70** (1948), 521–526.

[8]   H. S. M. Coxeter and W. O. J. Moser, *Generators and Relations for Discrete Groups* (Springer-Verlag, Berlin-Heidelberg-New York, 1980).

[9]   J. A. Green, On the structure of semigroups, *Ann. of Math.* **54** (1951), 163–172.

[10]   J. H. Remmers, On the geometry of semigroup presentations, *Adv. Math.* **36** (1980), 283–296.

[11]   R. M. Thomas, The Fibonacci groups revisited, *in* C. M. Campbell and E. F. Robertson (eds.), *Groups St Andrews 1989, vol. 2* (London Math. Soc. Lecture Note Series **160**, Cambridge University Press, 1991), 445–454.

[12]   T. G. Walker, *Semigroup Enumeration - Computer Implementation and Applications* (Ph.D. Thesis, University of St Andrews, 1992).

C M Campbell, E F Robertson   R M Thomas
and N Ruškuc   Department of Mathematics
Mathematical Institute   and Computer Science
University of St Andrews   University of Leicester
St Andrews, Fife KY16 9SS   Leicester LE1 7RH
Scotland   England

Generalised trees and Λ-trees

I. M. CHISWELL

1. The idea of a Λ-tree, where Λ is an ordered abelian group, was introduced in [9]. We shall reproduce the definition shortly, but for an account of the basic theory of Λ-trees we refer to [1]. In the special case Λ = ℤ, ℤ-trees are closely related to simplicial trees (trees in the ordinary graph-theoretic sense). The connection is spelt out in Lemma 4 below, which shows that Λ-trees may be viewed as generalisations of simplicial trees. However, there are other notions of generalised tree in the literature, and our purpose here is to consider two of these, and their relation to Λ-trees.

Firstly there is what we call an order tree. This is a partially ordered set $(P, \leq)$ such that the set of predecessors of any element is linearly ordered, that is, for all $x, y, z \in P$, if $x \leq z$ and $y \leq z$, then either $x \leq y$ or $y \leq x$. It is also convenient to assume that P has a least element (this can always be arranged just by adding one). By choosing a point in a Λ-tree, it is possible to make the Λ-tree into an order tree. We shall show that, conversely, any order tree $(P, \leq)$ can be embedded in a Λ-tree for some suitable Λ, so that the ordering on P is induced from the ordering on the Λ-tree defined by the (image of) the least element of P. Order trees occur in set theory (see, for example, [7]), but our interest in them stems from the fact that that they occur in the theory of pregroups ([6], [10]).

The other notion of generalised tree we shall consider was introduced by Herrlich in [5], and is obtained by writing down axioms satisfied by the set of segments in a Λ-tree. We shall show that any tree in the Herrlich sense can be embedded in a Λ-tree in such a way that segments are preserved (we shall give a precise statement later).

Our construction makes use of ultraproducts, and for a description of these and their relevant properties, see §3 in [11] or Ch. 5 in [2]. We also show that a countable order tree can be embedded in an ℝ-tree, using more conventional limiting processes, and there is a corresponding result for Herrlich trees.

2. In this section we collect together the results we shall need to prove the embedding results. We begin by giving the definition of a Λ-tree. If Λ is a (totally) ordered abelian group, a Λ-metric on a set X is a mapping $d : X \times X \to \Lambda$ satisfying the usual axioms for a metric with values in ℝ, and given such a metric the pair (X, d) is called a Λ-metric space. The mapping $\Lambda \times \Lambda \to \Lambda$ given by $(a, b) \mapsto |a - b|$, where $|x| = \max\{x, -x\}$, makes Λ itself into a Λ-metric space. A segment in an arbitrary Λ-metric space (X, d) is the image of an isometry $\alpha : [a, b] \to X$, where $[a, b] = \{x \in \Lambda; a \leq x \leq b\}$ (and $a \leq b$). The endpoints of the segment are $\alpha(a)$ and $\alpha(b)$. A Λ-metric space

(X, d) is *geodesic* if, for all $x, y \in X$, there is a segment in X with endpoints x and y.

Definition. A Λ-metric space (X, d) is a Λ-tree if

(a) it is geodesic

(b) the intersection of two segments with a common endpoint is a segment

(c) if two segments intersect in a single point, which is an endpoint of both, then their union is a segment.

This is the definition as given in [8]. It is a consequence of (b) that given x, y in X, there is a unique segment in X whose set of endpoints is $\{x, y\}$. A Λ-metric space with this property will be called geodesically linear, and the unique segment with set of endpoints $\{x, y\}$ will be denoted by $[x, y]$. The next result is well-known, and the proof is left as an exercise. We shall only use it in the case that (X, d) is a Λ-tree, a special case proved in Cor. 2.12 of [1].

Lemma 1. *Let (X, d) be a geodesically linear Λ-metric space, and let $x, y \in X$. Then a point $z \in X$ is in the segment $[x, y]$ if and only if $d(x, y) = d(x, z) + d(z, y)$.*

It will be convenient to use another characterisation of Λ-trees. Let (X, d) be a Λ-tree, and let $\delta \in \Lambda$, with $\delta \geq 0$. Choose a basepoint $p \in X$, and define, for $x, y \in X$,

$$x \cdot y = \frac{1}{2}(d(x, p) + d(y, p) - d(x, y)),$$

an element of $\frac{1}{2}\Lambda$. Following Gromov, we call (X, d) δ-hyperbolic with respect to p if, for all x, y and $z \in X$, $x \cdot y \geq \min\{x \cdot z, y \cdot z\} - \delta$. If (X, d) is δ-hyperbolic with respect to one point, then it is 2δ-hyperbolic with respect to any other point (see Proposition 1.2 in [4]). Consequently it makes sense to speak of a 0-hyperbolic space without reference to a basepoint. Also, (X, d) is δ-hyperbolic with respect to every point in X if and only if, for all $x, y, z, t \in X$,

$$d(x, y) + d(z, t) \leq \max\{d(x, z) + d(y, t), \ d(x, t) + d(y, z)\} + 2\delta.$$

See Proposition 1.6 in [4]. Consequently, (X, d) is 0-hyperbolic if and only if it satisfies this condition with $\delta = 0$ (the so-called four-point condition). A geometric interpretation of this is given by the "H-Proposition" (2.15 in [1]).

Λ-trees can be characterised as follows [1; Theorem 3.17]. It can be checked directly that, if condition (ii) in the the lemma below holds for one choice of basepoint, then it holds for any other choice of basepoint.

Lemma 2. *The Λ-metric space (X, d) can be isometrically embedded in a Λ-tree if and only if*

(i) *(X, d) is 0-hyperbolic.*

(ii) *For all $x, y \in X$, $x \cdot y \in \Lambda$.*

Further, (X, d) is a Λ-tree if and only if (i) and (ii) hold, and additionally

(iii) *(X, d) is geodesic.*

Let T be a simplicial tree, Λ an ordered abelian group, and suppose there is a mapping w assigning to every edge e of T an element $w(e) \in \Lambda$ with $w(e) > 0$. Then (T, w) is called a weighted tree. (We may view w as defined on unoriented edges; alternatively, if we view unoriented edges as consisting of two oppositely oriented edges $e, \bar{e}$, we require $w(e) = w(\bar{e})$.) Given vertices u, v in T, let $e_1, \ldots, e_n$ be the edges of the unique reduced path in T from u to v, and define

$$d(u, v) = \sum_{i=1}^{n} w(e_i)$$

(and $d(u, u) = 0$). Then it is easily verified that d is a Λ-metric.

Lemma 3. *Let X be the set of vertices of a weighted tree T. Then with the metric d just defined, (X, d) satisfies (i) and (ii) of Lemma 2, so can be isometrically embedded in a Λ-tree.*

Proof. Let p, x, y be vertices of T. The reduced paths from p to x and from p to y intersect in a reduced path, say $e_1, \ldots, e_k$, and with respect to p as basepoint, $x \cdot y = \sum_{i=1}^{k} w(e_i)$. The lemma follows easily from this.

In the special case $\Lambda = \mathbb{Z}$, and $w(e) = 1$ for all edges e, the metric d is called the path metric on T. The next result is Theorem 10 of [8], to which we refer for hints on the proof (but note that it is Axiom (c) for a Λ-tree, not (b), which implies there are no loops in the tree, indeed Axiom (b) is redundant when $\Lambda = \mathbb{Z}$).

Lemma 4. *Let (X, d) be a $\mathbb{Z}$-metric space. Then (X, d) is a $\mathbb{Z}$-tree if and only if there is a simplicial tree T such that X is the set of vertices of T, and d is the path metric on X.*

NOTE. The proof of Lemma 3 shows that for any $x, y \in X$, the set of vertices of T which belong to the segment $[x, y]$ in X is the set of all vertices of T on the reduced path in T from x to y. For in the proof, the point p lies in $[x, y]$ if and only if $x \cdot y = 0$, by Lemma 1. This happens if and only if $k = 0$, which is equivalent to saying that p lies on the reduced path between x and y. This applies in particular to the path metric in Lemma 4.

The final result we need is essentially that an ultraproduct of Λ-trees is a
*Λ-tree, where *Λ is an ultrapower of Λ. This is all we need, but with a view
to possible further applications, we shall be rather more general. Let I be an
index set and let $\mathcal{D}$ be an ultrafilter in $\mathcal{P}(I)$, the Boolean algebra of all subsets
of I. If $\{X_i; i \in I\}$ is a family of sets, we shall denote the equivalence class
of an element $(x_i)_{i \in I}$ of $\prod_{i \in I} X_i$ in the ultraproduct $\prod_{i \in I} X_i / \mathcal{D}$ by $\langle x_i \rangle_{i \in I}$.

For every $i \in I$ let G_i be a group acting as isometries on a Λ_i-tree (X_i, d_i),
where Λ_i is an ordered abelian group. Put

$$G = \prod_{i \in I} G_i / \mathcal{D}, \ \ \Lambda = \prod_{i \in I} \Lambda_i / \mathcal{D}, \ \ X = \prod_{i \in I} X_i / \mathcal{D}$$

so that G is a group, Λ is an ordered abelian group and X is a Λ-metric space
with metric $d = \prod_{i \in I} d_i / \mathcal{D}$, that is,

$$d(\langle x_i \rangle_{i \in I}, \langle y_i \rangle_{i \in I}) = \langle d_i(x_i, y_i) \rangle_{i \in I}.$$

These statements can be verified directly, and are manifestations of Łoš's
Theorem (see [2] or [11]). Further, the actions of G_i on X_i induce an action
of G on X as isometries. Let ℓ_i denote the hyperbolic length function for the
action of G_i on X_i (see §6 and Prop. 7.1 in [1]). We recall that an action of
a group as isometries on a Λ-tree is called free if every non-identity element
of the group is hyperbolic, that is, has positive hyperbolic length.

Lemma 5. *In the situation just described, (X, d) is a Λ-tree, and if ℓ is the
hyperbolic length function for the action of G on X, then $\ell = \prod_{i \in I} \ell_i / \mathcal{D}$. In
particular, if G_i acts freely on X_i for almost all i (i.e. the set of i for which
G_i acts freely belongs to $\mathcal{D}$), then G acts freely on X.*

Proof. It is easily checked that properties (i)-(iii) of Lemma 2 are inherited by
X from the X_i, so X is a Λ-tree. (For a hint on the proof that X satisfies (iii),
see the proof of Lemma 3(i) in [3]). Choose a basepoint p_i in X_i, and take
$p = \langle p_i \rangle_{i \in I}$ as basepoint in X. Then the Lyndon length function $L_i = L_{p_i}$ (see
5.1 and 5.3 in [1]) is given by $L_i(g_i) = d_i(p_i, g_i p_i)$ for $g_i \in G_i$. Consequently,
if $g = \langle g_i \rangle_{i \in I}$ is in G, then the Lyndon length function L_p is given by $L_p(g) =
\langle L_i(g_i) \rangle_{i \in I}$. Hence, by Prop. 7.1(c) of [1],

$$\begin{aligned} \ell(g) &= \max\{L_p(g^2) - L_p(g), 0\} \\ &= \langle \max\{L_i(g_i^2) - L_i(g_i), 0\} \rangle_{i \in I} \\ &= \langle \ell_i(g_i) \rangle_{i \in I} \end{aligned}$$

that is, $\ell = \prod_{i \in I} \ell_i / \mathcal{D}$ as claimed. The last part of the lemma follows immediately.

3. We now prove our embedding results for order trees. Let (X, d) be a
Λ-tree and choose a basepoint $x_0 \in X$. We define a binary relation $\le$ on X
by: $x \le y$ if and only if $x \in [x_0, y]$. It is not difficult to check that this makes
$(X, \le)$ into an order tree with least element x_0. We shall not give the details
since we shall prove a more general result later. We call this the ordering on
X relative to x_0.

If $(P, \leq)$ is an order tree with P finite, and with least element p, then P can be made into the vertex set of a simplicial tree, joining two elements x, y of P by an edge when $x < y$, but there is no element $z \in P$ such that $x < z < y$, where $<$ is the strict partial order corresponding to $\leq$ (see §2 of [10]). This defines a $\mathbb{Z}$-tree (P, d) by Lemma 4, and it is easily checked that the ordering $\leq$ is the ordering obtained from this $\mathbb{Z}$-tree as described in the previous paragraph with $x_0 = p$. We are now in a position to prove our first main result.

Theorem 1. *Let $(P, \leq)$ be an order tree with least element p. Then there exist an ordered abelian group Λ, a Λ-tree (X, d) and a $1-1$ mapping $\phi : P \to X$ such that, if X is given the ordering relative to $\phi(p)$, then $x \leq y$ if and only if $\phi(x) \leq \phi(y)$, for all $x, y \in P$.*

Proof. Let Σ be the set of all finite subsets of P which contain p. If $Q \in \Sigma$, restricting the ordering of P makes Q into a finite order tree, and as above, Q becomes a $\mathbb{Z}$-tree, with metric d_Q, say, and the ordering is the ordering on this $\mathbb{Z}$-tree relative to p.

For $Q \in \Sigma$, let $a_Q = \{R \in \Sigma; Q \subseteq R\}$. Then if $Q' \in \Sigma$, $a_{Q \cup Q'} \subseteq a_Q \cap a_{Q'}$, and $a_Q \neq \emptyset$ since $Q \in a_Q$. Therefore there is an ultrafilter $\mathcal{D}$ in $\mathcal{P}(\Sigma)$ such that $a_Q \in \mathcal{D}$ for all $Q \in \Sigma$. (See, for example, Cor. 3.5, Ch. 1 in [2]).

Let $X = \prod_{Q \in \Sigma} Q/\mathcal{D}$ with metric $d = \prod_{Q \in \Sigma} d_Q/\mathcal{D}$. Then by Lemma 5, (X, d) is a Λ-tree, where $\Lambda = \mathbb{Z}^\Sigma/\mathcal{D}$. For $Q \in \Sigma$, define $\phi_Q : P \to Q$ by

$$\phi_Q(x) = \begin{cases} x, & \text{if } x \in Q \\ p, & \text{otherwise} \end{cases}$$

and define $\phi : P \to X$ by $\phi(x) = \langle \phi_Q(x) \rangle_{Q \in \Sigma}$. Since $\phi_Q(x) = x$ for all $Q \in a_{\{p,x\}}$, and $a_{\{p,x\}} \in \mathcal{D}$, we have $\phi(x) = \langle y_Q \rangle_{Q \in \Sigma}$ if and only if $y_Q = x$ for almost all Q (that is, the set of Q for which $y_Q = x$ belongs to $\mathcal{D}$). It follows that ϕ is $1 - 1$.

Suppose that $x, y \in P$ and $x \leq y$. It follows by Lemma 1 that, for all $Q \in a_{\{p,x,y\}}$,

$$d_Q(p, y) = d_Q(p, x) + d_Q(x, y)$$

that is,

$$d_Q(\phi_Q(p), \phi_Q(y)) = d_Q(\phi_Q(p), \phi_Q(x)) + d_Q(\phi_Q(x), \phi_Q(y))$$

because $x, y, p \in Q$. Since $a_{\{p,x,y\}} \in \mathcal{D}$,

$$d(\phi(p), \phi(y)) = d(\phi(p), \phi(x)) + d(\phi(x), \phi(y))$$

hence $\phi(x) \leq \phi(y)$ in the ordering of X relative to $\phi(p)$, again by Lemma 1.

Conversely, if $\phi(x) \leq \phi(y)$, we have

$$d(\phi(p), \phi(y)) = d(\phi(p), \phi(x)) + d(\phi(x), \phi(y))$$

which means that, for some $A \in \mathcal{D}$,

$$d_Q(\phi_Q(p), \phi_Q(y)) = d_Q(\phi_Q(p), \phi_Q(x)) + d_Q(\phi_Q(x), \phi_Q(y))$$

for all $Q \in A$. Now $A \cap a_{\{p,x,y\}} \neq \emptyset$ since it belongs to $\mathcal{D}$, so if $Q \in A \cap a_{\{p,x,y\}}$, we have $d_Q(p, y) = d_Q(p, x) + d_Q(x, y)$, and again by Lemma 1, $x \leq y$.

In the case that P is countable, we can show, by a different argument, that we may take $\Lambda = \mathbb{R}$ in Theorem 1.

Theorem 2. *Let $(P, \leq)$ be an order tree with least element p, and assume P is countable. Then there exist an $\mathbb{R}$-tree (X, d) and a 1–1 mapping $\phi : P \to X$ such that, if X is given the ordering relative to $\phi(p)$, then $x \leq y$ if and only if $\phi(x) \leq \phi(y)$, for all x, $y \in P$.*

Proof. Enumerate the elements of P as $x_0, x_1, x_2, \ldots$, where $p = x_0$. We shall define recursively a sequence of weighted simplicial trees (X_n, w_n) for $n = 0, 1, \ldots$, where the weight function w_n takes values in $\Lambda = \mathbb{Z}[1/2]$. By Lemma 3, the set of vertices $V(X_n)$ of X_n has a Λ-metric d_n and can be isometrically embedded in a Λ-tree, giving an induced ordering $\leq_n$ on $V(X_n)$. Explicitly, by the note after Lemma 4, $x \leq_n y$ if and only if x is a vertex on the reduced path from p to y in X_n. This sequence of trees will have the following properties:

(i) $V(X_n) = \{x_0, \ldots, x_n\}$

(ii) the ordering $\leq_n$ is the restriction of the ordering on P to $V(X_n)$

(iii) for all edges e of X_n, $w_n(e) \geq 1/2^{n-1}$

(iv) for u, $v \in V(X_{n-1})$, $d_{n-1}(u, v) \leq d_n(u, v) \leq d_{n-1}(u, v) + (1/2^{n-1})$.

Define X_0 to be the tree with one vertex x_0 and no edges, and X_1 to have two vertices x_0, x_1 with a single edge e joining them, with $w_1(e) = 1$. Suppose that X_{n-1} has been defined and satisfies (i)-(iv). Let x be the largest element of $\{x_0, \ldots x_{n-1}\}$ such that $x \leq x_n$. Let $y_1, \ldots, y_m$ be the minimal elements of $\{z \in V(X_{n-1}); x_n \leq z\}$ (note that $m = 0$ is possible). Then X_{n-1} has a subgraph of the following form:

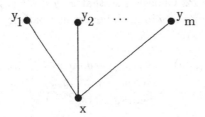

Let $\alpha_i = w_{n-1}(e_i)$, where e_i is the edge joining x and y_i. Modify X_{n-1} to obtain X_n by "stretching" the vertex x to an edge e joining x to x_n, so that e_i now joins y_i to x_n. Put $w_n(e_i) = \alpha_i$ and $w_n(e) = 1/2^{n-1}$, leaving X_{n-1} unchanged otherwise. Thus X_n has a subgraph of the form:

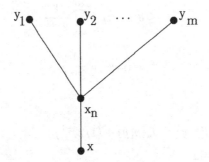

It is easily checked that properties (i)-(iv) are satisfied by X_n.

If $u, v \in P$, then $d_n(u, v)$ is defined for sufficiently large n, and by Property (iv), $(d_n(u, v))$ is a monotone increasing Cauchy sequence. Define $d(u, v) = \lim_{n \to \infty} d_n(u, v)$, so $d(u, v) \geq d_n(u, v)$. By Property (iii), if $u \neq v$ and $u, v \in V(X_n)$, then $d_n(u, v) \geq 1/2^{n-1} > 0$, so $d(u, v) > 0$. The other axioms for a metric are easily checked and (P, d) is an $\mathbb{R}$-metric space. Further, if $x, y, z, t \in P$, they all belong to $V(X_n)$ for sufficiently large n, and by Lemma 3,

$$d_n(x, y) + d_n(z, t) \leq \max\{d_n(x, z) + d_n(y, t), \; d_n(x, t) + d_n(y, z)\}$$

and taking limits as $n \to \infty$, we see that (P, d) is 0-hyperbolic, hence by Lemma 2 there is an isometric embedding of P into an $\mathbb{R}$-tree, say $\phi : P \to X$. For the rest of the proof we shall identify x with $\phi(x)$, for $x \in P$, and denote the metric on X by d. We give X the ordering relative to $x_0 = p$.

Suppose $x \leq y$ in P. For sufficiently large n we have $x, y \in V(X_n)$. By Property (ii) and Lemma 1, $d_n(x_0, y) = d_n(x_0, x) + d_n(x, y)$, and taking limits shows that $x \leq y$ in X, again by Lemma 1.

Suppose that $x \not\leq y$ in P. Choose n so that $x, y \in V(X_n)$. The reduced paths from p to x and from p to y in X_n intersect in a path from p to a vertex q, say, and $x \neq q$. By definition of d_n,

$$
\begin{aligned}
d_n(p, x) + d_n(x, y) &= d_n(p, q) + d_n(q, y) + 2d_n(q, x) \\
&= d_n(p, y) + 2d_n(q, x) \\
&\geq d_n(p, y) + (2/2^{n-1})
\end{aligned}
$$

by Property (iii). That is

$$
d_n(p, y) \leq d_n(p, x) + d_n(x, y) - (1/2^{n-2}) \tag{1}
$$

By Property (iv), for $k \geq 1$,

$$
\begin{aligned}
d_{n+k}(p, y) &\leq d_n(p, y) + \sum_{i=0}^{k-1} \frac{1}{2^{n+i}} \\
&\leq d_n(p, y) + \sum_{i=0}^{\infty} \frac{1}{2^{n+i}} \\
&= d_n(p, y) + \frac{1}{2^{n-1}}.
\end{aligned}
$$

Let $k \to \infty$ to get

$$
d(p, y) \leq d_n(p, y) + (1/2^{n-1}) \tag{2}
$$

From (1) and (2) we obtain

$$
\begin{aligned}
d(p, y) &\leq d_n(p, x) + d_n(x, y) - (1/2^{n-1}) \\
&\leq d(p, x) + d(x, y) - (1/2^{n-1}) \\
&< d(p, x) + d(x, y).
\end{aligned}
$$

By Lemma 1, $x \not\leq y$ in X. This proves the theorem.

4. It remains to prove our embedding results for trees in the sense of Herrlich. We recall the definition [5; (2.1)].

Definition. A Herrlich tree is a pair (V, S) where V is a set and $S : V \times V \to \mathcal{P}(V)$ is a mapping satisfying:

(i) $S(a, a) = \{a\}$ for all $a \in V$;

(ii) $\{a, b\} \subseteq S(a, b)$ for all $a, b \in V$;

(iii) $S(a, b) = S(b, a)$ for all $a, b \in V$;

(iv) for all a, b and $c \in V$, the set $S(a, b) \cap S(b, c) \cap S(c, a)$ has exactly one member, denoted by $\mu(a, b, c)$;

(v) for all $a, b \in V$, if $c \in S(a, b)$ then $S(a, b) = S(a, c) \cup S(c, b)$.

Any Λ-tree (X, d) gives rise to a Herrlich tree (V, S) with $V = X$ and $S(a, b) = [a, b]$. Clearly Axioms (i)-(iii) are satisfied, (iv) follows from 2.12 in [1] (where $\mu(a, b, c)$ is denoted by $Y(a, b, c)$), and Axiom (v) follows from 2.13 in [1].

Given a Herrlich tree (V, S), choose a basepoint $p \in V$, and define a binary relation $\leq$ on V by: $x \leq y$ if and only if $x \in S(p, y)$. We shall show this makes $(V, \leq)$ into an order tree. This generalises the ordering on a Λ-tree defined at the start of §3.

Firstly, $x \leq x$ for all $x \in V$ by Axiom (ii). Secondly, if $x \leq y$ and $y \leq x$, then by the remark after Axiom (v) in [5], $S(p, x) \cap S(x, y) = \{x\}$. But $y \in S(p, x)$ by assumption and $y \in S(x, y)$ by Axiom (ii), hence $x = y$. Suppose $x \leq y$ and $y \leq z$. Then $x \in S(p, y)$ and $y \in S(p, z)$, and by Axiom (v), $S(p, z) = S(p, y) \cup S(y, z)$, so $x \in S(p, z)$, i.e. $x \leq z$. Thus $\leq$ is a partial ordering on V.

Now suppose $x, y \leq z$, so $x, y \in S(p, z)$, and assume $x \not\leq y$ and $y \not\leq x$. By Axiom (v), $S(p, z) = S(p, x) \cup S(x, z)$, so $y \in S(x, z)$. Similarly $x \in S(y, z)$. Hence by Axiom (ii), $x, y \in S(x, z) \cap S(x, y) \cap S(y, z)$, so by Axioms (iii) and (iv), $x = y$, a contradiction. Thus either $x \leq y$ or $y \leq x$.

We could now obtain an embedding result for Herrlich trees by appealing to Theorem 1. However, we prefer to give a direct argument, similar to that of Theorem 1, which does not depend on a choice of basepoint. First we need to prove two lemmas.

Lemma 6. *Let (V, S) be a Herrlich tree, suppose that $x, y \in V$ and $u \in S(x, y)$. Then for any $z \in V$, either $S(u, z) \subseteq S(x, z)$ or $S(u, z) \subseteq S(y, z)$.*

Proof. Take $z \in V$, and let $v = \mu(x, y, z)$. Since $v \in S(x, y)$, we have $S(x, y) = S(x, v) \cup S(y, v)$ by Axiom (v), so either $u \in S(x, v)$ or $u \in S(y, v)$.

If $u \in S(x, v)$ then since $v \in S(x, z)$, $S(x, z) = S(x, v) \cup S(v, z)$ by Axiom (v), so $u \in S(x, z)$. Hence, again by Axiom (v), $S(x, z) = S(x, u) \cup S(u, z)$, so $S(u, z) \subseteq S(x, z)$. Similarly, if $u \in S(y, v)$ then $S(u, z) \subseteq S(y, z)$.

In the next lemma we use the term subtree as defined in (2.2) of [5].

Lemma 7. *Let (V, S) be a Herrlich tree and let X be a finite subset of V. Then there is a subtree (V', S') of (V, S) with V' finite and $X \subseteq V'$.*

Proof. Define $V' = \{\mu(x, y, z); x, y, z \in X\}$. Plainly V' is finite, and $X \subseteq V'$ since $x = \mu(x, x, x)$. Let u, v, $w \in V'$. Then for some elements x_i, y_i, z_i $(1 \leq i \leq 3)$ of X, we have $u = \mu(x_1, x_2, x_3)$, $v = \mu(y_1, y_2, y_3)$ and $w = \mu(z_1, z_2, z_3)$. We claim that, after suitable renumbering,

$$S(u, v) \cap S(v, w) \cap S(w, u) \subseteq S(x_1, y_1) \cap S(y_1, z_1) \cap S(z_1, x_1)$$

hence by Axiom (iv) $\mu(u, v, w) = \mu(x_1, y_1, z_1)$, in particular $\mu(u, v, w) \in V'$.

Now $v \in S(y_i, y_j)$ for $i \neq j$, so by Lemma 6, $S(u, v) \subseteq S(u, y_i)$ for at least two values of i. Similarly $S(v, w) \subseteq S(y_i, w)$ for at least two values of i. Since there are only three possible values of i, after renumbering the y_i, we may assume that $S(u, v) \subseteq S(u, y_1)$ and $S(v, w) \subseteq S(y_1, w)$. Again by Lemma 6, since $u \in S(x_i, x_j)$ for $i \neq j$, we have $S(u, y_1) \subseteq S(x_i, y_1)$ for at least two values of i. Similarly, $S(y_1, w) \subseteq S(y_1, z_i)$ for at least two values of i. Therefore, after renumbering the x_i and the z_i, we may assume that $S(u, v) \subseteq S(x_1, y_1) \cap S(x_2, y_1)$ and $S(v, w) \subseteq S(y_1, z_1) \cap S(y_1, z_2)$.

Once again by Lemma 6, $S(u, w) \subseteq S(x_i, w)$ for at least two values of i, so renumbering x_1 and x_2 if necessary we can assume that $S(u, w) \subseteq S(x_1, w)$. Also, $S(x_1, w) \subseteq S(x_1, z_i)$ for at least two values of i, and renumbering z_1, z_2 if necessary, we can assume that $S(u, w) \subseteq S(x_1, z_1)$. This establishes the claim.

By the remark after Definition 2.2 in [5], we obtain a subtree (V', S') by defining $S'(a, b) = S(a, b) \cap V'$, for a, $b \in V'$.

REMARK. If (V', S') is a subtree of a Herrlich tree (V, S), and a, $b \in V'$, then $S'(a, b) = S(a, b) \cap V'$. For if $v \in S(a, b) \cap V'$, then

$$S'(a, b) \cap S'(a, v) \cap S'(v, b) \subseteq S(a, b) \cap S(a, v) \cap S(v, b) = \{v\}$$

and since (V', S') satisfies Axiom (iv), $v \in S'(a, b)$, so $S(a, b) \cap V' \subseteq S'(a, b)$, and the reverse inclusion is immediate from the definition of subtree. Thus if (V', S') is a subtree, then S' is determined by V'.

We can now prove our main result on Herrlich trees.

Theorem 3. *Let (V, S) be a Herrlich tree. Then there exist an ordered abelian group Λ, a Λ-tree (X, d) and a $1 - 1$ mapping $\phi : V \to X$ such that, for all a, $b \in V$, $\phi(S(a, b)) = [\phi(a), \phi(b)] \cap \phi(V)$.*

Proof. Let Σ be the set of subtrees (V', S') of (V, S) with V' finite. For every finite subset X of V, let

$$a_X = \{(V', S') \in \Sigma; X \subseteq V'\}.$$

Then if Y is also a finite subset of V, $a_{X \cup Y} \subseteq a_X \cap a_Y$, and $a_X \neq \emptyset$ by Lemma 7. As in Theorem 1, there is an ultrafilter $\mathcal{D}$ in $\mathcal{P}(\Sigma)$ such that $a_X \in \mathcal{D}$ for all finite subsets X of V.

For $(V', S') \in \Sigma$, let $d_{V'}$ be the path metric on V' derived from its structure as a simplicial tree, as described in Lemma (2.1)(b) of [5]. By Lemma 4, $(V', d_{V'})$ is a $\mathbb{Z}$-tree.

Put $X = \prod_{(V',S')\in\Sigma} V'/\mathcal{D}$ with metric $d = \prod_{(V',S')\in\Sigma} d_{V'}/\mathcal{D}$. By Lemma 5, (X, d) is a Λ-tree, where $\Lambda = \mathbb{Z}^{\Sigma}/\mathcal{D}$. Let $(V', S') \in \Sigma$, and choose $x_{V'} \in V'$. Define $\phi_{V'} : V \to V'$ by

$$\phi_{V'}(x) = \begin{cases} x, & \text{if } x \in V' \\ x_{V'}, & \text{otherwise} \end{cases}$$

and define $\phi : V \to X$ by $\phi(x) = \langle \phi_{V'}(x)\rangle_{(V',S')\in\Sigma}$. Now for $x \in V$ and $(V', S') \in \Sigma$, $x \in V'$ if and only if $(V', S') \in a_{\{x\}}$, hence $\phi_{V'}(x) = x$ for all $(V', S') \in a_{\{x\}}$. Since $a_{\{x\}} \in \mathcal{D}$, we have $\phi(x) = \langle y_{(V',S')}\rangle_{(V',S')\in\Sigma}$ if and only if $y_{(V',S')} = x$ for almost all (V', S'). It follows that ϕ is $1 - 1$, and is independent of the choice of the points $x_{V'}$.

Suppose that $z \in S(x, y)$, where $x, y \in V$. Then for $(V', S') \in a_{\{x,y,z\}}$, $z \in S'(x, y)$ by the remark preceding the theorem. Also, by Lemma 2.1 in [5] and the note after Lemma 4, $S'(x, y)$ is the segment $[x, y]$ in the $\mathbb{Z}$-tree $(V', d_{V'})$, hence by Lemma 1, for all $(V', S') \in a_{\{x,y,z\}}$,

$$d_{V'}(x, y) = d_{V'}(x, z) + d_{V'}(z, y).$$

Since $x, y, z \in V'$ this is the same as

$$d_{V'}(\phi_{V'}(x), \phi_{V'}(y)) = d_{V'}(\phi_{V'}(x), \phi_{V'}(z)) + d_{V'}(\phi_{V'}(z), \phi_{V'}(y)).$$

Since $a_{\{x,y,z\}} \in \mathcal{D}$, it follows that

$$d(\phi(x), \phi(y)) = d(\phi(x), \phi(z)) + d(\phi(z), \phi(y))$$

and by Lemma 1, $\phi(z) \in [\phi(x), \phi(y)]$.
Conversely, suppose $\phi(z) \in [\phi(x), \phi(y)]$. By Lemma 1,

$$d(\phi(x), \phi(y)) = d(\phi(x), \phi(z)) + d(\phi(z), \phi(y))$$

which means that, for some $A \in \mathcal{D}$,

$$d_{V'}(\phi_{V'}(x), \phi_{V'}(y)) = d_{V'}(\phi_{V'}(x), \phi_{V'}(z)) + d_{V'}(\phi_{V'}(z), \phi_{V'}(y))$$

for all $(V', S') \in A$. Now if $(V', S') \in A \cap a_{\{p,x,y\}}$ (a non-empty set since it belongs to $\mathcal{D}$), we have $d_{V'}(x, y) = d_{V'}(x, z) + d_{V'}(z, y)$. By Lemma 1, $z \in S'(x, y) \subseteq S(x, y)$. This completes the proof.

To prove an analogue of Theorem 2, it seems easiest to make use of Theorem 2, but we first need another observation. If (V, S) is a finite Herrlich tree, and $p \in V$, let $(V, \leq)$ be the corresponding order tree as defined at the beginning of this section. Then this ordering makes V into the vertex set of a simplicial tree, as described just before Theorem 1. We note that this is the same simplicial tree structure as that given by Lemma 2.1 in [5].

For suppose $S(x,y) = \{x,y\}$ and $x \neq y$. Then $\mu(p,x,y)$ is either x or y, say x. Then $x \leq y$. If $x \leq z \leq y$, then $z \in S(p,y) = S(p,x) \cup S(x,y)$ by Axiom (v). If $z \in S(p,x)$ then $z \leq x$, so $z = x$, while if $z \in S(x,y)$ then z is either x or y. Thus there is no z such that $x < z < y$.

Conversely, suppose $x < y$ and there is no z such that $x < z < y$. Let $u \in S(x,y)$. By Axiom (v) $u \in S(p,y)$, that is, $u \leq y$. Again by Axiom (v), $x \in S(p,y) = S(p,u) \cup S(u,y)$, so either $x \leq u$ or $x \in S(u,y)$. If $x \in S(u,y)$ then x, $u \in S(u,y) \cap S(u,x) \cap S(x,y)$, so $x = u$ by Axiom (iv). In any case, $x \leq u \leq y$, so either $x = u$ or $u = y$, and $S(x,y) = \{x,y\}$ has two elements.

Theorem 4. *Let (V,S) be a Herrlich tree with V countably infinite. Then there exist an $\mathbb{R}$-tree (X,d) and a $1-1$ mapping $\phi : V \to X$ such that, for all a, $b \in V$, $\phi(S(a,b)) = [\phi(a),\phi(b)] \cap \phi(V)$.*

Proof. Choose a basepoint $p \in V$ and enumerate V as $V = \{x_0, x_1, \ldots\}$, where $x_0 = p$. View V as an order tree as at the beginning of §4 and carry out the construction in the proof of Theorem 2, to obtain an embedding of V in an $\mathbb{R}$-tree (X,d). We use the notation of the proof of Theorem 2, and write V_n for $V(X_n)$. Using Lemma 7, we may choose the enumeration so that there is a subsequence $V_{n_1}, V_{n_2}, \ldots$ such that there is a subtree (V_{n_i}, S_{n_i}) of (V,S) for $i \geq 1$. As in Theorem 2, we identify x and $\phi(x)$ for $x \in V$.

Suppose $z \in S(x,y)$, where x, $y \in V$. For sufficiently large i we have x, y, $z \in V_{n_i}$. By the remarks preceding the theorem, using Property (ii) in the proof of Theorem 2, and Lemma 1, $d_{n_i}(x,y) = d_{n_i}(x,z) + d_{n_i}(z,y)$, and taking limits as $i \to \infty$ shows that z belongs to the segment $[x,y]$ in X, again by Lemma 1.

Suppose $z \in V$ and $z \notin S(x,y)$. We can choose n such that x, y, $z \in V_n$ and $n = n_i$ for some i. Then by a calculation like that at the end of the proof of Theorem 2, replacing p, x, y by x, z, y, we see that $z \notin [x,y]$. This finishes the proof.

In the setting of Theorem 1, if $\gamma : P \to P$ is an order preserving injective map, one might hope that γ could be extended to an isometry from X into X, or at least that this is true when X is replaced by the subtree spanned by $\phi(P)$. However, it is easy to see that no such result can hold in general. For take $P = \mathbb{R} \cup \{-\infty\}$ with the usual ordering, and let γ be the the translation $x \mapsto x + 1$ on $\mathbb{R}$, with $\gamma(-\infty) = -\infty$. Then (identifying x with $\phi(x)$ for $x \in P$),

$$d(\gamma(-\infty), \gamma(0)) = d(-\infty, 0) + d(0,1) > d(-\infty, 0).$$

Also, $\mathbb{R} \cup \{-\infty\}$ is a Herrlich tree, where $S(x,y)$ is the segment $[x,y]$ in the usual sense, and the mapping γ defined above is a morphism of Herrlich trees (see 2.2 in [5]). Thus in Theorem 3, a morphism $f : (V,S) \to (V,S)$ will not, in general, extend to an isometry of X into X.

References

1 R.C. Alperin and H. Bass, "Length functions of group actions on Λ-trees". In: Combinatorial group theory and topology (ed. S.M.Gersten and J.R.Stallings), Annals of Mathematics Studies **111**, pp 265-378. Princeton: University Press 1987.

2 J.L. Bell and A.B. Slomson, *Models and ultraproducts*. Amsterdam: North-Holland 1971.

3 I.M. Chiswell,"Non-standard analysis and the Morgan-Shalen compactification", *Quart. J. Math. Oxford (2)* **42** (1991), 257-270.

4 M. Coornaert, T. Delzant and A.Papadopoulos, *Géométrie et théorie des groupes*, Lecture Notes in Mathematics 1441. Berlin, Heidelberg, Springer 1990.

5 F. Herrlich, "Moduli for stable marked trees of projective lines", *Math. Ann.* **291** (1991), 643-661.

6 A.H.M.Hoare, "Pregroups and length functions", *Math. Proc. Cambridge Philos. Soc.* **104** (1988), 21-30.

7 T.J. Jech, "Trees", *J. Symbolic Logic* **36** (1971), 1-14.

8 J.W. Morgan, "Λ-trees and their applications", *Bull. Amer. Math. Soc.* **26** (1992), 87-112.

9 J.Morgan and P.B.Shalen, "Valuations, trees and degenerations of hyperbolic structures: I", *Annals of Math.* **122** (1985), 398-476.

10 F.S. Rimlinger, *Pregroups and Bass-Serre Theory*, Mem. Amer. Math. Soc. no. 361 (Providence, American Mathematical Society 1987).

11 L.van den Dries and A. Wilkie, "Gromov's Theorem on groups of polynomial growth and elementary logic", *J. Algebra* **89** (1984), 349-374.

School of Mathematical Sciences
Queen Mary and Westfield College
University of London
Mile End Road
London E1 4NS.

The mathematician who had little wisdom: a story and some mathematics

DANIEL E. COHEN

1. The Story

If any of you who read this volume do not like stories, then I am sorry for you. Stories are the thread from which the fabric of the world is woven, and to dislike stories is to dislike life. But, to any such people, I would also say that if you read this story you will also learn some mathematics.

Once there was a mathematician who had little wisdom. One spring he attended a group theory conference in Scotland, which may or may not have been wise of him. Since the conference was long, he decided to take a couple of days off, which was certainly wise. He had heard much about the beauty of Scotland's rivers, and the fine salmon that swam in them, so he decided to go salmon-fishing. He did not think of the need for a licence, nor that a large charge is made for the right to fish for salmon in most places; indeed, he had not even checked whether there were salmon in the rivers at that time of year. This may seem foolish of him, but turned out not to be so.

He went to the Tweed, which was running sweetly. He saw many people fishing for salmon, but found a pool where no-one was. Not thinking that this might be because that was not a good place for salmon (for he had little wisdom, though, as we shall see, he was also lucky), he began fishing under a bright spring sky. And, after he had been fishing for some time, he caught a fine salmon, and, after a struggle, landed it. Great was his surprise when the salmon spoke to him. "Put me back" it said; "put me back in the river before I die."

The mathematician was wise enough to do as the salmon asked. I have heard some suggest that he should have kept the salmon in a tank, and made much money by exhibiting it as a talking fish. Even on the everyday level, this idea is foolish, since he had no way to compel the salmon to talk. But there are deeper reasons why this thought was foolish. Many say that our treatment of the world as something to be exploited for financial profit, rather than as something sacred, has led us to the brink of disaster. Be that as it may, to treat the creatures of the inner worlds in such a way is a certain recipe for disaster.

The salmon, now returned to the river, spoke again. "I am the Salmon of Wisdom. I have fed on the Nuts of Knowledge that fall from the hazel tree which leans over the Well at the World's End. Since you have set me free, I will grant you a gift."

Now the mathematician had been thinking hard about a certain finite presentation of a group. "What I want", said he, "is a machine that will tell me whether or not a word in the generators equals one in the group I have recently been considering."

The salmon replied "I must warn you that if you ask me for something impossible I will vanish, and what little wisdom you have may become even less."

"That's no problem" answered the mathematician. "I know that there are some finite presentations for which no such machine exists, but I have shown that in this case it can be done."

So the salmon swam deeper into the river, and returned carrying a little machine. It looked very delicate and attractive, and the mathematician was delighted by it. He took it home, and used it with great pleasure for several weeks. He enjoyed looking at it, for it was indeed a work of art, and his eyes kept following the tracery of lights that showed the machine was operating, finishing with either a green light or a red one. Since this machine was not of the ordinary world, it never ran out of memory in which to perform the computation.

But after a while the mathematician, who had little wisdom, became dissatisfied, and decided to ask the salmon for a more powerful machine. He went back, with the machine, to the pool where he had first caught the salmon. The river was running high, and the skies were grey, but he did not notice that. "Salmon," he called, "Great Salmon of Wisdom, I desire a gift from you".

The salmon came up to the surface and spoke. "What is it this time?"

"This machine is very nice, but I would like a machine that does more. Instead of one that just shows a green light when the word equals one, I would like one that actually tells me how to write that word as a product of conjugates of the relators and their inverses."

"Remember, if you ask for the impossible, I will vanish and you may lose even what little wisdom you have."

"I feel quite sure that, since you have given me one machine, you can also give me the more powerful one I desire."

"You are right" said the salmon, and swam deeper into the river. He emerged carrying a heavy and ugly machine. The mathematician dragged it home. For several weeks he was happy using it.

But after a while, since he had little wisdom, he became dissatisfied, and decided to ask the salmon for a more powerful machine. With an effort, he dragged the machine back to the pool where he had first seen the salmon. It was a hard journey, under a black sky, with thunder in the distance, and showers of hail. When he arrived, the river was in flood, and lapped almost to his feet. Since he had little wisdom, he did not see what this had to do with him, and once more called out "Salmon, Great Salmon of Wisdom, I desire a gift from you."

The salmon came up, and asked what he wanted this time. "This machine does what I asked for, but it is so slow. There have been many occasions when a word for which the first machine answered my question in a few seconds, has taken days on this new one. What I want now is a machine which will do what the second machine does, but will do it as fast as the first machine."

"I'm getting tired of your demands" said the salmon crossly. "Is that what you really want?" The mathematician, who had little wisdom, did not notice the warning in the salmon's tone, and said, firmly, "Yes, that's what I want."

At that, the salmon leapt into air, cried out "What you ask is impossible!", and fell back into the river with a splash that drenched the mathematician. When he recovered from being soaked, the salmon and the machine were gone.

As he began to start homewards, the salmon appeared briefly, and said "Read computer science journals, as well as mathematics journals. Then, it may be that instead of losing what little wisdom you have, you may even gain some."

When the mathematician returned to his home university, he did as the salmon had suggested, and started to look at computer science journals, including back issues. He discovered, to his surprise, that they contained a significant number of interesting results in group theory, and he learned why his last request was impossible, since there are presentations (of which his was one) for which the word problem has an easy solution, but for which any solution of the special kind asked for is difficult.

The last I heard of him, he was still looking at these journals, and also at logic journals, as well as those more directly concerned with his own interests. And, by doing so, his wisdom had increased.

2. The Mathematics of the Story

This section is based on work of Madlener and Otto [4], and its slight extension by Cohen, Madlener, and Otto [2]; see also [5]. It will explain the mathematics behind what happened to the mathematician of my story.

Let $\langle X; R \rangle$ be a finite presentation of a group G. A *solution to the word problem for* $\langle X; R \rangle$ is a computable function which, on input w, answers "No" if a word w of the free group $F(X)$ on X does not equal 1 in G and "Yes" if it does equal 1.

Here there is no need to make the definition of 'computable function' precise. All reasonable definitions lead to the same class of functions.

It is well known that there are presentations for which the word problem has no solution. Since, given two finite presentations of G, it is easy to see that the word problem for one presentation has a solution iff the word problem for the other has a solution, we then say that the word problem for G has a solution.

If w equals 1 in G then w is a product $\prod_1^k (u_i^{-1} r_i^{\epsilon_i} u_i)$ for some k and some elements $r_i \in R, u_i \in F(X)$, and $\epsilon_i = \pm 1$. A *pseudo-natural solution to the word problem* is a computable function which, as before, says "No" if w does not equal 1 in G, and if w does equal 1 it explicitly gives the elements $r_i, u_i,$ and e_i. This is not the definition used in [4], but it is pointed out in [2] that it is equivalent to the earlier definition.

If the word problem has a solution, then it has a pseudo-natural solution. For we first get the answer "Yes" or "No", and, if the answer is "Yes" we then proceed to search systematically through all possible choices of k, and $r_i, u_i,$ and e_i until we find one such that $w = \prod_1^k (u_i^{-1} r_i^{e_i} u_i)$. This systematic search can be done, for instance, by searching in order of magnitude of $k + \sum |u_i|$ and in some convenient order among those finitely many possibilities for which this sum has a given value.

However, this is a 'wait-and-see' approach, in which there is no way of telling in advance how long the procedure will take. We just have to wait until it gives an answer. A function constructed in this way is recursive, but is usually not primitive recursive.

To look at the question further, we need to consider the complexity of primitive recursive functions. The Grzegorczyk hierarchy is a collection E_n of primitive recursive functions for all n such that every primitive recursive function is in some E_n, $E_n \subseteq E_{n+1}$, and $E_n \neq E_{n+1}$. To understand the theorem below, this is all one needs to know, but it seems worthwhile to be more explicit about the hierarchy. See [1, 7] for details.

One definition uses programs. We need the programs to operate on infinite sequences of natural numbers, all but a finite number of which are zero. We start with some simple programs, such as those which add 1 to or subtract 1 from some entry. We build new programs from old ones by two procedures. If P and Q are programs then PQ is a program which performs first P and then Q. Also LOOP k P is a program which performs P x_k times where x_k is the k-th entry in the sequence. The class of all functions computed by such LOOP programs is exactly the class of primitive recursive functions. Note that we know in advance how many times the program LOOP k P performs P. If we modify our programs so that we can repeat P until a certain condition is satisfied then we get the larger class of partial recursive (or computable, but not necessarily total) functions. The class E_n can be defined as the class of those functions which can be computed by LOOP programs in which the depth of nesting the LOOPs is n.

Another definition uses Ackermann functions. There are a number of variants of these functions, and there may be a dimension-shift of 1 in the results, depending on which function we take. The two-variable function A is defined by certain initial conditions together with the rule $A(m+1, n+1) = A(m, A(m+1, n))$. Thus, if a_m is defined by $a_m(n) = A(m, n)$, then a_{m+1} is defined by iterating a_m, and each a_m is primitive recursive. However, A is recursive but not primitive recursive.

One of the nice properties of the class E_n is that a function is in E_n iff it can be computed by a LOOP program in a time which is itself an E_n function of the input.

We say that a function f is defined from functions g, h and k by *bounded primitive recursion* if f is given by $f(x_1, \ldots, x_r, 0) = g(x_1, \ldots, x_r)$ and $f(x_1, \ldots, x_r, y+1) = h(x_1, \ldots, x_r, y, f(x_1, \ldots, x_r, y))$ — these two conditions just say that f comes by primitive recursion from g and h — and, in addition, $f(x_1, \ldots, x_r, y) \leq k(x_1, \ldots, x_r, y)$. Then the class E_m can be defined to be the smallest class containing both a_m and some simple initial functions and closed under composition and bounded primitive recursion. Most functions of importance are in E_3 (which is sometimes called the class of *elementary functions*. By contrast, a_5, and even a_4, while computable in theory, grow too quickly to be computable in practice.

If we want to look at the complexity of functions defined on words (or more general objects), the obvious way to do so is to code these up into numbers. There are usually several ways of doing this, which give the same notion of the class E_n provided $n \geq 3$. Also, when we change presentations, the complexity classes of solutions and of pseudo-natural solutions remain unchanged if this class is at least 3. It is for this reason that we only look at complexity at least 3 in our theorem.

Theorem   *There is a presentation for which the word problem has a solution in E_3 but which has no primitive recursive pseudo-natural solution.*

More generally, if $3 \leq m \leq n$, there is a presentation for which the word problem has a solution in E_m but no solution in E_{m-1}, there is no pseudo-natural solution in E_{n-1}, and some pseudo-natural solution is in E_n.

This result is surprising at first sight, since it is hard to imagine how one could solve the word problem without using a pseudo-natural solution. The situation is clarified by looking at the similar situation for machines (equivalently, programs). A *solution for the halting problem* of a machine (or program) will be a computable function which, on a given input, answers "Yes" or "No" according to whether the machine does or does not halt on the given input. Given a machine, we can construct a corresponding group, and we obtain a group with unsolvable word problem from a machine with unsolvable halting problem (such machines exist, by general theory). A more precise analysis shows that, for $n \geq 3$, the word problem has a solution in E_n iff the corresponding machine has a halting problem with solution in E_n.

A *pseudo-natural solution to the halting problem* will, on an input from which the machine halts, give the computation of the machine from that input. Note that among the pseudo-natural solutions there is one natural solution; namely, run the original machine and make a note of its computation. Again we find that, for $n \geq 3$, the group has a pseudo-natural solution in E_n iff the machine has a pseudo-natural solution in E_n. Also, this holds iff the run-time of the machine (that is, the number of steps taken to halt if the machine does halt on that input, and 0 if the machine does not halt) is in E_n.

Now let f be a function which is recursive but not primitive recursive. Then it can be computed by a machine, which necessarily halts when started on a configuration corresponding to a natural number. If we start the machine on an arbitrary configuration, however, it may not halt. But it can be shown [6, 3] that we can alter the machine so that it halts when started on any configuration. In this case, the halting problem has the very easy solution which always answers "Yes". But, since the function is not primitive recursive, its run-time (and hence, any pseudo-natural solution) cannot be primitive recursive.

2. Computer Science Journals and Group Theory

Many group-theorists will know of the *International Journal of Algebra and Computing*, whose first issue came out a couple of years ago. This always contains material of interest, and is as much a journal of mathematics as it is one of computer science. The *Journal of Symbolic Computation* is primarily a computer science journal, but regularly has material of interest to group-theorists and other algebraists. In particular, the issues 5–6 of volume 9 (1990) and 4–5 of volume 12 (1991) are on computational group theory.

But other computer science journals do have relevant papers sufficiently often to be worth glancing at regularly, as do the *Springer Lecture Notes in Computer Science*. These include *Theoretical Computer Science, Acta Informatica, Journal of Computer and System Science*, and *Journal of Algorithms*, all of which have contained more than one paper of interest. Other computer science journals also contain relevant material occasionally. I have prepared a survey (not comprehensive, but based on personal preference) of some thirty papers. This is too long to include here, but may be obtained from me by mail (School of Mathematical Sciences, Queen Mary and Westfield College, Mile End Rd., London E1 4NS) or e-mail me (D.E.Cohen@maths.qmw.ac.uk).

References

[1] E. Börger, *Computabilty, Complexity, Logic.* North-Holland, Amsterdam and New York, (1989).

[2] D.E. Cohen, K. Madlener and F. Otto, 'Separating the intrinsic and the derivational complexity for the word problem of finitely presented groups'. *Mathematical Logic Quarterly* 39 (1993), 143-157.

[3] G. T. Herman, 'Strong computability and variants of the uniform halting problem.' *Zeitschrift Mathematische Logik Grundlagen Mathematik* 17 (1971) 115-131.

[4] K. Madlener and F. Otto, 'Pseudo-natural algorithms for the word problem for finitely presented monoids and groups'. *J. Symbolic Computation* 1 (1985), 383-418.

[5] K. Madlener and F. Otto, 'Pseudo-natural algorithms for finitely generated presentations of monoids and groups'. *J. Symbolic Computation* 5
(1988), 339-358.

[6] J.C. Shepherdson, 'Machine configurations and word problems of given
degree of unsolvabillity'. *Zeitschrift Mathematische Logik Grundlagen
Mathematik* 11 (1965), 149-175.

[7] G.J. Tourlakis, *Computability*. Reston Publishing Co., Boston, (1984).

School of Mathematical Sciences
Queen Mary & Westfield College
Mile End Road
London E1 4NS
U.K.
E-mail: D.E.Cohen@qmw.ac.uk

Palindromic automorphisms of free groups

DONALD J. COLLINS

1 Introduction

Let F be a free group with a given finite basis X. Let w be a word in X, say

$$w \equiv x_1^{\varepsilon_1} x_2^{\varepsilon_2} \ldots x_n^{\varepsilon_n}.$$

The *reverse* of w is the word

$$\overline{w} \equiv x_n^{\varepsilon_n} x_{n-1}^{\varepsilon_{n-1}} \ldots x_1^{\varepsilon_1}.$$

A (reduced) word w in X will be called a *palindrome* if it coincides with its reverse. An automorphism α of F will be called *palindromic* (relative to the basis X) if , for every $x \in X$, the image $x\alpha$ is (represented by) a palindrome. It is not hard to show that the set of all palindromic automorphisms of F is a subgroup of the full group $\mathrm{Aut}(F)$ of automorphisms of F; we shall denote it by $\Pi A(F)$.

Subgroups of $\mathrm{Aut}(F)$ are not, in general, well understood; a notable example is the lack of any known set of defining relations for the subgroup $IA(F)$, the kernel of the map to $\mathrm{Aut}(F/F')$. The group $\Pi A(F)$ of palindromic automorphisms lies towards the opposite extreme to $IA(F)$ being almost a supplement to the latter – the image of $\Pi A(F)$ in $\mathrm{Aut}(F/F')$ is the group generated by the squares of the transvections together with the permutations and inversions of the basis elements. The intersection of $\Pi A(F)$ with $IA(F)$ is, however, non-trivial, containing, for instance in the case that F has basis $\{x, y, z\}$ the automorphism whose action is $x \mapsto zyz^{-1}y^{-1}xy^{-1}z^{-1}yz$, $y \mapsto y$, $z \mapsto z$.

In this paper we show that the group $\Pi A(F)$ is generated by a finite set of 'elementary' palindromic automorphisms subject to a finite set of defining relations. An easy argument (passing to the quotient by $F'F^2$) shows that if α is palindromic, then $x\alpha$ has odd length, for every $x \in X$, and may be written in the form $\bar{u}_x(x\tau)^{\varepsilon_x}u_x$, $\varepsilon_x = \pm 1$, where τ is a permutation of the basis X. Since any automorphism of F which permutes the basis X is clearly palindromic, the essential part of the discussion is that concerned with the *pure palindromic* automorphisms, i.e. those automorphisms α for which $x\alpha = \bar{u}_x x^{\varepsilon_x} u_x$.

We use the following notation and terminology. By a *factor automorphism* σ of F is meant an automorphism which inverts some of the elements of the basis X. For each pair x, y of distinct elements of X, the automorphism $(x\|y)$ of F given by

$$(x\|y) \quad : \begin{cases} x \to yxy \\ z \to z & \text{if } z \neq x \end{cases}$$

63

will be called *elementary* palindromic. The factor automorphism σ_y which inverts only the single basis element y will also be called *elementary*.

Theorem *The group* $P\Pi A(F)$ *of pure palindromic automorphisms of the free group* $F(X)$ *of finite rank is the semidirect product of the group* $E\Pi A(F)$, *generated by the set*

$$\{(x\|y) : x,\ y \in X,\ x \neq y\}$$

of elementary palindromic automorphisms, and the group Φ *of factor automorphisms. The relations*

$$(x\|z)(y\|z) = (y\|z)(x\|z) \tag{1}$$

$$(x\|z)(y\|t) = (y\|t)(x\|z) \tag{2}$$

$$(x\|z)(y\|z)(x\|y) = (x\|y)(y\|z)(x\|z)^{-1} \tag{3}$$

where distinct letters denote distinct elements of X *and a relation occurs for every possible choice of* x, y, z *and* t, *form a set of defining relations for* $E\Pi A(F)$. *The group* Φ *is an elementary abelian 2-group generated by the automorphisms* $\{\sigma_x : x \in X\}$ *and the action of* Φ *on* $E\Pi A(F)$ *is given by* $\sigma_y^{-1}(x\|y)\sigma_y = (x\|y)$.

The presentation of $E\Pi A(F)$ is similar to the presentation of the group $P\Sigma A(F)$ of pure symmetric automorphisms studied by the author in [C]. An automorphism of $F(X)$ is *pure symmetric* if , for each $x \in X$, the image $x\alpha = u_x^{-1}xu_x$, for some word u_x, i.e. $x\alpha$ is conjugate to x. An *elementary* symmetric automorphism of $F(X)$ is an automorphism, denoted by $(x|y)$ with x and y distinct elements of the basis X, whose action is

$$(x|y) \quad : \begin{cases} x \to y^{-1}xy \\ z \to z \end{cases} \quad \text{if } z \neq x.$$

Such automorphisms generate $P\Sigma A(F)$ subject to the defining relations

$$(x|z)(y|z) = (y|z)(x|z) \tag{1}$$

$$(x|z)(y|t) = (y|t)(x|z) \tag{2}$$

$$(x|z)(y|z)(x|y) = (x|y)(y|z)(x|z) \tag{3}$$

with the same conventions on the use of distinct letters as in the theorem above. (This is implicit in [FR] and is spelled out in detail in [G].)

This parallel raises questions about the group $E\Pi A(F)$. For example can an alternative argument for obtaining a presentation be given using Whitehead automorphisms? If so this would suggest that one could calculate the virtual cohomological dimension of $E\Pi A(F)$ by employing the method of [CV] as was done in [C] for $P\Sigma A(F)$ - it also may be the case that $E\Pi A(F)$ is torsion-free since $P\Sigma A(F)$ is torsion-free.

2 Generation

2.1 Lemma *For any word* w, $\overline{w}^{-1} \equiv \overline{w^{-1}}$.

2.2 Proposition *The group* $P\Pi A(F)$ *of pure palindromic automorphisms is generated by the elementary palindromic and factor automorphisms.*

Proof. For $\alpha \in P\Pi A(F)$, define $L(\alpha) = \sum_{x \in X} |x\alpha|$ where $|x\alpha|$ is the length of $x\alpha$ relative to the basis X.

The argument is a "Nielsen cancellation argument" by induction on $L(\alpha)$. The case $L(\alpha) = 0$ is clearly trivial; so suppose $L(\alpha) > 0$. By Proposition 2.13 of [LS] the set $\{\overline{u}_x x^{\varepsilon_x} u_x : x \in X\}$ is not Nielsen reduced. Since all the words $\overline{u}_x x^{\varepsilon_x} u_x$ have odd length it follows that condition $(N2)$ [LS,p.6] is not relevant. Hence, without loss of generality, there exist distinct elements x and y of X and η_x, $\eta_y = \pm 1$ such that in cancelling the product $(\overline{u}_x x^{\varepsilon_x} u_x)^{\eta_x} (\overline{u}_y y^{\varepsilon_y} u_y)^{\eta_y}$ the displayed $x^{\varepsilon_x \eta_x}$ is cancelled by a letter of $\overline{u}_y$ (if $\eta_y = 1$) or of u_y^{-1} (if $\eta_y = -1$). Routine calculations show that one of the following then holds:

(a) $\eta_x = \eta_y = 1$, $L((y||x)\alpha) < L(\alpha)$;

(b) $\eta_x = -1, \ \eta_y = 1, \ L((y||x)\sigma_y \alpha) < L(\alpha)$;

(c) $\eta_x = 1, \ \eta_y = -1, \ L((y||x)^{-1}\alpha) < L(\alpha)$;

(d) $\eta_x = \eta_y = -1, \ L((y||x)^{-1}\sigma_y \alpha) < L(\alpha)$.

The result thus follows.

3 Defining Relations

To show that the relations displayed in the Theorem form a set of defining relations we shall give an argument which parallels that in [FR]. For this we consider products $\gamma_1 \gamma_2 \cdots \gamma_s$ of elementary palindromic automorphisms and their inverses. The *diagram* of such a product is the sequence $(l_1, l_2, \ldots, l_s)$ where $l_i = L(\gamma_i \cdots \gamma_s)$. The diagram is *monotone* if $l_1 > l_2 > \ldots > l_s$ or $s = 0$.

3.1 Proposition *Every product* $\gamma_1 \gamma_2 \cdots \gamma_s$ *is equivalent modulo the relations* $(1) - (3)$ *of the theorem to a product with a monotone diagram.*

Once the proposition is established the theorem is essentially immediate. Firstly, the relations giving the action of the factor automorphisms enable one to write any word θ in the generators as a product $\theta = \psi \gamma_1 \gamma_2 \cdots \gamma_s$ where ψ is a factor automorphism and $\gamma_1 \gamma_2 \cdots \gamma_s$ is a product of elementary palindromic automorphisms and their inverses. The proposition enables one to assume that the diagram of $\gamma_1 \gamma_2 \cdots \gamma_s$ is monotone. If θ represents the identity automorphism, it follows that $L(\gamma_1 \gamma_2 \cdots \gamma_s) = 0$, since clearly application of a factor automorphism will not alter the value obtained by applying

L. Therefore $\gamma_1\gamma_2\cdots\gamma_s$ is the empty product. It is obvious that the group Φ is elementary abelian of exponent 2 and there is nothing further to prove - except, of course, to check by direct calculation that the relations (1)-(3) are valid. In the process we have shown that the claimed semidirect product structure occurs.

It is convenient to extend our notation $(x||y)$ to include the the possibility that x and y are inverses of elements of the basis X. When we do this we observe that $(x^{-1}||y^{-1}) = (x||y)$ and that $(x||y)^{-1} = (x||y^{-1}) = (x^{-1}||y)$. Formally we have enlarged our presentation by the use of Tietze transformations and when we do this, simple calculations show that we can extend the relations of the type (1) - (3) so that the letters are interpreted as standing for elements of the set $X \cup X^{-1}$. In what follows we shall use letters such as x, y, z, t to stand for elements of $X \cup X^{-1}$.

3.2 Lemma *Any product $\gamma_1\gamma_2$ has monotone diagram, unless $\gamma_2 = \gamma_1^{-1}$.*

Proof. The diagram is of the form $(l, 1)$ where $l = L(\gamma_1\gamma_2)$. Let $\gamma_1\gamma_2 = (x||y)(z||t)$, making no assumptions on x, y, z and t other than are necessary so that the notation is well-defined. Various cases occur.

Case 1. $x = z^{\pm 1}$; without loss of generality $x = z$. Then $x\gamma_1\gamma_2 = ytxty$ while $z\gamma_1\gamma_2 = z$ so that $L(\gamma_1\gamma_2) = 2$, since $y \neq t^{-1}$.

Case 2. $x \neq z^{\pm 1}$, $y = t^{\pm 1}$. This time $x\gamma_1\gamma_2 = yxy$ while $z\gamma_1\gamma_2 = tzt$ and again $L(\gamma_1\gamma_2) = 2$.

Case 3. $x \neq z^{\pm 1}$, $y \neq t^{\pm 1}$. Here $y = z^{\pm 1}$ is possible and while $z\gamma_1\gamma_2 = tzt$, there are two possibilities for $x\gamma_1\gamma_2$, namely yxy (if $y \neq z^{\pm 1}$) and $t^{\pm 1}yt^{\pm 1}xt^{\pm 1}yt^{\pm 1}$ (if $y = z^{\pm 1}$). In the latter case a little cancellation is possible but still $L(\gamma_1\gamma_2) \geq 2$ as required.

We now proceed to the proof of the Proposition, which is effectively an induction with the initial case now having been disposed of. Formally we are given a product $\gamma_1\gamma_2\cdots\gamma_s$ with diagram $(l_1, l_2, \ldots, l_s)$. If this is not monotone then, using Lemma 3.2 it follows that there must exist r, $1 < r < s$ such that $l_{r-1} \leq l_r > l_{r+1} > \ldots > l_s$. We show that by using (the extended) relations (1)-(3), we can write $\gamma_{r-1}\gamma_r = \delta_1\cdots\delta_q$ so that $L(\delta_j\cdots\delta_q\gamma_{r+1}\cdots\gamma_s) < L(\gamma_r\cdots\gamma_s)$, for $2 \leq j \leq q$. If we use the pair $(s - r, l_r)$ with lexicographic ordering as an inductive measure, then replacing $\gamma_{r-1}\gamma_r$ by $\delta_1\cdots\delta_q$ reduces this measure as required (possibly after eliminating inverse pairs in the new product).

The problem to be solved is therefore the following: suppose we are given arbitrary ζ, $(x||y)$ and $(z||t)$ such that

$$L((x||y)(z||t)\zeta) \leq L((z||t)\zeta) > L(\zeta),$$

can we then find a product $\delta_1\cdots\delta_q$ equivalent modulo (1)-(3) to $(x||y)(z||t)$ such that

$$L(\delta_j\cdots\delta_q\zeta) < L((z||t)\zeta)$$

for $2 \leq j \leq q$.

We fix the following notation:

$$x\zeta \equiv \overline{u}_x x^{\varepsilon_x} u_x, \quad y\zeta \equiv \overline{u}_y y^{\varepsilon_y} u_y, \quad z\zeta \equiv \overline{u}_z z^{\varepsilon_z} u_z, \quad t\zeta \equiv \overline{u}_t t^{\varepsilon_t} u_t.$$

The argument proceeds by case analysis, the different cases being dependent upon the coincidences among the sets $\{x, x^{-1}\}$, $\{y, y^{-1}\}$, $\{z, z^{-1}\}$, $\{t, t^{-1}\}$.

Case I. Suppose that either there are no coincidences or that $\{y, y^{-1}\} = \{t, t^{-1}\}$, but this is the only coincidence. We take $\delta_1 \delta_2 = (z||t)(x||y)$, using (1)-(2). It suffices to show $L((x||y)\zeta) \leq L(\zeta)$. However this is immediate since the difference between $L((x||y)(z||t)\zeta)$ and $L((z||t)\zeta)$ arises solely from the difference between the respective images of x and we see that

$$x(x||y)(z||t)\zeta = x(x||y)\zeta, \quad x(z||t)\zeta = x\zeta.$$

Case II. Suppose that $\{z, z^{-1}\} = \{x, x^{-1}\}$; then $\{y, y^{-1}\}$ and $\{t, t^{-1}\}$ must be distinct from $\{x, x^{-1}\}$ but we allow the possibility that they coincide. Without loss of generality , $x = z$. So we have

$$L((x||y)(x||t)\zeta) \leq L((x||t)\zeta) > L(\zeta).$$

The inequality $L((x||t)\zeta) > L(\zeta)$ is equivalent to $|(txt)\zeta| > |x\zeta|$, i.e. to

$$|\overline{u}_t t^{\varepsilon_t} u_t \overline{u}_x x^{\varepsilon_x} u_x \overline{u}_t t^{\varepsilon_t} u_t| > |\overline{u}_x x^{\varepsilon_x} u_x|. \qquad (II.0.1)$$

The displayed occurrences of t^{ε_t} are not therefore cancelled in the process of reducing to normal form. If p is the normal form of $u_x \overline{u}_t$ then the reduced form of $x(x||t)\zeta = (txt)\zeta$ is

(A) $$\overline{u}_t t^{\varepsilon_t} \overline{p} x^{\varepsilon_x} p t^{\varepsilon_t} u_t$$

or

(A′) $$\overline{u}_t t^{\varepsilon_t} \overline{q} x^{-\varepsilon_x} q t^{\varepsilon_t} u_t.$$

In turn a partially reduced form of $x(x||y)(x||t)\zeta$ is

(B) $$\overline{u}_y y^{\varepsilon_y} u_y \overline{u}_t t^{\varepsilon_t} \overline{p} x^{\varepsilon_x} p t^{\varepsilon_t} u_t \overline{u}_y y^{\varepsilon_y} u_y$$

or

(B′) $$\overline{u}_y y^{\varepsilon_y} u_y u_t t^{\varepsilon_t} \overline{q} x^{-\varepsilon_x} q t^{\varepsilon_t} u_t \overline{u}_y y^{\varepsilon_y} u_y$$

with the latter case occurring whenever $p \equiv x^{-\varepsilon_x} q$ and $\overline{u}_t \equiv u_x^{-1} p$. Therefore the inequality $L((x||y)(x||t)\zeta) \leq L((x||t)\zeta)$ is equivalent to

$$|(B)| \leq |(A)| \quad \text{or, if appropriate,} \quad |(B')| \leq |(A')| \qquad (II.0.2),$$

where $|(B)|$ denotes the length of the fully reduced form of (B).

Subcase II.1. Suppose $|u_t| > |u_y|$; to achieve $(II.0.2)$, $u_t \equiv vy^{-\varepsilon_y} \overline{u}_y^{-1}$, for some v, and the inequality must actually be strict. The relation (3) on this occasion yields

$$\begin{aligned} (x||y)(x||t) &= (t||y^{-1})(x||t)(t||y)(x||y^{-1}) \\ &= \delta_1 \delta_2 \delta_3 \delta_4. \end{aligned}$$

We need, therefore, to establish the inequalities

$$L((x||y^{-1})\zeta) \;<\; L((x||t)\zeta) \tag{II.1.1}$$
$$L((t||y)(x||y^{-1})\zeta) \;<\; L((x||t)\zeta) \tag{II.1.2}$$
$$L((x||t)(t||y)(x||y^{-1})\zeta) \;<\; L((x||t)\zeta). \tag{II.1.3}$$

Since $p \equiv u_x \overline{u}_t \equiv u_x u_y^{-1} y^{-\varepsilon_y} \overline{v}$, $(II.1.1)$ is equivalent to

$$|u_y^{-1} v^{-1} \overline{p} x^{\varepsilon_x} p \overline{v}^{-1} \overline{u}_y^{-1}| < |u_y^{-1} y^{-\varepsilon_y} \overline{v} t^{\varepsilon_t} \overline{p} x^{\varepsilon_x} p t^{\varepsilon_t} v y^{-\varepsilon_y} \overline{u}_y^{-1}| \tag{II.1.4}$$

which holds regardless of whether the right hand side refers to (A) or (A').

For $II.1.2$ we have to examine the various images of t and x. Noting that $t\delta_3\delta_4 = t\delta_3$, $x\delta_3\delta_4 = x\delta_4$ and $t(x||t) = t$, $II.1.2$ turns out to be equivalent to

$$|x(x||y^{-1})\zeta| + |t(t||y)\zeta| < |x(x||t)\zeta| + |t\zeta|. \tag{II.1.5}$$

However $II.1.1$, which depends only on the images of x, actually implies that it suffices to prove that

$$|t(t||y)\zeta| < |t\zeta|. \tag{II.1.6}$$

which amounts to proving

$$|\overline{u}_y \overline{v} t v u_y| < |\overline{u}_y^{-1} y^{-\varepsilon_y} \overline{v} t v y^{-\varepsilon_y} \overline{u}_y^{-1}|.$$

This, however, is immediate since the word on the right hand side is reduced. Finally $II.1.3$ reduces firstly to

$$|x(x||t)(t||y)(x||y^{-1})| + |t(t||y)\zeta| < |x(x||t)\zeta| + |t\zeta|$$

and then, by $II.1.6$, to $|x(x||t)(t||y)(x||y^{-1})| < |x(x||t)\zeta|$. In the light of the previous calculations, this amounts to showing

$$|\overline{u}_y \overline{v} t^{\varepsilon_t} \overline{p} x^{\varepsilon_x} p t^{\varepsilon_t} v u_y| < |u_y^{-1} y^{-\varepsilon_y} \overline{v} t^{\varepsilon_t} \overline{p} x^{\varepsilon_x} p t^{\varepsilon_t} v y^{-\varepsilon_y} \overline{u}_y^{-1}|$$

which is immediate, regardless of whether the right hand side refers to (A) or (A').

Subcase II.2. Suppose that $|u_t| < |u_y|$. From $II.0.2$ we must have (at least) $u_y \equiv v t^{-\varepsilon_t} \overline{u}_t^{-1}$, for some v. From an instance of the relation (3) we obtain

$$\begin{aligned}
(x||y)(x||t) &= (y||t^{-1})(x||t^{-1})(x||y)(y||t) \\
&= (x||t^{-1})(y||t^{-1})(x||y)(y||t) \\
&= \delta_1\delta_2\delta_3\delta_4.
\end{aligned}$$

We again have three inequalities to establish, namely

$$L((y||t)\zeta) \;<\; L((x||t)\zeta) \tag{II.2.1}$$
$$L((x||y)(y||t)\zeta) \;<\; L((x||t)\zeta) \tag{II.2.2}$$
$$L((y||t^{-1})(x||y)(y||t)\zeta) \;<\; L((x||t)\zeta). \tag{II.2.3}$$

Now $II.2.1$ is equivalent to $|x\zeta| + |(tyt)\zeta| < |(txt)\zeta| + |y\zeta|$. By $II.0.1$ it suffices to prove that $|(tyt)\zeta| < |y\zeta|$, i.e. that

$$|\overline{u}_t\overline{v}y^{\varepsilon_y}vu_t| < |u_t^{-1}t^{-\varepsilon_t}\overline{v}y^{\varepsilon_y}vt^{\varepsilon_t}\overline{u}_t^{-1}|. \qquad (II.2.4)$$

Since the word on the right hand side is already in normal form this is immediate.

Next we observe that $II.2.2$ is equivalent to

$$|(tytxtyt)\zeta| + |(tyt)\zeta| < |(txt)\zeta| + |y\zeta|.$$

Having established $II.2.4$, it suffices to prove $|(tytxtyt)\zeta| \leq |(txt)\zeta|$, i.e. that

$$|\overline{u}_t\overline{v}y^{\varepsilon_y}v\overline{p}x^{\varepsilon_x}p\overline{v}y^{\varepsilon_y}vu_t| \leq |\overline{u}_tt^{\varepsilon_t}\overline{p}x^{\varepsilon_x}pt^{\varepsilon_t}u_t| \qquad (II.2.5)$$

where p is the reduced form of of $u_x\overline{u}_t$. We know that $v = u_y\overline{u}_tt^{\varepsilon_t}$ so that the inequality $II.0.2$ becomes

$$|u_t^{-1}t^{-\varepsilon_t}\overline{v}y^{\varepsilon_y}v\overline{p}x^{\varepsilon_x}p\overline{v}y^{\varepsilon_y}vt^{-\varepsilon_t}\overline{u}_t^{-1}| \leq |\overline{u}_tt^{\varepsilon_t}\overline{p}x^{\varepsilon_x}pt^{\varepsilon_t}u_t|,$$

where at most one cancellation is possible in $\overline{p}x^{\varepsilon_x}p$ It therefore follows that

$$|\overline{v}y^{\varepsilon_y}v\overline{p}x^{\varepsilon_x}p\overline{v}y^{\varepsilon_y}v| \leq |\overline{p}x^{\varepsilon_x}p|$$

whence $II.2.5$ is obtained. Finally, since $y(y||t^{-1})(x||y)(y||t) = y$, $II.2.3$ is equivalent to $II.2.5$ which we have just established.

Subcase II.3. Suppose that $|u_t| = |u_y|$; then $II.0.2$ forces, at least, $\overline{u}_t \equiv u_y^{-1}$ and we obtain

$$|\overline{u}_yy^{\varepsilon_y}t^{\varepsilon_t}\overline{p}x^{\varepsilon_x}pt^{\varepsilon_t}y^{\varepsilon_y}u_y| \leq |\overline{u}_tt^{\varepsilon_t}\overline{p}x^{\varepsilon_x}pt^{\varepsilon_t}y^{\varepsilon_y}u_y|$$

which is patently false and so this case does not arise, even when $\{y, y^{-1}\} = \{t, t^{-1}\}$ since certainly $t \neq y^{-1}$

Case III. Suppose that $\{y, y^{-1}\} = \{z, z^{-1}\}$ and $\{x, x^{-1}\} \neq \{t, t^{-1}\}$. Without loss of generality we may assume that $y = z$. From the relations (3) we obtain

$$(x||y)(y||t) = (y||t)(x||t)(x||y)(x||t) = \delta_1\delta_2\delta_3\delta_4.$$

We know $L((y||t)\zeta) > L(\zeta)$ which is equivalent to

$$|\overline{u}_tt^{\varepsilon_t}u_t\overline{u}_yy^{\varepsilon_y}u_y\overline{u}_tt^{\varepsilon_t}u_t| > |\overline{u}_yy^{\varepsilon_y}u_y| \qquad (III.1)$$

and $L((x||y)(y||t)\zeta) \leq L(x\zeta)$, which is equivalent to

$$|\overline{u}_tt^{\varepsilon_t}u_t\overline{u}_yy^{\varepsilon_y}u_y\overline{u}_tt^{\varepsilon_t}u_t\overline{u}_xx^{\varepsilon_x}u_x\overline{u}_tt^{\varepsilon_t}u_t\overline{u}_yy^{\varepsilon_y}u_y\overline{u}_tt^{\varepsilon_t}u_t| \leq |\overline{u}_xx^{\varepsilon_x}u_x|. \qquad (III.2)$$

From $III.1$ we observe that the displayed occurrences of t^{ε_t} in the word on the left hand side must survive the process of cancellation to normal form so that if p is the reduced form of $u_y\overline{u}_t$, then $\overline{u}_t t^{\varepsilon_t} \overline{p} y^{\varepsilon_y} p t^{\varepsilon_t} u_t$ allows at most one cancellation and

$$|\overline{u}_t t^{\varepsilon_t} \overline{p} y^{\varepsilon_y} p t^{\varepsilon_t} u_t| > |\overline{u}_y y^{\varepsilon_y} u_y|. \qquad (III.3)$$

Then $III.2$ becomes

$$|\overline{u}_t t^{\varepsilon_t} \overline{p} y^{\varepsilon_y} p t^{\varepsilon_t} u_t \overline{u}_x x^{\varepsilon_x} u_x \overline{u}_t t^{\varepsilon_t} \overline{p} y^{\varepsilon_y} p t^{\varepsilon_t} u_t| \leq |\overline{u}_x x^{\varepsilon_x} u_x|. \qquad (III.4)$$

whence the displayed occurrences of y^{ε_y} must be cancelled in reducing to normal form. Hence $u_x \equiv v y^{-\varepsilon_y} \overline{p}^{-1} t^{-\varepsilon_t} \overline{u}_t^{-1}$, for some v and $III.4$ becomes transparent in the form

$$|\overline{u}_t t^{\varepsilon_t} \overline{p}\, \overline{v} x^{\varepsilon_x} v p t^{\varepsilon_t} u_t| \leq |u_t^{-1} t^{-\varepsilon_t} p^{-1} y^{-\varepsilon_y} \overline{v} x^{\varepsilon_x} v y^{-\varepsilon_y} \overline{p}^{-1} t^{-\varepsilon_t} \overline{u}_t^{-1}|. \qquad (III.5)$$

The rewriting process requires three inequalities namely

$$\begin{aligned} L((x||t)\zeta) &< L((y||t)\zeta) & (III.6) \\ L((x||y)(x||t)\zeta) &< L((y||t)\zeta) & (III.7) \\ L((x||t)(x||y)(x||t)\zeta) &< L((y||t)\zeta). & (III.8) \end{aligned}$$

To obtain $III.6$, it suffices, using $III.1$, to show that $|(txt)\zeta| < |x\zeta|$, that is

$$|\overline{u}_t t^{\varepsilon_t} u_t \overline{u}_x x^{\varepsilon_x} u_x \overline{u}_t t^{\varepsilon_t} u_t| < |\overline{u}_x x^{\varepsilon_x} u_x|$$

which is immediate, given that the analysis subsequent to $III.4$ shows that u_x ends with $t^{-\varepsilon_t} \overline{u}_t^{-1}$. For $III.7$, another use of $III.1$ shows that it is enough to prove that $|(ytxty)\zeta| < |x\zeta|$, that is, using the form of u_x derived after $III.4$,

$$|\overline{u}_y \overline{v} x^{\varepsilon_x} v u_y| < |u_t^{-1} t^{-\varepsilon_t} p^{-1} y^{-\varepsilon_y} \overline{v} x^{\varepsilon_x} v y^{-\varepsilon_y} \overline{p}^{-1} t^{-\varepsilon_t} \overline{u}_t^{-1}|. \qquad (III.9)$$

This will follow if $|u_y| < y^{-\varepsilon_y} \overline{p}^{-1} t^{-\varepsilon_t} \overline{u}_t^{-1}|$ which is immediate if $|u_y| \leq |u_t|$ and otherwise follows from the fact that $|p| \geq |u_y| - |u_t|$.

Lastly $III.8$ is immediate from $III.1$ and $III.2$

Case IV. Suppose that $\{x, x^{-1}\} = \{t, t^{-1}\}$ and $\{y, y^{-1}\} \neq \{z, z^{-1}\}$. As usual, without loss of generality we may assume that $x = t$. This time we are given $L((x||y)(z||x)\zeta) \leq L((z||x)\zeta) > L(\zeta)$. Using relations (3) again, we obtain

$$(x||y)(z||x) = (z||y^{-1})(z||x)(x||y)(z||y^{-1}) = \delta_1 \delta_2 \delta_3 \delta_4.$$

From the given inequalities we firstly obtain, writing p for the reduced form of $u_z \overline{u}_x$,

$$|\overline{u}_x x^{\varepsilon_x} \overline{p} z^{\varepsilon_z} p x^{\varepsilon_x} u_x| > |\overline{u}_z z^{\varepsilon_z} u_z| \qquad (IV.1)$$

and, secondly, observing that necessarily $u_x \equiv v y^{-\varepsilon_y} \overline{u}_y^{-1}$, for some v, we obtain the strict(!) inequality

$$|\overline{u}_y \overline{v} x^{\varepsilon_x} v u_y| < |u_y^{-1} y^{-\varepsilon_y} \overline{v} x^{\varepsilon_x} v y^{-\varepsilon_y} \overline{u}_y^{-1}|. \qquad (IV.2)$$

The rewriting process requires

$$L((z||y^{-1})\zeta) \quad < \quad L((z||x)\zeta) \qquad\qquad (IV.3)$$
$$L((x||y)(z||y^{-1})\zeta) \quad < \quad L((z||x)\zeta) \qquad\qquad (IV.4)$$
$$L((z||x)(x||y)(z||y^{-1})\zeta) \quad < \quad L((z||x)\zeta). \qquad\qquad (IV.5)$$

The procedure is much as before. Firstly $IV.3$ is equivalent to

$$|u_y^{-1}y^{-\varepsilon_y}\overline{u}_y^{-1}\overline{u}_z z^{\varepsilon_z}u_z u_y^{-1}y^{-\varepsilon_y}\overline{u}_y^{-1}| < |\overline{u}_x x^{\varepsilon_x}\overline{p}z^{\varepsilon_z}px^{\varepsilon_x}u_x|. \qquad (IV.6)$$

Since we know $u_z = p\overline{u}_x^{-1} = p\overline{v}^{-1}y^{\varepsilon_y}u_y$, $IV.6$ is equivalent to

$$|u_y^{-1}v^{-1}\overline{p}z^{\varepsilon_z}p\overline{v}^{-1}\overline{u}_y^{-1}| < |\overline{u}_y^{-1}y^{-\varepsilon_y}\overline{v}x^{\varepsilon_x}\overline{p}z^{\varepsilon_z}px^{\varepsilon_x}vy^{-\varepsilon_y}\overline{u}_y^{-1}|. \qquad (IV.7)$$

However $IV.7$ follows since $IV.2$ implies that either the word in the right hand side is reduced or $\overline{u}_x \equiv u_z p \equiv uz^{-\varepsilon_z}q$.

Secondly $IV.4$ is implied by $IV.6$ and $IV.1$. Finally $IV.5$ reduces, via $IV.7$ and the observations which define p and v, to

$$|\overline{u}_y\overline{v}x^{\varepsilon_x}\overline{p}z^{\varepsilon_z}px^{\varepsilon_x}vu_y| < |u_y^{-1}y^{-\varepsilon_y}\overline{v}x^{\varepsilon_x}\overline{p}z^{\varepsilon_z}px^{\varepsilon_x}vy^{-\varepsilon_y}\overline{u}_y^{-1}|.$$

Like $IV.7$ this follows from $IV.1$.

Case V. Suppose that $\{x,\ x^{-1}\} = \{t,\ t^{-1}\}$ and that $\{y,\ y^{-1}\} = \{z,\ z^{-1}\}$. Without loss of generality $y = z$. We shall show that this case cannot arise, that is , the inequalities

$$L((x||y)(y||x^{\pm 1})\zeta) \leq L((y||x^{\pm 1})\zeta) > L(\zeta)$$

are inconsistent.

So suppose that these inequalities hold. The second one yields

$$|(\overline{u}_x x^{\varepsilon_x}u_x)^{\pm 1}\overline{u}_y y^{\varepsilon_y}u_y(\overline{u}_x x^{\varepsilon_x}u_x)^{\pm 1}| > |\overline{u}_y y^{\varepsilon_y}u_y|. \qquad (V.1)$$

This means that the displayed occurrences of $x^{\pm\varepsilon_x}$ cannot be cancelled when the left hand word is reduced to normal form.

Subcase V.1. Suppose that $t = x$. So if p is the reduced form of $u_y\overline{u}_x$, then $V.1$ becomes

$$|\overline{u}_x x^{\varepsilon_x}\overline{p}y^{\varepsilon_y}px^{\varepsilon_x}u_x)^{\pm 1}| > |\overline{u}_y y^{\varepsilon_y}u_y|, \qquad (V.2)$$

and most one further cancellation is possible, occurring in $\overline{p}y^{\varepsilon_y}p$.

In this case the first inequality is equivalent to $|(xyxxxyx)\zeta| \leq |x\zeta|$, which becomes

$$|\overline{u}_x x^{\varepsilon_x}\overline{p}y^{\varepsilon_y}px^{\varepsilon_x}u_x\overline{u}_x x^{\varepsilon_x}u_x\overline{u}_x x^{\varepsilon_x}\overline{p}y^{\varepsilon_y}px^{\varepsilon_x}u_x| \leq |\overline{u}_x x^{\varepsilon_x}u_x|$$

which is manifestly impossible since $u_x\overline{u}_x$ is freely reduced and there is no cancellation between any x^{ε_x} and the reduced form of $\overline{p}y^{\varepsilon_y}p$.

Subcase V.2. Suppose that $t = x^{-1}$. This time the second of the two initial inequalities shows that the displayed occurrences of $x^{-\varepsilon_x}$ cannot be cancelled and we obtain

$$|u_x^{-1}x^{-\varepsilon_x}\overline{p}y^{\varepsilon_y}px^{-\varepsilon_x}\overline{u}_x^{-1}| > |\overline{u}_y y^{\varepsilon_y}u_y|$$

where, this time, p is the reduced form of $u_y u_x^{-1}$, with no cancellation between either $x^{-\varepsilon_x}$ and the reduced form of $\bar{p} y^{\varepsilon_y} p$.

Now the first of the initial inequalitites is equivalent to $|(x^{-1}yx^{-1}yx^{-1})\zeta| \leq |x\zeta|$ which becomes

$$|u_x^{-1} x^{-\varepsilon_x} \bar{p} y^{\varepsilon_y} p x^{-\varepsilon_x} \bar{p} y^{\varepsilon_y} p x^{-\varepsilon_x} \bar{u}_x^{-1}| \leq |\bar{u}_x x^{\varepsilon_x} u_x|.$$

As written, the number of letters in the word on the left hand side is $2|u_x| + 4|p| + 5$ while the length of the right hand word is $2|u_x| + 1$. The maximum possible loss due to cancellation is 4, caused by the single possible cancellation in each of the two occurrences of $\bar{p} y^{\varepsilon_y} p$. If $|p| \geq 1$, the result is immediate but also if $|p| = 0$ then no cancellation occurs and again the desired result follows.

It remains only to observe that systematic enumeration of possibilities demonstrates that all possible cases have now been examined and the proof of Proposition 3.1 is complete.

References

[C] Collins, D.J., Cohomological dimension and symmetric automorphisms of a free group, Comment. Math. Helvetici, 64 (1989), 44-61.

[CV] Culler, M. and Vogtmann, K., Moduli of graphs and automorphisms of free groups, Invent. Math., 84 (1986), 91-119.

[FR] Fouxe-Rabinowitsch, D.I. Ueber die Automorphismengruppen der freien Produkte, I, Mat. Sb., 8 (1940), 265-276.

[G] Gilbert, N.D., Automorphisms of free products, Ph.D Thesis, University of London, 1985.

[LS] Lyndon, R.C. and Schupp, P.E., *Combinatorial Group Theory*, Berlin-Heidelberg-New York, Springer 1977.

School of Mathematical Sciences
Queen Mary and Westfield College
Mile End Road
London E1 4NS, UK
email d.j.collins@qmw.ac.uk

A Freiheitssatz for certain one-relator amalgamated products

BENJAMIN FINE, FRANK ROEHL AND GERHARD ROSENBERGER

1. Introduction

Suppose $G = \langle x_1, ..., x_n; R = 1 \rangle$ is a one-relator group with R a cyclically reduced word in the free group on $\{x_1, ..., x_n\}$ which involves all the generators. The classical *Freiheitssatz* or *independence theorem* of Magnus (see [25]) asserts that the subgroup generated by any proper subset of the generators is free on those generators. More generally suppose X and Y are disjoint sets of generators and suppose that the group A has the presentation $A = \langle X; rel(X) \rangle$ and that the group G has the presentation $G = \langle X, Y; rel(X), Rel(X, Y) \rangle$. Then we say that G satisfies a *Freiheitssatz* relative to A if $< X >_G = A$, that is the subgroup of G generated by X is isomorphic to A. In this more general language Magnus' result says that a one-relator group satisfies a Freiheitssatz relative to the free group on any proper subset of generators.

A great deal of work has gone into proving the Freiheitssatz for the class of *one-relator products*. These are groups of the form $G = (A * B)/N(R)$ where R is a non-trivial,cyclically reduced word in the free product $A * B$ of syllable length at least two. A and B are called the *factors* and R is called the *relator*. In this case the Freiheitssatz means that A and B inject into G. For one-relator products there have been two approaches to the Freiheitssatz: to either impose conditions on the factors or on the relator. B.Baumslag[1], Brodskii[4], Howie[20] and Short[31] have shown that the Freiheitssatz holds in a one-relator product, for any relator, if the factors are locally indicable. In the other direction Howie [20],[21] has proved that if $R = S^m$ with S a non-trivial,cyclically reduced word in the free product $A * B$ of syllable length at least two and not a proper power, then the Freiheitssatz holds if $m \geq 4$. (The case $m \geq 7$ actually follows from small cancellation theory - see [5],[24]). Fine, Howie and Rosenberger[8] proved the Freiheitssatz for $m \geq 2$ for one-relator products of cyclics. This last result used representation techniques which were extended in [8],[16],[17] to provide versions of the Freiheitssatz for other related classes of groups, specifically groups of special NEC type and generalized tetrahderon groups. One-relator products of cyclics and the two classes of groups mentioned above can be considered as algebraic generalizations of discrete groups (see [10] and [15]). General surveys of the Freiheitssatz and its extensions can be found in [7],[14],[24].

In this paper we prove a Freiheitssatz for certain *one-relator amalgamated products.* By this we mean groups of the form

$$G = (A *_C B)/N(R)$$

where R is a non-trivial, cyclically reduced element of length at least two in the amalgamated product $A *_C B$. In particular suppose that $G = (A *_C B)/N(S^m)$ with C cyclic and $m \geq 2$. Then if $A *_C B$ admits a complex two-dimensional representation which is faithful on A and B then under certain conditions this representation can be extended to a representation $\rho : G \to PSL_2(\mathbb{C})$ which is also faithful on A and B. In particular then A and B both inject into G. Our proof is a generalization of results in [3],[8],[10],[27] and [30] and is related to a result of Vinberg [33]. Once we establish this Freiheitssatz we can prove a series of results about these one-relator amalgamated products.

2. The Freiheitssatz and Some Consequences

We first give our main Freiheitssatz for one-relator amalgamated products. If $A, B \in PSL_2(\mathbb{C})$ then we say that the pair $\{A, B\}$ is *irreducible* if A, B, regarded as linear fractional transformations, have no common fixed point, that is, $\operatorname{tr}[A, B] \neq 2$.

Theorem 1.   The Freiheitssatz. *Suppose $H = H_1 *_A H_2$ with $A = \langle a \rangle$ cyclic. Let $R \in H \setminus A$ be given in a reduced form $R = a_1 b_1 ... a_k b_k$ with $k \geq 1$ and $a_i \in H_1 \setminus A, b_i \in H_2 \setminus A$ for $i = 1,\ldots,k$. Assume that there exists a representation $\phi : H \to PSL_2(\mathbb{C})$ such that $\phi|_{H_1}$ and $\phi|_{H_2}$ are faithful and the pairs $\{\phi(a_i), \phi(a)\}$ and $\{\phi(b_i), \phi(a)\}$ are irreducible for $i = 1,\ldots,k$.*

Then the group $G = H/N(R^m)$, $m \geq 2$ admits a representation $\rho : G \to PSL_2(\mathbb{C})$ such that $H_1 \to G \xrightarrow{\rho} PSL_2(\mathbb{C})$ and $H_2 \to G \xrightarrow{\rho} PSL_2(\mathbb{C})$ are faithful and $\rho(R)$ has order m. In particular $H_1 \to G$ and $H_2 \to G$ are injective.

Proof. Let $\phi : H \to PSL_2(\mathbb{C})$ be the given representation of H such that $\phi|_{H_1}$ and $\phi|_{H_2}$ are faithful and the pairs $\{\phi(a_i), \phi(a)\}$ and $\{\phi(b_i), \phi(a)\}$ are irreducible for $i = 1,\ldots,k$.

We may assume that $\phi(a)$ has the form $\phi(a) = \begin{pmatrix} s & 0 \\ 0 & s^{-1} \end{pmatrix}$ or $\phi(a) = \begin{pmatrix} 1 & 1 \\ 0 & 1 \end{pmatrix}$. Suppose first that $\phi(a) = \begin{pmatrix} s & 0 \\ 0 & s^{-1} \end{pmatrix}$ and let $T = \begin{pmatrix} t & 0 \\ 0 & t^{-1} \end{pmatrix}$ with t a variable whose value in $\mathbb{C}$ is to be determined. Define

1. $\rho(h_1) = \phi(h_1)$ for $h_1 \in H_1$ and

2. $\rho(h_2) = T\phi(h_2)T^{-1}$ for $h_2 \in H_2$.

Since T commutes with $\phi(a)$ for any t the map $\rho : G \to PSL_2(\mathbb{C})$ will define a representation with the desired properties if there exists a value of t such that

$\rho(R)$ has order m. Recall that a complex projective matrix A in $PSL_2(\mathbb{C})$ will have finite order $m \geq 2$ if $\operatorname{tr} A = \pm 2\cos(\pi/m)$.

As in the statement of the theorem assume $R = a_1 b_1 ... a_k b_k$ with $k \geq 1$ and $a_i \in H_1 \setminus A, b_i \in H_2 \setminus A$ for $i = 1, ..., k$ and assume that the pairs $\{\phi(a_i), \phi(a)\}$ and $\{\phi(b_i), \phi(a)\}$ are irreducible for $i = 1, ..., k$. Define

$$f(t) = \operatorname{tr} \phi(a_1)T\phi(b_1)T^{-1}....\phi(a_k)T\phi(b_k)T^{-1}.$$

$f(t)$ is a Laurent polynomial in t of degree $2k$ in both t and t^{-1}. The coefficients of t^{2k} and t^{-2k} are non-zero because the pairs $\{\phi(a_i), \phi(a)\}$ and $\{\phi(b_i), \phi(a)\}$ are irreducible for $i = 1, ..., k$. Therefore by the fundamental theorem of algebra there exists a t_0 with $f(t_0) = 2\cos(\pi/m)$. With this choice of t_0 we have $\operatorname{tr} \rho(R) = 2\cos(\pi/m)$ and thus $\rho(R)$ has order m. Therefore ρ is a representation with the desired properties.

Now assume $\phi(a) = \begin{pmatrix} 1 & 1 \\ 0 & 1 \end{pmatrix}$. In this case define $T = \begin{pmatrix} 1 & t \\ 0 & 1 \end{pmatrix}$ with t again a variable. Again T commutes with $\phi(a)$ and the proof goes through as above giving the desired representation.

We now state some consequences of the Freiheitssatz. These consequences are related to similar properties for one-relator products of cyclics (see [9]). In particular the existence of a complex representation as constructed in theorem 1 leads to many linear properties – specifically being virtually torsion-free, satisfying the Tits Alternative, being SQ-universal and having a free product with amalgamation decomposition. In what follows we let R always be of the allowed form $R = a_1 b_1 ... a_k b_k$ as in theorem 1. Note that if H_i, $i = 1 \text{or} 2$ has a free subgroup of rank 2 then G does also.

Recall that a *group of F-type* is a group H with a presentation of the form

$$H = \langle a_1, .., a_n; a_1^{e_1} = ... = a_n^{e_n} = UV = 1 \rangle \tag{1}$$

where $n \geq 2, e_i = 0$ or $e_i \geq 2$ for $i = 1, ..., n$, $U = U(a_1, ..., a_p)$, $1 \leq p \leq n-1$ is a cyclically reduced word in the free product on $a_1, ..., a_p$ which is of infinite order and $V = V(a_{p+1}, ..., a_n)$ is a cyclically reduced word in the free product on $a_{p+1}, ..., a_n$ which is of infinite order. A group of F-type is free product with amalgamation of the groups

$$H_1 = \langle a_1, ..., a_p; a_1^{e_1} = ... = a_p^{e_p} = 1 \rangle$$

$$H_2 = \langle a_{p+1}, ..., a_n; a_{p+1}^{e_{p+1}} = ... = a_n^{e_n} = 1 \rangle$$

with the cyclic amalgamation $A = \langle U \rangle = \langle V^{-1} \rangle$. Groups of F-type are natural algebraic generalizations of Fuchsian groups (see [10] and [15]).

Corollary 1. *Suppose H is a group of F- type, that is H has a presentation of form (1) above. Assume further that $n \geq 4, 2 \leq p \leq n-2$ and neither $U = U(a_1, ..., a_p)$ nor $V = V(a_{p+1}, ..., a_n)$ is a proper power in the free product on*

the generators they involve. Assume also that UV involves all the generators.
Let $R = c_1 d_1 ... c_k d_k, k \geq 1$ with $c_i \in \langle a_1, ..., a_p \rangle \setminus \langle U \rangle, d_i \in \langle a_{p+1}, ..., a_n \rangle \setminus \langle V \rangle$
for $i = 1, ..., k$ and let $m \geq 2$.

Then the conclusions of theorem 1 hold for $G = H/N(R^m)$ with $H_1 =$
$\langle a_1, ..., a_p; a_1^{e_1} = ... = a_p^{e_p} = 1 \rangle, H_2 = \langle a_{p+1}, ..., a_n; a_{p+1}^{e_{p+1}} = ... = a_n^{e_n} = 1 \rangle$ and
$A = \langle U \rangle = \langle V^{-1} \rangle$.

Proof. As remarked earlier H is a free product of H_1 and H_2 with a cyclic
amalgamation. From [26] under the conditions in the corollary H admits a
faithful representation ϕ in $PSL_2(\mathbb{C})$ such that the pairs $\{\phi(c_i), \phi(U)\}$ and
$\{\phi(d_i), \phi(V)\}$ are irreducible for $i = 1, ..., k$. Therefore the theorem applies.

In [10] it was shown that groups of F-type satisfy many linearity properties.
The existence of the extended representation ρ to a one-relator quotient of the
special groups of F-type in corollary 1 allows us also to extend these linearity
properties. We first need the following ideas.

Recall that if $\mathcal{P}$ is a group property then a group G is virtually $\mathcal{P}$ if G contains a subgroup of finite index which satisfies $\mathcal{P}$. If $\mathcal{P}$ is a subgroup-inherited
property such as torsion-freeness or solvability then the intersection of conjugates of a subgroup H of finite index satisfying $\mathcal{P}$ will be a normal subgroup
of finite index satisfying $\mathcal{P}$. Thus being virtually $\mathcal{P}$ in these cases implies the
existence of a normal subgroup of finite index satisfying $\mathcal{P}$. Further being
virtually torsion-free can be considered a linear property since from a result
of Selberg [29] all finitely generated linear groups are virtually torsion-free.

Suppose G is a group with the presentation

$$G = \langle a_1, ..., a_n; a_1^{e_1} = \cdots = a_n^{e_n} = R_1^{m_1} = \cdots = R_k^{m_k} = 1 \rangle$$

where $e_i = 0$ or $e_i \geq 2$ for $i = 1 ... n$, $m_j \geq 1$ for $j = 1, ..., k$ and each
R_i is a cyclically reduced word in the free product on $\{a_1, ..., a_n\}$ of syllable
length at least two. A representation $\rho : G \rightarrow$ Linear Group is an *essential
representation* if $\rho(a_i)$ has infinite order if $e_i = 0$ or exact order e_i if $e_i \geq 2$ and
$\rho(R_j)$ has order m_j. Essential representations have proved crucial in the study
of one-relator products of cyclics. For any group G a linear representation ρ
is an *essentially faithful representation* if ρ is finite dimensional with torsion-
free kernel. In [17] the following straightforward result was proved which had
been used previously (see [9],[11],[12] and the references there).

Theorem. *[17] Let G be a finitely generated group. Then G admits an
essentially faithful representation if and only if G is virtually torsion-free.*

Corollary 2.   *Let $G = H/N(R^m)$ with H a special group of F-type and
R the relator as in corollary 1. Suppose that $m \geq 8$. Then G is virtually
torsion-free.*

Proof. In H any element of finite order is conjugate to a power of a generator. Since $m \geq 8$ from the torsion theorem for small cancellation products (see [24] or [5]) then any element of finite order in G is conjugate to a power of a generator or a power of R. Since $\rho(R)$ has exact order m it follows that ρ is essentially faithful and therefore G is virtually torsion-free.

Recall that a group G is *SQ-universal* if every countable group can be embedded in a quotient of G. If G has an SQ-universal quotient then G is itself SQ-universal and similarly if G has a SQ-universal subgroup of finite index then G is itself SQ-universal (see [24]). From the work of Higman, Neumann and Neumann a free group of rank 2, and hence any non-abelian free group, is SQ-universal. In particular a group is SQ-universal if it contains a subgroup of finite index which maps homomorphically onto a free group of rank 2. Baumslag and Pride [2] have proved that if a group G has a deficiency of 2 or greater then G contains a subgroup of finite index mapping onto a free group of rank 2 while if the deficiency is 1 then G has a subgroup of finite index mapping onto $\mathbb{Z}$ and is thus infinite.

Corollary 3. *Let $G = H/N(R^m)$ with H a special group of F-type and R the relator as in corollary 1. For $i = 1,...,n$ let $\alpha_i = 0$ if $e_i = 0$ and $\alpha_i = (1/e_i)$ if $e_i \geq 2$. Then*

1. *if $\sum_{i=1}^{n} \alpha_i + (1/m) \leq n-2$ then G has a subgroup of finite index mapping homomorphically on $\mathbb{Z}$. In particular G is infinite;*

2. *if $\sum_{i=1}^{n} \alpha_i + (1/m) < n-2$ then G has a subgroup of finite index mapping homomorphically onto a free group of rank 2. In particular G is SQ-universal.*

Proof. Suppose $G = H/N(R^m)$ with H the special group of F-type and R the relator as in corollary 1.

From corollary 1 the group G admits an essential representation ρ into $PSL_2(\mathbb{C})$. From Selberg's theorem there exists a normal torsion-free subgroup K of finite index in $\rho(G)$. Thus the composition of maps (where π is the canonical map) $G \overset{\rho}{\to} \rho(G) \overset{\pi}{\to} \rho(G)/K$ gives an essential representation ϕ of G onto a finite group.

Let $X = \langle a_1, \ldots, a_n; a_1^{e_1} = \cdots = a_n^{e_n} = 1 \rangle$ be the free product on $a_1, \ldots, a_n$. There is a canonical epimorphism β from X onto G. We therefore have the sequence $X \overset{\beta}{\to} G \overset{\phi}{\to} \rho(G)/K$. Let $Y = \ker(\phi \circ \beta)$. Then Y is a normal subgroup of finite index in X and since ϕ is essential, Y is torsion-free. Since X is a free product of cyclics and Y is torsion-free, Y is a free group of finite rank r.

Suppose $|X : Y| = j$. Regard X as a Fuchsian group with finite hyperbolic area $\mu(X)$. (Every f.g. free product of more than two cyclics can be faithfully represented as such a Fuchsian group.) From the Riemann–Hurwitz formula, $j\mu(X) = \mu(Y)$ where $\mu(Y) = r - 1$ and

$$\mu(X) = (n - 1) - (\alpha_1 + ... + \alpha_n).$$

Equating these we obtain

$$r = 1 - j(\alpha_1 + ... + \alpha_n - (n - 1)).$$

Now G is obtained from X by adjoining the relations UV, R^m and so $G = X/M$, where M is the normal closure of UV, R^m, and M is contained in Y since $\rho(R)$ has order m. Therefore the quotient Y/M can be considered as a subgroup of finite index in G. Applying the Reidemeister–Schreier process or repeated applications of Corollary 3 in [2], Y/M can be defined on r generators subject to $j + (j/m)$ relations. The deficiency d of this presentation for Y/M is then

$$d = r - j - \frac{j}{m} = 1 - j(\sum_{i=1}^{n} \alpha_i + (1/m) - (n - 2)).$$

If $\sum_{i=1}^{n} \alpha_i + (1/m) \leq n - 2$ then $d \geq 1$ and from the work of Baumslag and Pride [2] Y/M, and hence G has a subgroup of finite index mapping onto $\mathbb{Z}$ and is thus infinite. If $\sum_{i=1}^{n} \alpha_i + (1/m) < n - 2$ then $d \geq 2$ and again from [2] Y/K, and hence G, has a subgroup of finite index mapping onto a free group of rank 2 and is thus SQ-universal.

From the proof it is clear that the deficiency argument would carry through for part (1) if $n \geq 5$ and would carry through for part (2) if either $n \geq 6$ or $n = 5$ and at least one of m or the e_i is not equal to 2.

We mention that the proof of corollary 3 essentially mirrors the proof given for groups of F-type in [10] and is related to work of Vinberg [33] and of Thomas [32]. In [17] we proved the following more general result.

Theorem. *[17] Suppose that G is a finitely presented group with a presentation of the form*

$$G = \langle a_1, \ldots, a_n; a_1^{e_1} = \ldots = a_n^{e_n} = R_1^{m_1} = \cdots = R_k^{m_k} = 1 \rangle.$$

Suppose that each $m_j \geq 2$ and that the extended deficiency $d* = n - k \geq 3$. Then if G admits an essential representation, G must contain a subgroup of finite index which maps onto a non-abelian free group. In particular G is SQ-universal and contains a non-abelian free subgroup. The result is still true if $d = 2$ but not all $m_j = 2$ or not all $e_i = 2$.*

Corollary 4.   Let H and G be as in corollary 1. Then G is a non-trivial free product with amalgamation.

Proof. We may assume that each $e_i \geq 2$. If not we introduce the relations $a_i^{f_i} = 1$ with $f_i \geq 2$ and f_i sufficiently large, for each a_i with $e_i = 0$ and write

$f_j = e_j$ if $e_j \geq 2$. The group G^* obtained by replacing each e_i by f_i is an epimorphic image of G. It is known that if $\rho : G_1 \to G_2$ is an epimorphism and G_2 is a non-trivial free product with amalgamation then G_1 is also, via ρ.

Then suppose $e_i \geq 2$ for all $i = 1, \ldots, n$: that is, G is generated by elements of finite order. From corollary 1 there exist an irreducible representation of G in $PSL_2(\mathbb{C})$ and therefore the space of representations of G as a subset of $(PSL_2(\mathbb{C}))^n$ is non-empty and contains a point related to an irreducible representation. From a result of Culler and Shalen [6] if the dimension of the space of characters of representations of a group in $PSL_2(\mathbb{C})$ is positive (as an affine variety) then the group admits a splitting as a non-trivial free product with amalgamation or an HNN group. From the construction of the representation in theorem 1 the dimension of the character space of G is $2n - 6 - 1 = 2n - 7$. This is positive if $n \geq 4$. Thus if $n \geq 4$, G splits as a non-trivial free product with amalgamation or an HNN group. However G is generated by elements of finite order so its abelianization G^{ab} is finite. Therefore G cannot be an HNN group.

Corollary 5. *Let H and G be as in corollary 1. If $n \geq 5$ or $m = 4$ and at least one of the e_i is not equal to 2 then G has a free subgroup of rank 2.*

Proof. From theorem 1 we know that either $\mathbb{Z} * \mathbb{Z}$, or $\mathbb{Z}_n * \mathbb{Z}_k$ with $k, n \geq 2, n + k \geq 5$, or $\mathbb{Z} * \mathbb{Z}_k$ with $k \geq 2$, or $\mathbb{Z}_2 * \mathbb{Z}_2 * \mathbb{Z}_2$ embed into G. Each of these has a free subgroup of rank 2 and hence G does also.

3. An Interesting Example

The Bianchi group Γ_3 is $PSL_2(\mathcal{O}_3)$ where $\mathcal{O}_3$ is the group of integers in the quadratic imaginary number field $\mathbb{Q}(\sqrt{-3})$. Using Tietze transformations on the standard presentation for this group [35] we obtain the following alternative presentation

$$\Gamma_3 = \langle a, b, c; a^2 = b^2 = c^3 = (ab)^3 = (bc)^3 = (abac)^3 = 1 \rangle.$$

This can be shown to be a one-relator amalgamated product. The Freiheitssatz of theorem 1 and the representation constructed in its proof allow us to perform an analysis of the class of groups in the following result related to the above example.

Theorem 2. *Let $G = \langle a, b, c; a^p = b^q = c^r = (ab)^k = (bc)^n = R^m(a, b, c) = 1 \rangle$, with $p = 0$ or $p \geq 2$, $q = 0$ or $q \geq 2$, $r = 0$ or $r \geq 2$, $k \geq 2$, $n \geq 2$, $m \geq 2$, such that $(1/k) + (1/p) + (1/q) \neq 1$ if $2 \leq p, q$, and $(1/n) + (1/q) + (1/r) \neq 1$ if $2 \leq q, r$, where $R(a, b, c) = a_1 b_1 \cdots a_t b_t$, with $t \geq 1$, and with $a_i \in \langle a, b; a^p = b^q = (ab)^k = 1 \rangle \setminus \langle b \rangle$, $b_i \in \langle b, c; b^q = c^r = (bc)^n \rangle \setminus < b >$. Then*

(a) There exists an essential representation $\rho : G \to PSL_2(\mathbb{C})$.

(b) *The orders are as given in the presentation.*

(c) *Let $\alpha = 0$ if $p = 0$, $\alpha = 1/p$ if $p \geq 2$, $\beta = 0$ if $q = 0$, $\beta = 1/p$ if $q \geq 2$, $\gamma = 0$ if $r = 0$ and $\gamma = 1/r$ if $r \geq 2$. If $\alpha + \beta + \gamma + (1/k)+(1/n)+(1/m) < 2$ then G has a subgroup of finite index mapping homomorphically on to a free group of rank 2 and hence G is SQ-universal.*

Proof. Let $H_1 = \langle a, b; a^p = b^q = (ab)^k = 1 \rangle$ and $H_2 = \langle b, c; b^q = c^r = (bc)^n = 1 \rangle$. Then H_1 and H_2 are ordinary triangle groups and therefore have faithful representations in $PSL_2(\mathbb{C})$. If $g \in H_1, g \notin \langle b \rangle$ then, regarded as linear fractional transformations, g and b have no common fixed points. Similarly for $h \in H_2, h \notin \langle b \rangle$. Therefore the free product with amalgamation $H_1 *_A H_2$ with $A = \langle b; b^k = 1 \rangle$ has a faithful representation into $PSL_2(\mathbb{C})$ using the arguments from [26] and [30]. Therefore theorem 1 can be applied to get part (a). Parts (b) and (c) follow as in the corollaries.

Corollary 6.   *Let G be as in theorem 2.*

(6.1) If $p = 0$ or $q = 0$ or $r = 0$ then G is infinite.

(6.2) G has a free subgroup of rank 2 under any of the following conditions:

(a) $p = 0$ and $q \neq 2$ or $k \neq 2$

(b) $q = 0$ and $p \neq 2$ or $k \neq 2$

(c) $q = 0$ and $r \neq 2$ or $n \neq 2$

(d) $r = 0$ and $q \neq 2$ or $n \neq 2$

(e) $2 \leq p, q$ and $(1/p) + (1/q) + (1/k) < 1$

(f) $2 \leq q, r$ and $(1/q) + (1/r) + (1/n) < 1$

(g) $p = 0$, $q = k = 2$, $r > 2$, $n > 2$

(h) $p = 0$ or $r = 0$ and $m \geq 5$

(i) $q = 0$ and at least one of p,r,k,n is not equal to 2

(j) $m \geq 6, p, q, r \geq 2$ and at least 2 of p,q,k or of q,r,n are not equal to 2

(k) $m \geq 6, p, q, r \geq 2$ and at least one of r,n and at least one of p,k is greater than 4

(l) $m \geq 6, p, q, r \geq 2$ and at least one of r,n and at least one of p,k is greater than 2 and $m = 2m_1$ is even

(m) $m \geq 8, p, q, r \geq 2$ and at least one of r, n and at least one of p, k is greater than 3

(6.3) If two of p, q, r are equal to 0 then G either has a free abelian group of rank 2 as a subgroup of index 2 or G is SQ-universal

(6.4) If $q = 0$ and $p = r = k = n = 2$ then G is infinite solvable for any $m \geq 2$. In this case G is a factor group of $H = \langle x, y, z; x^2 = y^2 = z^2 = (xyz)^2 = 1 \rangle$

Proof. As remarked in the proof of Theorem 2, H_1 and H_2 as defined there are ordinary triangle groups and embed into G. Most of the parts of corollary 6 follow easily from the fact that they are true in H_1 or H_2. We show the details for two cases where this is not so obvious. The others follow in the same manner. For more details on this type of argument see [13].

(6.3) If two of p, q, r are equal to zero then after applying Nielsen transformations ($a \to ab$ if p = 0 for example) we may write G as a one-relator product of cyclics. The SQ-universality in this case then follows from the SQ-universality of one-relator products of cyclics. [28].

(6.2)(h) Suppose $p = 0$. Then by (6.2)(a) and (6.2)(g) we may asume that $q = k = 2$ and $r = 2 orn = 2$. Assume $r = 2$, the case $n = 2$ being analagous. If $m = n \geq 5$ then $(1/m) + (1/n) < 1/2$ and hence G has a free subgroup of rank 2 from theorem 2 part (c). So now suppose $m \neq n$.

From the construction in Theorem 1 we have an irreducible, essential representation $\rho : G \to PSL_2(\mathbb{C}), \rho(G) = \overline{G}, \rho(a) = A, \rho(b) = B, \rho(c) = C$, such that $G_1 = \langle a, b; b^2 = (ab)^2 = 1 \rangle$ and $G_2 = \langle b, c; b^2 = c^2 = (bc)^n = 1 \rangle$ embed into $\overline{G}$. Assume that G does not have a free subgroup of rank 2. Assume first that $n \geq 3$. From [27,Theorem 2.3] it follows that A and BC necessarily commute, that is , $BCA = ABC = BA^{-1}C$ and hence $A^{-1} = CAC$. From this we get that $R(A, B, C) = A^\alpha (BC)^\gamma B^\epsilon$ for some $\alpha \in \mathbb{N}, 0 \leq \gamma < n, 0 \leq \epsilon \leq 1$. Since $m \geq 3$ we must have $\alpha = 0 = \epsilon$. Otherwise if $\epsilon = 1$ then $R^2(A, B, C) = (BC)^\gamma A^\alpha B A^\alpha B (BC)^{-\gamma} = 1$ contradicting $m \geq 3$; while if $\epsilon = 0, \alpha \neq 0$ then $1 = R^{nm}(A, B, C) = A^{nm\alpha}(BC)^{nm\gamma} = A^{nm\alpha}$ which also gives a contradiction.

Therefore the above gives $R(A, B, C) = (BC)^\gamma$. Then we get that $m = n/d$ where $d = \gcd(\gamma, n)$ and, especially, $5 \leq m \leq n$ and $(1/m) + (1/n) < 1/2$ which contradicts theorem 2 part (c). Hence if $n \geq 3$, then G has a free subgroup of rank 2.

Now suppose $n = 2$. From [27 , Theorem 2.3] we have two possibilities: $AC = CA$ or $CAC = A^{-1}$. If $CAC = A^{-1}$ then A and BC commute and we may argue as above. Now let $AC = CA$. From this we get that $R(A, B, C) = A^\alpha, R(A, B, C) = A^\alpha C, R(A, B, C) = A^\alpha B$ or $R(A, B, C) = A^\alpha BC$ for some $\alpha \in \mathbb{Z}$ (recall that n = 2). In the first two cases we must have $\alpha = 0$ which contradicts $m \geq 3$. In the last two cases we have $R^2(A, B, C) = 1$ which also contradicts $m \geq 3$. Hence G also has a free subgroup of rank 2 if $n = 2$.

The other parts of the corollary follow in essentially the same manner.

Relative to the groups considered in the corollary we make the following conjecture:

Conjecture.   *Let G, H be as in Theorem 2. Suppose $p = 0$ or $q = 0$. Then if $m \geq 3$, G has a free subgroup of rank 2.*

Suppose $p = 0$. Then from the proof of (6.3)(h) above we see that the conjecture is true except for the following possible exceptions: $q = k = 2, 3 \leq m \leq 4$ and $r = 2$ or $n = 2$. Without loss of generality assume $q = k = r = 2$ and $3 \leq m \leq 4$. Then the proof of (6.3)(h) shows that the conjecture is true with the possible exceptions:

(i) $m = n = 3$,

(ii) $m = 3, n = 6$,

(iii) $m = n = 4$.

The corollaries after theorem 1 provide examples for the main theorem for the case when H is a special group of F-type. Theorem 2 provides another example when the factors in the free product with amalgamation are ordinary triangle groups. For other example of free products with amalgamation $H = H_1 *_A H_2$ where H_1 and H_2 admit faithful representations into $PSL_2(\mathbb{C})$ we may take for H_1 and H_2 any non-elementary Kleinian groups or combinations of such groups via the Klein-Maskit combination theorems.

4. Some Generalizations of the Main Theorem

Exactly the same proof as that used for theorem 1 can be employed to prove the following two generalizations.

Theorem 3.   *Let $H = H_1 *_A H_2$ with $A = H_1 \cap H_2$. Let $R \in H \setminus A$ be given in reduced form $R = a_1 b_1 \ldots a_k b_k$ with $k \geq 1$ and $a_i \in H_1 \setminus A$, $b_i \in H_2 \setminus A$ for $i = 1, \ldots, k$. Assume that there exists a representation $\phi : H \to PSL_2(\mathbb{C})$ with the properties that $\phi|_{H_1}$ and $\phi|_{H_2}$ are faithful, for each $a \in A$ there exists $\gamma_a \in \mathbb{C}$, such that $\phi(a) = \begin{pmatrix} \gamma_a & 0 \\ 0 & \gamma_a^{-1} \end{pmatrix}$ or for each $a \in A$ there exists $\gamma_a \in \mathbb{C}$ such that $\phi(a) = \begin{pmatrix} 1 & \gamma_a \\ 0 & 1 \end{pmatrix}$ and for each $i = 1, \ldots, k$ and for each $a \in A$ the pairs $\{\phi(a_i), \phi(a)\}$ and $\{\phi(b_i), \phi(a)\}$ are irreducible.*

Then the group $G = H/N(R^m), (m \geq 2)$ admits a representation $\rho : G \to PSL_2(\mathbb{C})$ such that $H_1 \to G \overset{\rho}{\to} PSL_2(\mathbb{C})$ and $H_2 \to G \overset{\rho}{\to} PSL_2(\mathbb{C})$ are faithful and $\rho(R)$ has order m. In particular $H_1 \to G$ and $H_2 \to G$ are injective.

Theorem 4. *Let $H = H_1 *_A H_2$ with $A = H_1 \cap H_2$. Let $R \in H \setminus A$ be given in reduced form $R = a_1 b_1 \cdots a_k b_k$ with $k \geq 1$ and $a_i \in H_1 \setminus A, b_i \in H_2 \setminus A$ for $i = 1, \ldots, k$. Assume that there exists a representation $\phi : H \to PSL_2(\mathbb{C})$ with the properties that $\phi|_{H_1}$ and $\phi|_{H_2}$ are faithful, for each $a \in A$ there exists $\gamma_a \in \mathbb{C}$, such that $\phi(a) = \begin{pmatrix} 1 & \gamma_a \\ 0 & 1 \end{pmatrix}$ and for each $i=1,\ldots,k$ there exists $x_i \neq 0, y_i \neq 0$ such that $\phi(a_i) = \begin{pmatrix} * & * \\ x_i & * \end{pmatrix}, \phi(b_i) = \begin{pmatrix} * & * \\ y_i & * \end{pmatrix}$.*

Then the group $G = H/N(R^m), m \geq 2$ admits a representation $\rho : G \to PSL_2(\mathbb{C})$ such that $H_1 \to G \xrightarrow{\rho} PSL_2(\mathbb{C})$ and $H_2 \to G \xrightarrow{\rho} PSL_2(\mathbb{C})$ are faithful and $\rho(R)$ has order m. In particular $H_1 \to G$ and $H_2 \to G$ are injective.

The proof of theorem 3 is exactly the same as the proof of theorem 1. The conditions on the images of the elements of the amalgamated subgroup again force the trace polynomial to be non-constant. In theorem 4 the form of the images of the elements which appear in the relator force these images to be irreducible in pairs with the images of the elements of the amalgamated subgroup. Again this forces the trace polynomial to be non-constant and the proof goes through as in theorem 1.

We close with a general question related to the above.

Question. *Suppose $H = H_1 *_A H_2$ with $A = H_1 \cap H_2$. Suppose there exist faithful representations $\phi_i : H_i \to PSL_2(\mathbb{C})$ for $i = 1, 2$. Under what conditions on H_1, H_2, A may we construct a representation $\phi : H \to PSL_2(\mathbb{C})$ such that $\phi|_{H_1}$ and $\phi|_{H_2}$ are faithful?*

We note that it is possible to construct such a representation in the following cases:

1. H_1, H_2 are free and A is maximal cyclic. [30]

2. H_1, H_2 are free products of cyclics and A is cyclic. [10],[26]

3. Each $H_i, i = 1, 2$ is either free, a free product of cyclics or a non-elementary Fuchsian group with positive dimensional Teichmuller space and A is cyclic.

4. A is finite cyclic

A general criterion to answer the question is given by P. Shalen in [30] and generalized by S.Litz [22] to $GL_2(\mathbb{C})$.

Finally we note that if H_1, H_2 embed into $PSL_2(\mathbb{C})$ then the free product $H_1 * H_2$ also embeds into $PSL_2(\mathbb{C})$

References

[1] B.Baumslag, Free products of locally indicable groups with a single relator, *Bull. Austr. Math. Soc.* **29** (1984) 401–404.

[2] B.Baumslag and S.Pride, Groups with two more generators than relators, *J. London Math. Soc.* **17** (1987) 425–426.

[3] G. Baumslag, J.Morgan and P.Shalen, Generalized triangle groups, *Math. Proc. Camb. Phil. Soc.* **102** (1987) 25–31.

[4] S. Brodskiĭ, Equations over groups and groups with a single defining relation, *Siberian Math. J.* **25** (1984) 231–251.

[5] D. Collins and F. Perraud, Cohomology and finite subgroups of small cancellation quotients of free products, *Math. Proc. Camb. Phil. Soc.* **37** (1985) 243–259.

[6] M. Culler and P. Shalen, Varieties of Group Representations and Splittings of Three Manifolds, *Ann. of Math.* **117** (1983) 109–147.

[7] A. J. Duncan and J. Howie, One-relator products with high powered relators, in: *Geometric Group Theory, Volume 1* (G. A. Niblo and M. A. Roller, eds.), London Mathematical Soc. Lecture Notes Series **181**, Cambridge University Press (1993) 48-74.

[8] B.Fine, J.Howie and G.Rosenberger, One-relator quotients and free products of cyclics, *Proc. Amer. Math. Soc.* **102** (1988) 1–6.

[9] B.Fine and G.Rosenberger, Complex Representations and One-Relator Products of Cyclics, in: *Geometry of Group Representations, Contemporary Math.* **74** (1987) 131–149.

[10] B.Fine and G.Rosenberger, Generalizing Algebraic Properties of Fuchsian Groups, in: *Proceedings of Groups St. Andrews 1989, Volume 1* (C. M. Campbell and E. F. Robertson, eds.), London Mathematical Soc. Lecture Notes Series **159**, Cambridge University Press, (1989) 124–148.

[11] B. Fine, F. Levin and G. Rosenberger, Free Subgroups and Decompositions of One-Relator Products of Cyclics: Part 1: The Tits Alternative, *Arch. Math.* **50** (1988) 97–109.

[12] B.Fine, F.Levin and G.Rosenberger, Free subgroups and decompositions of one-relator products of cyclics; Part 2 : Normal Torsion-Free Subgroups and FPA Decompositions *J. Indian Math. Soc.* **49** (1985) 237–247.

[13] B.Fine and G.Rosenberger, Note on Generalized Triangle Groups *Abh. Math. Sem.Univ. Hamburg* **56** (1986) 233–244.

[14] B.Fine and G.Rosenberger, The Freiheitssatz of Magnus and its Extensions, in: *Groups, Geometries and Special Functions:The Legacy of Wilhelm Magnus.* Contemporary Mathematics, American Mathematical Society (to appear)

[15] B.Fine, A. Hempel and G.Rosenberger, Algebraic Generalizations of Certain Discrete Groups, *Amalele Universitatii din Timisora* (1993) 57-91.

[16] B.Fine and G.Rosenberger, On Groups of Special NEC Type, Contemporary Mathematics, *Proceedings of the Haifa Topology Conference,* 1992, Contemporary Mathematics (to appear).

[17] B.Fine and G.Rosenberger, Groups which admit essentially faithful representations, (to appear)

[18] B.Fine, F. Levin, F. Roehl and G.Rosenberger, Generalized Tetrahedron Groups, *Proceedings of the Ohio State Geometric Group Theory Conference,* 1992, World Scientific Press (to appear).

[19] J. Howie, One-Relator Products of Groups, in: *Proceedings of Groups St. Andrews - 1985* (C. M. Campbell and E. F. Robertson, eds.), Camb. Univ. Press (1986) 216-219.

[20] J. Howie, Cohomology of One-Relator Products of Locally Indicable Groups, *J.London Math. Soc.* **30** (1984) 419-430.

[21] J. Howie, The quotient of a free product of groups by a single high-powered relator. I. Pictures. Fifth and higher powers, *Proc. London Math. Soc.* **59** (1989) 507-540.

[22] J. Howie, The quotient of a free product of groups by a single high-powered relator. II. Fourth powers, *Proc. London Math. Soc.* **61** (1990) 33-62.

[23] S. Litz, Darstellungen gewisser freier Produkte mit Amalgam in die $GL_2(\mathbb{C})$, *Diplomarbeit,* Dortmund, 1990.

[24] R.C. Lyndon and P.E. Schupp, *Combinatorial Group Theory,* Springer-Verlag 1977.

[25] W. Magnus, A. Karass and D. Solitar, *Combinatorial Group Theory,* Wiley Interscience, New York 1968.

[26] G. Rosenberger, Faithful linear representations and residual finiteness of certain one-relator product of cyclics, *J. Siberian Math. Soc.* 1990.

[27] G. Rosenberger, On Free Subgroups of Generalized Triangle Groups, *Alg. i Logika* **28** 1989 227-240.

[28] G. Rosenberger, The *SQ*-universality of one-relator products of cyclics, *Results in Math.* **21** 1992 396–402.

[29] A. Selberg, On discontinuous groups in higher dimensional symmetric spaces, *Int. Colloq. Function Theory* , Tata Institute, Bombay (1960) 147–164.

[30] P. Shalen, Linear representations of certain amalgamated products, *J. Pure and Applied Algebra* **15** 1979 187–197.

[31] H. Short, Topological Methods in Group Theory: The Adjunction Problem, Ph.D. Thesis, University of Warwick, 1984.

[32] R.Thomas, Cayley Graphs and Group Presentations, *Math. Proc. Cambridge Phil. Soc.* **103** 1988 385–387.

[33] E.B. Vinberg On Groups Generated by Periodic Relations, (to appear)

[34] H. Zieschang, On decompositions of discontinuous groups of the plane, *Math. Z.* **151** (1976), 165-168.

[35] B. Fine, *The Algebraic Theory of the Bianchi Groups,* Marcel Dekker 1989.

Benjamin Fine,
Department of Mathematics,
Fairfield University,
Fairfield, Connecticut 06430,
United States

Frank Roehl,
Department of Mathematics,
University of Alabama,
Tuscaloosa, Alabama
United States

Gerhard Rosenberger,
Fachbereich Mathematik Universitat,
Dortmund,
44221 Dortmund,
Germany.

Isoperimetric functions of groups and exotic cohomology

S. M. GERSTEN

Abstract

This is an expanded version of the lecture given at the ICMS conference in Edinburgh, March 1993. The complexity of the word problem is formulated in terms of isoperimetric functions. Exotic cohomology theories are defined to give lower bounds for isoperimetric functions. Invariance of cohomological dimension under quasi-isometries for groups possessing finite Eilenberg-MacLane spaces is discussed along with several consequences.

1. Let $\mathcal{P} = \langle x_1, x_2, \ldots, x_k \mid R_1, \ldots, R_m \rangle$ be a finite presentation for the group $G = G(\mathcal{P}) = F(\mathcal{P})/\langle\langle \vec{R} \rangle\rangle$, where $F = F(\mathcal{P})$ is the free group on the generators and $\langle\langle \vec{R} \rangle\rangle$ is the normal closure of the relators. The word problem for $\mathcal{P}$ asks whether there is an algorithm to determine whether or not a general word in the generators (and their inverses) represents the identity in G. It is unsolvable in general, as we know from the work of Novikov, Boone, and Higman (see [28]). It is now possible to give a measure of complexity to the word problem. In fact there are several such, but the most useful have been isoperimetric functions, which were introduced by Gromov in his fundamental paper, "Hyperbolic groups" [21]. A word w represents 1 in G (written $w \equiv_G 1$) if $w =_F \prod_{j=1}^n R_{i_j}^{\pm u_j}$. We set $\text{Area}_{\mathcal{P}}(w)$ to be the minimum n among all such expressions for w in the free group F. Geometrically, this means that w can be thought of as the label of the boundary circuit of a singular disc (a so-called "van Kampen diagram") mapped into the 2-complex $K(\mathcal{P})$ canonically associated to $\mathcal{P}$; here $K(\mathcal{P})$ has one vertex, a geometric edge for each free generator x_i, and a 2-cell with attaching map spelling out R_i for each relator R_i. Then an *isoperimetric function* $f : \mathbb{N} \to \mathbb{N}$ for $\mathcal{P}$ is any function satisfying

$$f(n) \geq \max\{\text{Area}_{\mathcal{P}}(w) \mid w \equiv_G 1, \ \ell(w) \leq n\},$$

where $\ell(w)$ is the length of the word w. The *Dehn function* of $\mathcal{P}$ is the minimal such isoperimetric function, defined by the right hand side of the inequality above.

Theorem 1 ([16]) *The group G has a solvable word problem iff there exists a recursive isoperimetric function for $\mathcal{P}$ iff the Dehn function for $\mathcal{P}$ is recursive.*

In order to discuss the invariance properties of isoperimetric functions, we need to bring in geometry.

Let G be a finitely generated group with finite set A of generators. The word metric on G is defined by

$$d(g',g) = \min\{n \mid g' = ga_1^{\pm 1}a_2^{\pm 1}\ldots a_n^{\pm 1}\},$$

where the $a_i \in A$. The Cayley graph Γ has vertex set G and has one geometric edge for each triple (g,a,g'), where $g,g' \in G$, $a \in A$, and $g' = ga$. It is a metric space where each edge has length 1 and where it is given the path metric; so the metric on Γ agrees with the word metric on vertices.

2. The appropriate equivalence relation on metric spaces is quasi-isometry. It represents how an astigmatic, far-sighted person would see things (so nearby things are not recognized and far-away things are distorted by a bounded distortion factor). Precisely, let $(X,d),(X',d')$ be metric spaces with metrics d,d'. A quasi-isometry is a pair of maps $f : X \to X', f' : X' \to X$ such that there exist constants $\lambda > 0$ and $\epsilon, C \geq 0$ so that

$$d'(f(x),f(y)) \leq \lambda d(x,y) + \epsilon \qquad\qquad d(f' \circ f(x),x) \leq C,$$
$$d(f'(x'),f'(y')) \leq \lambda d'(x',y') + \epsilon, \qquad d'(f \circ f'(x'),x') \leq C.$$

Note that these functions f, f' need not be continuous, because of the ϵ.

Examples:

a. A finitely generated group G is quasi-isometric to its Cayley graph and all Cayley graphs for G are mutually quasi-isometric.

b. If M is a closed Riemannian manifold, then the universal cover $\widetilde{M}$ is quasi-isometric to $\pi_1(M)$ [29][40].

Theorem 2 [1][38] *Let $\mathcal{P},\mathcal{P}'$ be finite presentations for quasi-isometric groups G, G'. Let f be an isoperimetric function for $\mathcal{P}$. Then there exist $A, B, C, D, E > 0$ so that the function $f'(n) = Af(Bn + C) + Dn + E$ is an isoperimetric function for $\mathcal{P}'$.*

3. We write $f' \preceq f$ if f', f are related as in the theorem, and we write $f' \sim f$ if both $f' \preceq f$ and $f \preceq f'$. This is an equivalence relation among functions, and, from the theorem, it makes sense to speak of an isoperimetric function (resp. Dehn function) for a finitely presented group being linear, quadratic, polynomial, exponential, recursive, etc.

Examples:

a. A finitely presented group G is word hyperbolic iff there is a linear isoperimetric function (recall that G is called *word hyperbolic* if there exists $\delta \geq 0$ so that for any geodesic triangle ABC in the Cayley graph and point $P \in [AB]$, the distance of P from the union of the other two sides is at most δ). This result is stated in a very general form in [21]2.3.F. The paper [30] contains a complete proof that a subquadratic isoperimetric inequality for finitely presented groups implies word hyperbolic. I recently received a very interesting paper [31], which verifies Gromov's original assertions in their most general form and which gives significant new insights into foundational questions of hyperbolic groups.

b. If G is automatic, then there exists a quadratic isoperimetric function [11]. Recall that the finitely generated group G is called automatic if one is given for each $g \in G$ an edge-path $p(g)$ from the identity to g in the Cayley graph Γ, so that

(i) the collection of labels of these paths is the language of a finite state automaton, and

(ii) there exists a constant $k > 0$ so that whenever $g, g' \in G$ are connected by an edge in Γ, the paths $p(g), p(g')$ remain uniformly k-close (the path $p(g)$ is parametrized by arc-length until it reaches its end point g, then is extended to the positive reals as the constant map at g).

c. Polycyclic groups have exponential isoperimetric functions [17].

d. Finitely generated nilpotent groups have polynomial isoperimetric functions [17][9]. Gromov argues in [22] that there is an isoperimetric polynomial of degree $c + 1$ if the class of nilpotence is c, and Pittet has confirmed this in the case of lattices in simply connected homogeneous nilpotent Lie groups of class c [32].

e. There is one definitive result for a restricted class of polycyclic groups:

Theorem 3 [4] *Let* $G = \mathbb{Z}^n \rtimes_\phi \mathbb{Z}$ *where* $\phi \in \mathrm{Gl}_n(\mathbb{Z})$. *Then the Dehn function for* G *is either polynomial or exponential. It is polynomial iff all eigenvalues of* ϕ *are roots of unity, in which case the degree of the polynomial is* $c + 1$, *where* c *is the size of the largest Jordan block of* ϕ.

For example, if $G = \mathbb{Z}^2 \rtimes_\phi \mathbb{Z}$ with $\phi = \begin{pmatrix} 1 & 1 \\ 0 & 1 \end{pmatrix}$, then G is the Heisenberg group, $c = 2$, and the Dehn function is a cubic polynomial.

It is also proved in [4] that the groups $\mathbb{Z}^n \rtimes_\phi \mathbb{Z}$ and $\mathbb{Z}^{n'} \rtimes_{\phi'} \mathbb{Z}$ with *unipotent* monodromies ϕ, ϕ' are quasi-isometric if and only if both $n = n'$ and ϕ and ϕ' have the same Jordan canonical forms. Lattices in the Lie group Sol (*e.g.* groups $\mathbb{Z}^2 \rtimes_\phi \mathbb{Z}$ where $\mathrm{Tr}(\phi^2) > 2$) give counterexamples to the analogous assertion in the general case, when ϕ is not unipotent.

f. Magnus showed that all 1-relator groups have a solvable word problem [28]. I calculated that Ackermann's function f_ω is an isoperimetric function for all 1-relator groups (unpublished). Recall that the Ackermann hierarchy [34] is defined by $f_1(n) = 2n$; $f_{r+1}(n) = f_r^{(n)}(1)$ (where $f^{(n)}$ means the n-fold iterate of f); and $f_\omega(n) = f_n(n)$. Thus f_2 grows exponentially and f_3 grows faster than any *iterated* exponential.

I showed that the group $\langle x, y \mid x^{(x^y)} = x^2 \rangle$ has Dehn function f_3 [13] (I learned at the Conference from Zelmanov's talk that f_3 is called the "tower function" and f_4 is called the "wowzer"). It is not known whether f_4 is possible as the Dehn function for a 1-relator group.

g. There exists a finitely presented solvable group G of derived length 3 $(D^3 G = 1)$ with an unsolvable word problem [26]. It would be of interest to understand better the geometry underlying such examples.

4. In calculating Dehn functions, there are two steps involving significantly different techniques. Upper bounds are derived from normal forms for groups elements, whereas lower bounds involve the techniques of algebraic topology. I want now to discuss results of several of my papers on calculating lower bounds [13][14][15]. There is some overlap with the results announced in [35] concerning the quasi-isometry invariance of the exotic real cohomology theories introduced below, as was pointed out by Shmuel Weinberger; the idea of applying these notions to finitely supported cohomology over $\mathbb{Z}$ is due to me.

Let X be a complex of type $K(G, 1)$. Consider the set $C^i_{(p)}(\widetilde{X}, \mathbb{R})$ of cellular i-cochains c on $\widetilde{X}$ which are p-summable,

$$\sum_{\widetilde{e}^{(i)}} |c(\widetilde{e}^{(i)})|^p < \infty$$

(the sum is over all i-cells $\widetilde{e}^{(i)}$ of $\widetilde{X}$). If $p = \infty$ the summability condition is replaced by the condition that there exists a constant M_c so that $|c(\widetilde{e}^{(i)})| \leq M_c$ for all $\widetilde{e}^{(i)}$.

If $X^{(n)}$ is a finite complex, then the subgroups $C^i_{(p)}(\widetilde{X}, \mathbb{R}) \subset C^i(\widetilde{X}, \mathbb{R})$ are stable under the coboundary for $i < n$, so we define

$$H^i_{(p)}(\widetilde{X}, \mathbb{R}) = Z^i_{(p)}/B^i_{(p)}$$

for $i \leq n$.

We say that a group G is of type $\mathcal{F}_n$ if it possesses an Eilenberg-MacLane space of type $K(G, 1)$ with a finite n-skeleton, and we say that G is of type $\mathcal{F}_\infty$ if it is of type $\mathcal{F}_n$ for all n. Gromov notes that the condition $\mathcal{F}_n$ is quasi-isometry invariant [22] 1.C$_2$. H. Meinert informed me at the conference that J. Alonso has found a homotopy theoretic proof of this fact.

Theorem 4 *Suppose that G, G' are quasi-isometric groups possessing Eilenberg-MacLane spaces X, X' with finite n-skeleta. Then*

$$H^n_{(p)}(\widetilde{X}, \mathbb{R}) \cong H^n_{(p)}(\widetilde{X'}, \mathbb{R}).$$

This result is proved in detail for $p = \infty$ in Section 11 of [14], where it is observed that the same argument works for all $p > 0$.

Consequently we define $H^*_{(p)}(G, \mathbb{R}) = H^*_{(p)}(\widetilde{X}, \mathbb{R})$.

Remark   If $p \geq 1$ (so $C^i_{(p)}(\widetilde{X}, \mathbb{R})$ is a normed linear space), we can also form the *reduced* cohomology groups $\overline{H}^i_{(p)}(\widetilde{X}, \mathbb{R}) = Z^i_{(p)}/\overline{B}^i_{(p)}$, where $\overline{B}^i_{(p)}$ is the closure of $B^i_{(p)}$ in the ℓ_p-norm. These groups are also quasi-isometry invariant, provided X has a finite n-skeleton and $i \leq n$. If $p = 2$, the reduced groups $\overline{H}^i_{(2)}(\widetilde{X}, \mathbb{R})$ are the ℓ_2-cohomology groups of [8]. The action of G on $\widetilde{X}$ by deck transformations makes both $H^*_{(p)}(\widetilde{X}, \mathbb{R})$ and $\overline{H}^*_{(p)}(\widetilde{X}, \mathbb{R})$ into G-modules and, when $p = 2$, there is the notion of von Neumann dimension for the latter modules. However the G-module structure is not quasi-isometry invariant, so the von Neumann dimensions are not invariant. But the condition of vanishing of the von Neumann dimension is quasi-isometry invariant, since it is equivalent to the vanishing of the corresponding cohomology group.

5.   An interesting example is given by taking $p = \infty$, $\star = 2$.

Theorem 5   *Let $z \in Z^2_{(\infty)}(\widetilde{X}, \mathbb{R})$, where $X^{(2)}$ is finite and X is an Eilenberg-MacLane space. Then there exists a constant $C > 0$ so that*

$$|\langle z, D_w \rangle| \leq C \text{ Area}(w)$$

for all circuits w in $\widetilde{X}^{(1)}$ and all van Kampen diagrams D_w for w.

Proof.   We first show that the left hand side of the inequality does not depend on D_w. If then D_w and D'_w are two van Kampen diagrams for w, we can combine them to obtain a spherical diagram $f : S^2 \to \widetilde{X}^{(2)}$, so

$$\langle z, S^2 \rangle = \langle z, D'_w \rangle - \langle z, D_w \rangle.$$

But $z = \delta b$ with $b \in C^1(\widetilde{X}, \mathbb{R})$, since $\widetilde{X}$ is contractible. It follows that

$$\langle z, S^2 \rangle = \langle \delta b, S^2 \rangle = \langle b, \partial S^2 \rangle = 0,$$

whence $\langle z, D'_w \rangle = \langle z, D_w \rangle$.

Thus we may pick D_w of minimal area $\text{Area}(w)$, and we obtain

$$|\langle z, D_w \rangle| \leq ||z||_\infty \text{Area}(w).$$

This completes the proof.

Remark   One must take care not to confuse H^*_b with $H^*_{(\infty)}$, where the former is Gromov's bounded cohomology [24]. There exist natural maps $H^*_b(G, \mathbb{R}) \to H^*(G, \mathbb{R})$ and $H^*(G, \mathbb{R}) \to H^*_{(\infty)}(G, \mathbb{R})$, and the composition of these maps is always zero [14] Proposition 10.3. For example, if G is the fundamental group of a closed orientable surface of genus g, then the map $H^2_b(G, \mathbb{R}) \to H^2(G, \mathbb{R})$ is surjective if $g > 1$, whereas the map $H^2(G, \mathbb{R}) \to H^2_{(\infty)}(G, \mathbb{R})$ is injective if $g = 1$, the last assertion following from Theorem 6 below.

For the preceding theorem to be useful, we need examples of classes in $H^2_{(\infty)}$. The next two results show how to produce such classes in two different situations.

Theorem 6   *If G is amenable and of type $\mathcal{F}_n$, then the canonical map $H^n(G, \mathbb{R}) \to H^n_{(\infty)}(G, \mathbb{R})$ is injective.*

This is proved in [14] Theorem 10.13.

Theorem 7   *Let $H < G$ where both H, G possess finite aspherical presentations. Then the restriction homomorphism $H^2_{(\infty)}(G, \mathbb{R}) \to H^2_{(\infty)}(H, \mathbb{R})$ is defined and is surjective.*

This is a restatement of Theorem C of [13] in terms of cohomology. Thus, if one knows classes for H, this result produces classes for G. As an application of this result, it follows that *if the group G in the last theorem contains a subgroup isomorphic to the group $\langle x, y \mid x^y = x^2 \rangle$, then the Dehn function of G must grow at least exponentially.* The reader will find other examples of lower bound estimates for isoperimetric functions in [14] §10.

6.   Although we have implicitly assumed above that p-summable meant $p > 0$, everything above makes sense for $p = 0$, provided ℓ_0 summability is interpreted as cochains of finite support on $\tilde{X}$. Now everything makes sense for all rings Λ, and in particular everything is defined for $\mathbb{Z}$. We obtain the next result for the finitely supported cohomology groups $H^*_f(\tilde{X}, \Lambda)$ (a proof is contained in the Appendix below).

Theorem 8   *Let the quasi-isometric groups G, G' admit Eilenberg-MacLane spaces X, X' respectively with finite n-skeleta. Then for all rings Λ we have $H^n_f(\tilde{X}, \Lambda) \cong H^n_f(\tilde{X'}, \Lambda)$.*

Remark   The group $H^n_f(\tilde{X}, \Lambda)$ is isomorphic to $H^n(G, \Lambda G)$ when X is a $K(G, 1)$ with a finite n-skeleton (*cf.* [2] p.209 Proposition 7.5).

Corollary 1   *The number of ends of a finitely generated group is a quasi-isometry invariant.*

Remark   This fact was proved previously for hyperbolic groups in [19] and in general in [6].

Proof of Corollary 1 Let G be a finitely generated group with an Eilenberg-MacLane space X having a finite 1-skeleton, and let k be a field. We have the short exact sequence of cochain complexes

$$0 \to C^*_f(\widetilde{X}, k) \to C^*(\widetilde{X}, k) \to \mathcal{Q} \to 0,$$

where $\mathcal{Q}$ is the quotient complex, and the corresponding exact cohomology sequence

$$0 \to H^0_f(\widetilde{X}, k) \to H^0(\widetilde{X}, k) \to H^0(\mathcal{Q}) \to H^1_f(\widetilde{X}, k) \to 0,$$

where we used $\widetilde{X}$ is contractible. It is shown in [12] that $\dim_k(H^0(\mathcal{Q}))$ is the number of ends of G; also, $\mathcal{Q} = 0$ if and only if G is finite. Thus we see that if G is finite, the number of ends is 0, whereas if G is infinite, then the number of ends is $1 + \dim_k(H^1_f(\widetilde{X}, k))$. The result now follows by applying Theorem 8.

7. In order to state the next corollary, we need to recall the definition of Poincaré duality group (resp. Bieri-Eckmann duality group). For our purposes we say that the group G is a Poincaré duality group (resp. Bieri-Eckmann duality group) if

(i) G is of type $\mathcal{F}_\infty$, so G has an Eilenberg-MacLane space X with $X^{(n)}$ finite for all n, and

(ii) G has finite cohomological dimension over $\mathbb{Z}$, and

(iii) in addition, $H^i_f(\widetilde{X}, \mathbb{Z})$ vanishes for all but one integer d where

$H^d_f(\widetilde{X}, \mathbb{Z}) \cong \mathbb{Z}$ (resp. $H^d_f(\widetilde{X}, \mathbb{Z})$ is a torsion free abelian group, in the case of Bieri-Eckmann duality).

For *finitely presented* groups, the definition just given for duality groups agrees with the more familiar definition in terms of cup product isomorphisms [2] p. 200 ff (for finitely presented groups, the condition FP in [2] is equivalent to the conjunction of (i) and (ii) above).

Corollary 2 *Let G, G' be quasi-isometric groups of type $\mathcal{F}_\infty$, both having finite cohomological dimension over $\mathbb{Z}$. Then G is a Poincaré duality group (resp. Bieri-Eckmann duality group) if and only if G' is a Poincaré duality group (resp. Bieri-Eckmann duality group).*

8. We recall now that Bieri showed that if X is a finite $K(G, 1)$, then $\mathrm{cd}(G) = \sup\{n \mid H^n_f(\widetilde{X}, \mathbb{Z}) \neq 0\}$ [2] p. 209. The next result follows from this observation and Theorem 8.

Theorem 9   *If the quasi-isometric groups G, G' both admit finite Eilenberg-MacLane spaces, then* $\mathrm{cd}(G) = \mathrm{cd}(G')$.

This result settles affirmatively a conjecture of Gromov's.

9.   I want next to give an application of the preceding result, which is a far-reaching generalization of Bieberbach's theorem; recall that Bieberbach's theorem states that if G is a discrete group of isometries of Euclidean space $\mathbb{R}^n$ acting properly discontinuously with compact quotient, then G contains a subgroup of finite index isomorphic to $\mathbb{Z}^n$. The argument below appears in my joint paper with Bridson [4], and only uses the invariance of cohomological dimension under quasi-isometries as applied to polycyclic groups, which is proved in our joint paper. Note that a polycyclic group is of type $\mathcal{F}_\infty$, as is clear from the facts that it contains a poly-$\mathbb{Z}$ subgroup of finite index and that a poly-$\mathbb{Z}$ group has a finite Eilenberg-MacLane space, as follows from the fact that that this is obtained by iterated fibrations over the circle (a good reference for polycyclic groups is [37]).

Corollary 1   *If a finitely generated group G is quasi-isometric to $\mathbb{Z}^n$, then G contains a subgroup of finite index isomorphic to $\mathbb{Z}^n$.*

Proof.   Since G has polynomial growth, it follows that G is virtually nilpotent [23]. Thus we may replace G by a finite index subgroup H, which is finitely generated torsion-free nilpotent.

The growth of H is polynomial of degree $\sum_{i=1}^{c} i\ \mathrm{rk}(H_i/H_{i+1})$ [3]; here, $H_1 = H$ and $H_{r+1} = [H, H_r]$, the lower central series of H, and c is the class of nilpotence of H. Since the growth of $\mathbb{Z}^n$ is polynomial of degree n, it follows that

$$\sum_{i=1}^{c} i\ \mathrm{rk}(H_i/H_{i+1}) = n.$$

On the other hand, $\mathrm{cd}(H) = \sum_{i=1}^{c} \mathrm{rk}(H_i/H_{i+1})$, so by the quasi-isometry invariance of cohomological dimension, we get

$$\sum_{i=1}^{c} \mathrm{rk}(H_i/H_{i+1}) = n.$$

It follows that

$$\sum_{i=2}^{c}(i-1)\ \mathrm{rk}(H_i/H_{i+1}) = 0,$$

whence H_2 is trivial. Thus H is abelian, and the proof is complete.

Corollary 2   *If G is a connected semi-simple real Lie group, then no uniform lattice in G is quasi-isometric to any nonuniform lattice in G.*

Proof. It is known that lattices in semi-simple Lie groups are of type $\mathcal{F}_\infty$ and have finite virtual cohomological dimensions [33]. The corollary follows from the fact that the virtual cohomological dimensions of uniform and nonuniform lattices are different.

Open question It is interesting to speculate whether there are partial converses for the preceding corollary. For example, we may ask whether the quasi-isometry type of a lattice in $SO(n, 1)$ is determined by its virtual cohomological dimension, or, equivalently, whether two nonuniform lattices in $SO(n, 1)$ are quasi-isometric. The fact that all such lattices satisfy the same isoperimetric inequalities as Euclidean spaces (a consequence of the fact that they are all automatic groups [11]) can be taken as positive evidence for this conjecture.

Another consequence of Theorem 9 is the fact that the q-rank of an arithmetic lattice Λ in the real semi-simple Lie group $\mathcal{G}$ is a quasi-isometry invariant; here one changes the $\mathbb{Q}$-structure, but fixes the group $\mathcal{G}$ of real points. This assertion follows from the formula

$$\operatorname{vcd}(\Lambda) = \dim(\mathcal{G}/K) - q - \operatorname{rank}(G);$$

here K is a maximal compact subgroup of $\mathcal{G}$, and G is an algebraic group defined over $\mathbb{Q}$ with (connected component of the) set of real points $G(\mathbb{R})$ equal to $\mathcal{G}$, and where Λ is commensurable with $G(\mathbb{Z})$, the set of integral points; the q-rank of G is the dimension of a maximal $\mathbb{Q}$-split torus.

10. Here is another application of these ideas.

Theorem 10 *Suppose that the finitely generated group G is quasi-isometric to a polycyclic group. If in addition either a) G is a subgroup of $\operatorname{Gl}_n(\mathbb{C})$ or b) G is solvable, then G is virtually polycyclic.*

Remark It is an open question whether one can omit the assumptions of linearity or solvability of G from the hypothesis of Theorem 10 (note that both conditions are necessary for G to be polycyclic). It is also open whether the class of finitely generated virtually solvable groups is closed under quasi-isometry (that is, whether the property of having a solvable subgroup of finite index is a so-called *geometric property* [20]).

Proof of Theorem 10 Assume first that $G < \operatorname{Gl}_n(\mathbb{C})$. Although it is not known whether virtual solvability is a geometric property, it is true that amenability is a geometric property. Since polycyclic groups are amenable, it follows that G is amenable. Since G is also linear, it follows from Tits's theorem [41] that G is virtually solvable. Let H be a torsion-free solvable subgroup of G of finite index; such a subgroup H exists by Selberg's lemma. The property $\mathcal{F}_n$ of groups is a geometric property [22], and since polycyclic groups enjoy this property, it follows that H has property $\mathcal{F}_n$ for all n. But Kropholler proved that a torsion-free linear group of type $\mathcal{F}_\infty$ has finite cohomological dimension [27] Corollary to Theorem A. It follows that H has finite cohomological dimension. But it follows from Corollary 2 to Theorem 8 that H is a Poincaré duality group. Finally, Bieri showed that every solvable Poincaré duality group is polycyclic [5] Theorem 9.23, whence H is polycyclic. This completes the proof of the theorem under the assumption that G is linear.

Let us assume now that G is finitely generated and solvable and quasi-isometric to a polycyclic group. As in the preceding paragraph, it follows that G has property $\mathcal{F}_\infty$. We may then invoke the proof of the corollary to Theorem B in Section 5 of [27] to deduce that G has a torsion-free subgroup H of finite index of type $\mathcal{F}_\infty$. It then follows from Theorem A of [27] that H has finite cohomological dimension over $\mathbb{Z}$. Then we can invoke Corollary 2 to Theorem 8 again to deduce that H is a Poincaré duality group, and it follows as before from Bieri's theorem that H is polycyclic. This completes the proof in the solvable case.

Acknowledgement   I am grateful to Peter Kropholler for having shown me the argument above in the case when G is solvable.

11.   As yet another application of Theorem 8, we have the following result.

Theorem 11   *If the finitely generated group G is quasi-isometric to the finitely generated free group F, then G is virtually free.*

Remark   A proof of this result for torsion-free groups G is given in [19] §7 Theorem 19, while a sketch is offered for the general case.

Proof of Theorem 11   Note that G is of type $\mathcal{F}_\infty$ and G is word hyperbolic, since both of these are geometric properties. Hence we have the Rips complex P for G [19] Chapter 4. We recall that P is a contractible finite dimensional cell complex on which G acts properly discontinuously and cellularly with compact quotient complex $G \backslash P$; the stabilizer G_σ of each cell σ of P is a finite subgroup of G. It follows that $\mathrm{cd}_\mathbb{Q}(G) < \infty$. The analog of Bieri's observation,

$$\mathrm{cd}_\mathbb{Q}(G) = \max\{n \mid H^n_f(\widetilde{X}, \mathbb{Q}) \neq 0\},$$

then holds, where X is an Eilenberg-MacLane space for G with $X^{(m)}$ finite for all numbers m. Since the groups $H^n_f(\widetilde{X}, \mathbb{Q})$ are quasi-isometry invariants by Theorem 8, it follows from the fact that $\mathrm{cd}(F) \leq 1$ that $\mathrm{cd}_\mathbb{Q}(G) \leq 1$. We can now apply Corollary 3.16 of [10] p. 114 to deduce that G has a free subgroup of finite index. This completes the proof.

12. It may be helpful to interpret these results in terms of a notion of dimension. Let X be an Eilenberg-MacLane space of type $K(G,1)$ with a finite n-skeleton for all n. If one considers only $\mathbb{Z}$-coefficients, then the groups $H_{(p)}^n(\widetilde{X}, \mathbb{Z})$ all reduce to $H_f^n(\widetilde{X}, \mathbb{Z})$ (the ℓ_0 cohomology) when $p < \infty$, so one has only $H_{(0)}^n(\widetilde{X}, \mathbb{Z})$ and $H_{(\infty)}^n(\widetilde{X}, \mathbb{Z})$. In fact, the groups $H_{(\infty)}^n(\widetilde{X}, \mathbb{Z})$ are closely related to $H_{(\infty)}^n(\widetilde{X}, \mathbb{R})$; in the first version of [14], I used integer cohomology, and only changed to real cohomology when I needed Theorem 6 above, whose proof uses the existence of means on amenable groups. This motivates the following

Definition   If G is a group of type $\mathcal{F}_\infty$ and X is a $K(G,1)$ with a finite n-skeleton for all n, then the ℓ_i-dimension of G is defined by

$$\ell_i\mathrm{dim}(G) = \sup\{n \mid H_{(i)}^n(\widetilde{X}, \mathbb{Z}) \neq 0\},$$

$i = 0, \infty$. Observe that $\ell_0\mathrm{dim}(G) = \sup\{n \mid H^n(G, \mathbb{Z}G) \neq 0\}$. These notions of dimension are quasi-isometry invariant. This contrasts with the notions of cohomological dimension and virtual cohomological dimension which are not quasi-isometry invariant in this generality. The following results hold:

a. If $G \in \mathcal{F}_\infty$ and $\mathrm{vcd}(G) < \infty$, then $\mathrm{vcd}(G) = \ell_0\mathrm{dim}(G)$.

b. If $G = \pi_1(M^n)$, where M^n is a closed smooth aspherical manifold of dimension n, then $\ell_0\mathrm{dim}(G) = n$; furthermore (i) if G is amenable, then $\ell_\infty\mathrm{dim}(G) = n$, and (ii) if G is non-amenable, then $\ell_\infty\mathrm{dim}(G) < n$. (This last result follows from [25] 4.2.)

c. The group F first considered by R. J. Thompson and studied by Brown and Geoghegan is of type $\mathcal{F}_\infty$, has $H^*(F, \mathbb{Z}F) = 0$, and has $H_n(F, \mathbb{Z}) \cong \mathbb{Z}^2$ for all $n \geq 1$ [7]. Hence $\mathrm{cd}(F) = \infty$ while $\ell_0\mathrm{dim}(F) = 0$.

It is not known what the ℓ_∞-cohomology of F is at the time of writing; if F were amenable, then $H^*(F, \mathbb{Z})$ would inject in $H_{(\infty)}^*(F, \mathbb{R})$ and in $\overline{H}_{(\infty)}^*(F, \mathbb{R})$, but amenability of F is an open problem.

Open question   If G and G' are torsion-free groups which are quasi-isometric, then is $\mathrm{cd}(G) = \mathrm{cd}(G')$?

This is true if G and G' are of type $\mathcal{F}_\infty$ and both groups have finite cohomological dimension. These conditions hold if the groups are of type $\mathcal{F}_\infty$ and both groups are either linear groups over $\mathbb{C}$ or solvable, by Kropholler's results [27].

13. I would like to add a word about producing examples of quasi-isometries of finitely generated groups. The basic result here, which includes the examples given previously of the Cayley graph and the universal cover of a closed Riemannian manifold, is the following.

Proposition 12 ([19] p. 60)   *Let X be a proper geodesic metric space and let G be a discrete subgroup of isometries acting properly discontinuously on X. If $G\backslash X$ is compact, then G is finitely generated and its Cayley graph for any finite set of generators is quasi-isometric to X.*

A consequence of this result is that if $N \lhd G$ with G a finitely generated group, then G is quasi-isometric to N if N is of finite index in G, and G is quasi-isometric to G/N if N is finite. Furthermore, it follows from Proposition 12 that two cocompact lattices in the same connected real Lie group $\mathcal{G}$ are each quasi-isometric to $\mathcal{G}$ and hence to each other.

There are however quasi-isometries arising from group theory which are somewhat exotic.

Theorem 13 ([18] Theorem 3.1)   *If we have a central extension*

$$1 \to \mathbb{Z} \to E \to G \to 1$$

where G is finitely generated and where the extension is defined by a bounded real 2-cocycle $G \times G \to \mathbb{R}$, then the group E is quasi-isometric to the product $\mathbb{Z} \times G$.

Nontrivial examples of Theorem 13 are given in [18]. For instance, if Λ, Λ' are uniform lattices in the Lie groups $SO(2,1) \times \mathbb{R}$ and $\widetilde{Sl}_2(\mathbb{R})$, respectively, then Λ is quasi-isometric to Λ', although Λ is never isomorphic to Λ' (as one sees by comparing the Seifert invariants of the corresponding orbifolds [36]).

One of the banes of the theory is the existence of quasi-isometries arising from geometry which appear to have nothing to do with group theory. One such is the bi-Lipschitz homeomorphism of $\mathbb{R}^2$ given by the formula in polar coordinates

$$f(r,\theta) = (r, \theta + \log(r+1)),$$

which maps rays from the origin into logarithmic spirals. The existence of examples such as this appears to preclude the existence of any reasonable boundary for general finitely generated groups, although Gromov views the "asymptotic cone" construction as a replacement for the boundary in general [22] §2. In the asymptotic cone construction, a logarithmic spiral in $\mathbb{R}^2$ gets identified with some ray from the origin; but it appears to be outside the realm of constructive mathematics to decide which ray it is identified with.

Appendix: Quasi-isometry Invariance of the Groups H_f^*

In this appendix I shall show that the groups $H_f^i(G, \Lambda)$ are quasi-isometry invariants if the group G is of type $\mathcal{F}_n$ and $i \leq n$ and Λ is any ring. The argument presented here is just a rewording of the argument of Section 11 of [14].

We shall assume that G, G' are groups and that X, X' are *simplicial complexes* of type $K(G,1)$, $K(G',1)$ respectively, each possessing a finite n-skeleton. It can be shown by standard techniques that if G has an Eilenberg-MacLane complex with a finite n-skeleton, then there is a simplicial complex of type $K(G,1)$ with a finite n-skeleton, so we have not lost any generality by our assumption. Furthermore we assume that a quasiisometry is given between G and G'. It is convenient for us to formulate this in terms the sets of vertices $V(\Gamma)$, $V(\Gamma')$ of the 1-skeletons Γ, Γ' of the universal covers $\widetilde{X}$, $\widetilde{X'}$ of X, X', where Γ, Γ' are equipped with their word metrics (so that each is a geodesic metric space, where each edge has length 1 and the metric is left invariant for the action of deck transformations) as follows. We assume that we are given maps $\alpha : V(\Gamma) \to V(\Gamma')$ and $\alpha' : V(\Gamma') \to V(\Gamma)$ such that there exist positive constants C, λ, ϵ with

$$|1_{V(\Gamma)} - \alpha'\alpha| \leq C,$$

$$|1_{V(\Gamma')} - \alpha\alpha'| \leq C,$$

$$|\alpha(x) - \alpha(y)| \leq \lambda|x - y| + \epsilon \qquad \text{for all } x, y \in V(\Gamma), \text{ and}$$

$$|\alpha'(x') - \alpha'(y')| \leq \lambda|x' - y'| + \epsilon \qquad \text{for all } x', y' \in V(\Gamma').$$

Here we have indiscriminately denoted all the word metrics by $|x - y|$.

Definition A.1 Let $f : K \to L$ be a simplicial map, where the domain is a triangulation of S^i or D^i. We call $\text{Area}_i f$ the number of i-simplexes of the domain. More generally, if K is a polyhedron which is homeomorphic to either S^i or D^i, if L is a simplicial complex and if $f : K \to L$ is a polyhedral map [39] which is simplicial for some triangulation of K *without subdividing* L, we say that $\text{Area}_i f \leq N$ if the number of i-simplexes of a subdivision of K which makes f simplicial is at most N.

Lemma A.2 *Let X be a simplicial complex of type $K(G,1)$ with a finite n-skeleton. For each number $1 \leq i \leq n$ and for each number $M > 0$ there is a number $N = N_i(M)$ with the property that if $f : K \to \widetilde{X}$ is a simplicial mapping of a simplicial complex K homeomorphic to S^i with $\text{Area}_i f \leq M$, then f extends to a simplicial mapping $F : K' \to \widetilde{X}$, where the pair (K', K) is homeomorphic to the pair (D^{i+1}, S^i) and where $\text{Area}_{i+1} F \leq N$.*

Proof. Projecting f down to X, we obtain a simplicial map $\bar{f} : K \to X$ with $\text{Area}_i f \leq M$ such that $\bar{f}$ is null homotopic. But there are only finitely many such maps $\bar{f}$, so extending each over the $(i + 1)$-disc simplicially, we obtain finitely many maps of $D^{i+1} \to X$. Let N be the maximum number of $(i + 1)$-simplices in the domains of these finitely many simplicial maps.

Then lifting these maps to $\widetilde{X}$, we obtain extensions of the type required and a bound on their Area_{i+1}.

Proposition A.3   *There is a polyhedral map $\alpha_* : \widetilde{X} \to \widetilde{X}'$ extending the map α on the zero skeleta and numbers N_i, $1 \leq i \leq n$ such that if Δ is an i-simplex of $\widetilde{X}$, then $\alpha|_\Delta$ is a polyhedral mapping with $\text{Area}_i(\alpha|_\Delta) \leq N_i$ (where one does not subdivide $\widetilde{X}'$, but only subdivides the domain of $\alpha|_\Delta$).*

Proof.   For $i = 1$, we define the extension of α to $\Gamma = \widetilde{X}^{(1)}$ as follows. If e is an edge of $\widetilde{X}$, then $|\alpha(\partial_1 e) - \alpha(\partial_0 e)| \leq \lambda + \epsilon$. We may thus choose a path in $\widetilde{X}'$ from $\alpha(\partial_0 e)$ to $\alpha(\partial_1 e)$ of length at most $N_1 = \lambda + \epsilon$, subdivide e and extend the map to send the subdivided edge to this path.

In general, having extended α over $\widetilde{X}^{(i)}$, $i < n$, so that for each i-simplex $\Delta^{(i)}$ of $\widetilde{X}$ we have $\text{Area}_i \alpha_*|_{\Delta^{(i)}} \leq N_i$, let Δ be an $(i+1)$-simplex of $\widetilde{X}$. The extension α_* of α restricted to $\partial\Delta$ has Area_i at most $(i+1)N_i$, so by Lemma A.2 we may extend over a subdivision of Δ simplicially so that Area_{i+1} of the extension is uniformly bounded by some number N_{i+1}.

Once the extension α_* over $\widetilde{X}^{(n)}$ has been defined, we extend polyhedrally over the rest of $\widetilde{X}$ without further concern for the areas.

We indicate now how to define $\alpha^* : H_f^i(\widetilde{X}', \Lambda) \to H_f^i(\widetilde{X}, \Lambda)$, where $i \leq n$. If $f' \in C_f^i(\widetilde{X}', \Lambda)$ and Δ is an i-simplex of $\widetilde{X}$, we set

$$\alpha^* f'(\Delta) = f'(\alpha_*(\Delta)),$$

where $\alpha_*(\Delta)$ is considered as an i-chain on $\widetilde{X}'$. Since the area of $\alpha_*|_\Delta$ is bounded by N_i, it follows that the diameter of the support of the chain $\alpha_*(\Delta)$ is also bounded by N_i. From this it follows that $\alpha^* f'$ takes nonzero values on only a finite subset of the i-simplexes of $\widetilde{X}$, and hence $\alpha^* f' \in C_f^i(\widetilde{X}, \Lambda)$. Since the mapping α_* on chains is induced by a mapping of spaces, it follows that α^* is a cochain mapping in the range of dimensions for which it is defined, and hence it induces homomorphisms $\alpha^* : H_f^i(G', \Lambda) \to H_f^i(G, \Lambda)$, $i \leq n$.

If we apply the same considerations to $\alpha' : X'^{(0)} \to X^{(0)}$, we get the induced map α'^* defined at the level of cochains $C_f^i(\widetilde{X}, \Lambda) \to C_f^i(\widetilde{X}', \Lambda)$ inducing homomorphisms on finite supported cohomology which we denote with the same symbol α'^*. Thus one has the compositions of maps $\alpha'^* \cdot \alpha^*$ and $\alpha^* \cdot \alpha'^*$ at the level of finitely supported valued cochains. We must show these compositions are the identity at the level of cohomology, in the range of dimensions where they are defined. Furthermore, in the preceding paragraph, we must show that we get the same map α^* at the level of cohomology if we had chosen a different extension of α. Both of these questions are handled by showing that there are geometrically defined homotopies with finiteness properties for Area_i.

In the first case, one uses the data that $|1_{V(\Gamma')} - \alpha\alpha'| \leq C$, $|1_{V(\Gamma)} - \alpha'\alpha| \leq C$ as the initial step of the construction of the required homotopy, together with Lemma A.3 to give a bound on areas in the inductive step. In the second case, if we had another extension of α to a polyhedral map $\widetilde{X} \to \widetilde{X'}$, the homotopy is constant at the map α on the zero skeleton and the extension proceeds by repeated application of Lemma A.3. The outcome of these considerations is the following result.

Theorem A.4 *Let G, G' be two groups each having an Eilenberg-MacLane space with a finite n-skeleton. Let (α, α') be a quasiisometry between them. Then there are well-defined inverse isomorphisms $\alpha^* : H_f^i(G', \Lambda) \to H_f^i(G, \Lambda)$, $\alpha'^* : H_f^i(G, \Lambda) \to H_f^i(G', \Lambda)$ for all $i \leq n$.*

References

[1] J. Alonso, Inégalités isopérimétriques et quasi-isométries, *C. R. Acad. Sci. Paris Série 1* **311** (1990), 761–764

[2] K. Brown, *Cohomology of Groups* Springer-Verlag, 1982

[3] H. Bass, The degree of polynomial growth of finitely generated nilpotent groups, *Proc. London Math. Soc.(3)* **25** (1972), 603–614

[4] M.R. Bridson and S. M. Gersten, The optimal isoperimetric inequality for $\mathbb{Z}^n \rtimes_A \mathbb{Z}$ Preprint, University of Utah 1992.

[5] R. Bieri, *Homological dimension of discrete groups,* Queen Mary College Notes, 1976

[6] S. Brick, Quasi-isometries and ends of groups, *Jour. Pure Appl. Alg* **86** (1993), 23–33

[7] K. S. Brown and R. Geoghegan, An infinite-dimensional torsion-free FP_∞ group, *Invent. Math.* **77** (1984), 367–381

[8] J. Cheeger and M. Gromov, L_2-cohomology and group cohomology *Topology* **25** (1986), 189–215

[9] G. R. Conner, Isoperimetric functions for central extensions, to appear in proceedings of conference on Geometric Group Theory held at Ohio State University, June 1992 (R. Charney, W. D. Neumann, M. Shapiro, eds.), de Gruyter Math. Series.

[10] W. Dicks and M. J. Dunwoody, *Groups acting on graphs* Cambridge University Press, 1989

[11] D. B. A. Epstein, J. W. Cannon, D. F. Holt, S. V. F. Levy, M. S. Paterson, and W. P. Thurston, *Word processing in groups* Bartlett and Jones, Boston, 1992

[12] D. B. A. Epstein, Ends, in: *Topology of 3-manifolds and related topics,*(M. K. Fort, Jr., ed.), Prentice-Hall, 1962

[13] S. M. Gersten, Dehn functions and ℓ_1-norms of finite presentations, in: *Algorithms and Classification in Combinatorial Group Theory* (G. Baumslag and C. F. Miller III, eds.), MSRI series **23** Springer-Verlag, 1991

[14] —— Bounded cohomology and combings of groups Preprint "version 5.5", University of Utah (1992)

[15] —— Quasi-isometry invariance of cohomological dimension *Comptes Rendus Acad. Sci. Paris Série 1 Math.* **316** (1993), 411–416

[16] —— Isoperimetric and isodiametric functions of finite presentations, in: *Geometric Group Theory, Volume 1* (G. A. Niblo and M. A. Roller, eds.), London Mathematical Society Lecture Notes series **181** Cambridge University Press, 1993, 79-96

[17] —— Isodiametric and isoperimetric inequalities in group extensions, Preprint, University of Utah (1991)

[18] —— Bounded cocycles and combings of groups *Intern. Jour. Alg. and Computation* **2** (1992), 307–326

[19] E. Ghys and P. de la Harpe, *Sur les groupes hyperboliques d'après Mikhael Gromov,* Progress in Math. **83** Birkhäuser, 1990

[20] E. Ghys, Les groupes hyperboliques, *Séminaire Bourbaki* no. 722 *Astérisque* **189–190** (1990)

[21] M. Gromov Hyperbolic groups, in: *Essays in group theory* (S. M. Gersten, ed.), MSRI series **8** Springer-Verlag, 1987

[22] —— Asymptotic invariants of infinite groups, in: *Geometric Group Theory, Volume 2* (G. A. Niblo and M. A. Roller, eds.), London Mathematical Society Lecture Notes series **182** Cambridge Univ. Press, 1993

[23] —— Groups of polynomial growth and expanding maps *IHES Publ. Math. Sci.* **53** (1981), 53–78

[24] —— Volume and bounded cohomology, *IHES Publ. Math* **56** (1982), 5–100

[25]   ———   , Hyperbolic manifolds, groups and actions, in: *Riemann surfaces and Related Topics, Proceedings of the 1978 Stony Brook Conference,* (I. Kra, ed.), Annals of Math. Study vol. 97, Princeton Univ. Press, 1980

[26] O. Kharlampovich, A finitely presented soluble group with insoluble word problem, *Izvestia Akad. Nauk Ser. Mat.* **45** (1981), 852–873

[27] P. H. Kropholler, On groups of type $(FP)_\infty$, Preprint, Queen Mary and Westfield College, University of London, 1993

[28] R. C. Lyndon and P. E. Schupp, *Combinatorial Group Theory,* Springer-Verlag, 1977

[29] J. Milnor, A note on curvature and fundamental group, *Jour. Diff. Geom.* **2** (1963), 1–7

[30] A. Yu Ol'shanskii, Hyperbolicity of groups with subquadratic isoperimetric inequality, *Intern. Jour. Alg. and Computation* **1** (1991), 281–290

[31] P. Papasoglu, Strongly geodesically automatic groups are hyperbolic, Preprint, Columbia University, 1993

[32] C. Pittet, Isoperimetric inequalities for homogeneous nilpotent groups, Preprint, Université de Genève, 1992

[33] M. S. Raghunathan, *Discrete subgroups of Lie groups,* Ergebnisse der Mathematik **68** Springer-Verlag, 1972

[34] J. Robbins, Large numbers and unprovable theorems, *Amer. Math. Monthly* **90** (1983), 669–675

[35] J. Roe, Exotic cohomology and index theory, *Bull. Amer. Math. Soc.* **23** (1990), 447–453

[36] P. Scott, The geometries of 3-manifolds, *Bull. London Math. Soc.* **15** (1983), 405–487

[37] D. Segal, *Polycyclic Groups,* Cambridge University Press, 1983

[38] H. Short, Groups and combings, Preprint, ENS Lyon, July 1990

[39] J. Stallings, Brick's quasi-simple filtrations for groups and 3-manifolds, in: *Geometric Group Theory, Volume 1* (G. A. Niblo and M. A. Roller, eds.), London Mathematical Society Lecture Notes series **181** Cambridge University Press, 1993, 188-203.

[40] A. A. Švarc, A volume invariant for coverings *Dokl. Akad. Nauk SSSR* **105** (1955), 32–34

[41] J. Tits, Free subgroups in linear groups, *J. Algebra* **20** (1972), 250–270

Department of Mathematics,
University of Utah,
Salt Lake City
UT 84112
USA
E-mail: gersten@math.utah.edu

Some embedding theorems and undecidability questions for groups

C. McA. Gordon

AUTHOR'S NOTE. This paper was written in 1980, but was not published at that time. The reference made to it in Miller's survey article [10], however, made me think that it might be worth publishing. It is unchanged except for the deletion of some remarks that either are outdated or no longer seem interesting. I am grateful to Professor Miller for his interest in the paper.

1. Introduction.

In the first part of this paper we give proofs of two embedding theorems for groups, and a version of the Adjan-Rabin construction for showing that many group-theoretic decision problems are unsolvable, which seem to be simpler than the standard ones. In the second part we consider some specific undecidability questions.

In [5], Higman, Neumann and Neumann proved the following theorem (see also [11]).

Theorem 1 *Every countable group G can be embedded in a 2-generator group H. If G is n-relator, then H may be taken to be n-relator.*

P. Hall (unpublished) proved that every countable group can be embedded in a finitely generated simple group. This was sharpened by Goryushkin [3] and Schupp [13] to

Theorem 2 *Every countable group can be embedded in a 2-generator simple group.*

(Compare [4] and [7], p.190.)

Following [13], we define an *Adjan-Rabin* (AR) *construction* to be an effective procedure for producing, from a finite presentation Π of a group G and a word w in the generators of Π, a finite presentation Π_w, of a group G_w, say, such that if $w = 1$ in G, then $G_w = \{1\}$, whereas if $w \neq 1$ in G, then G embeds in G_w. Such constructions are described in [1] and [12], where they are used to deduce from the unsolvability of the word problem, the unsolvability of many other decision problems concerning finite presentations of groups. (See also the version of [12] given in [8] and [7], p.192.) Schupp showed [13] that in fact one may take Π_w to be a 2-generator presentation. Thus we may state

Theorem 3   AR *constructions exist. Moreover,* Π_w *may be taken to have* 2 *generators.*

We shall give proofs of Theorems 1, 2 and 3 which seem to be both simpler and more elementary than those referred to above. For example, we make no use of HNN extensions or small cancellation theory, but only free products with amalgamation.

A consequence of our AR construction is that the triviality problem for finitely presented groups is unsolvable at level 3 with respect to free products with amalgamation in the sense of [9].

Next we consider the second homology group (Schur multiplier) $H_2(G)$ of a group G. If G is finitely presented, then $H_2(G)$ is a finitely generated abelian group. However, we show that there is no algorithm for computing $H_2(G)$ from a finite presentation of G. (I am grateful to Andrew Casson for bringing this problem to my attention.) More precisely, we have

Theorem 4   *There is no algorithm for deciding, given a finite presentation of a group* G, *whether or not* $H_2(G) = 0$.

Closely related to this is

Theorem 5   *There is no algorithm for computing, from a finite presentation of a group* G, *the deficiency of* G.

Our proof of Theorem 4 also has the following consequence.

Theorem 6   *There is no algorithm for deciding, given a finite presentation of a group* G, *whether or not* G *is a higher-dimensional knot group.*

2. Embedding theorems and an AR construction.

The proofs of Theorems 1, 2 and 3 will depend on choosing elements in a free group, or free product, which form a basis for a free subgroup. The fact that the stated elements do have the desired property will follow from standard normal form arguments. More precisely, one shows that any product corresponding to a reduced non-trivial word in the elements represents a non-trivial element in the free product, by showing that it has positive length when expressed in normal form. We suppress the details, but choose elements such that the possibilities of cancellation in the juxtaposition of normal forms are sufficiently restricted that these may be readily supplied.

Finally, $[x,y]$ means $x^{-1}y^{-1}xy$, and for $1 \leq m \leq \infty$, we let I_m denote $\{1, 2, \ldots, m\}$ if $m < \infty$, and $\{1, 2, \ldots\}$ if $m = \infty$.

Proof of Theorem 1.   Let G be a given countable group, and let x_i, $i \in I_m$ (for some m, $1 \leq m \leq \infty$), be any set of generators of G. Let $F(a, \alpha)$, $F(b, \beta)$ be the free groups on a, α and b, β respectively. Let H be the free product with amalgamation $(G * F(a, \alpha)) *_C F(b, \beta)$, where C is the free group corresponding to the amalgamating relations

(i) $a\alpha a^{-1} = b^2$

(ii) $\alpha a \alpha^{-1} = b\beta b^{-1}$

(iii) $a^{2i}x_i\alpha^{2i} = \beta^i b\beta^{-i}$, $\qquad i \in I_m$.

Then G is embedded in H.

Let S be the subgroup of H generated by a and b. Then $a, b \in S$ implies $\alpha \in S$ (by (i)); $a, \alpha, b \in S$ implies $\beta \in S$ (by (ii)); and $a, \alpha, b, \beta \in S$ implies $x_i \in S$, $i \in I_m$, (by (iii)). Thus H is generated by a and b.

Finally, if G has a presentation $(x_i : r_1, \ldots, r_n)$, then in the presentation of H obtained by adjoining additional generators a, α, b, β, and relations (i), (ii), and (iii), we may successively express α, β, and the x_i in terms of a and b, using (i), (ii), and (iii), which yields for H the 2-generator, n-relator presentation $(a, b : \bar{r}_1, \ldots, \bar{r}_n)$, where $\bar{r}_j$ is obtained from r_j by substituting for each x_i the corresponding word in a and b.

Proof of Theorem 2. Let $u_i, v_i \in F(b, \beta)$ be defined by

$$\begin{aligned} u_i &= \beta^{2i}b\beta^{-2i}, & i = 1, 2, \ldots, \\ v_i &= \beta^{2i-1}b\beta b^{-1}\beta^{-(2i-1)}, & i = 1, 2, \ldots . \end{aligned}$$

Let g_i, $1 \leq i < |G|$, be the set of all distinct, non-trivial elements of G. Let H be the free product with amalgamation $(G * F(a, \alpha)) *_D F(b, \beta)$, where D is a free group and the amalgamating relations are

(i) $a\alpha a^{-1} = b^2$

(ii) $\alpha a \alpha^{-1} = b\beta b^{-1}$

(iii) $a^{2i}g_i\alpha^{2i} = u_{2i-1}$, $\qquad 1 \leq i < |G|$

(iv) $[g_i, a] = u_{2i}$, $\qquad 1 \leq i < |G|$

(v) $[g_i, \alpha] = v_i$, $\qquad 1 \leq i < |G|$.

As before, (i), (ii) and (iii) show that H is generated by a and b.

Now note that setting $g_i = 1$ for some i implies $u_{2i} = 1$ (by (iv)), hence $b = 1$; similarly, $v_i = 1$ (by (v)), hence $\beta = 1$, and hence $a = 1$ (by (ii)). In other words, if N is any normal subgroup of H with $N \cap G \neq \{1\}$, then $N = H$. The rest of the argument is now the usual one. Let M be a maximal normal subgroup of H such that $M \cap G = \{1\}$. Then $Q = H/M$ is a simple 2-generator group, and the composition $G \subset H \to Q$ is an embedding.

Proof of Theorem 3. Let $\Pi = (x_1, \ldots, x_m : r_1, \ldots, r_n)$ be a finite presentation of G, and let w be a word in $x_1, \ldots, x_m$. Let Π'_w be the presentation obtained from Π by adjoining additional generators a, α, b, β, and additional relations

(i) $a\alpha a^{-1} = b^2$

(ii) $\alpha a \alpha^{-1} = b\beta b^{-1}$

(iii) $a^{2i} x_i \alpha^{2i} = u_{i+1}$, $\qquad 1 \le i \le m$

(iv) $[w, a] = u_1$

(v) $[w, \alpha] = v_1$.

As before, Π'_w is Tietze equivalent to a presentation Π_w on the 2 generators a and b.

Let G_w be the group presented by Π_w. If $w = 1$ in G, then, in G_w, we have $u_1 = 1$, $v_1 = 1$ (by (iv) and (v)), hence $b = 1$, $\beta = 1$; hence $a = 1$ (by (ii)), and hence $G_w = \{1\}$. On the other hand, if $w \ne 1$ in G, then G_w is a free product with amalgamation $(G * F(a, \alpha)) *_E F(b, \beta)$, where E is a free group of rank $m + 4$, and therefore G is embedded in G_w.

Recall [9] that a group is *at level n with respect to free products with amalgamation* if it can be expressed as a free product with amalgamation of two groups at level $n - 1$, the groups at level 0 being the free groups. Since the triviality problem for $\{\Pi_w\}$ is equivalent to the word problem for Π, and since there exist finitely presented G with unsolvable word problem at level 2 with respect to free products with amalgamation (see [9], p.514), it follows easily from the above construction that the triviality problem for finitely presented groups is unsolvable at level 3.

3. Some undecidability questions.

Let G have a finite presentation $\Pi = (x_1, \ldots, x_m : r_1, \ldots, r_n)$, with corresponding presentation sequence $1 \to R \to F \to G \to 1$. Then $H_2(G)$ is the finitely generated abelian group $R \cap [F, F]/[R, F]$. An equivalent description is $H_2(G) \cong H_2(K)/h\pi_2(K)$, where K is the finite 2-complex (with one 0-cell, m 1-cells, and n 2-cells) associated with the presentation Π, and $h : \pi_2(K) \to H_2(K)$ is the Hurewicz map. The cellular chain groups $C_1(K)$, $C_2(K)$ are the free abelian groups on $\bar{x}_1, \ldots, \bar{x}_m$ and $\bar{r}_1, \ldots, \bar{r}_n$, say, respectively, and $H_2(K) = \ker(\partial_2 : C_2(K) \to C_1(K))$, where ∂_2 is given by $\partial_2(\bar{r}_j) = \sum_{i=1}^m e_{ij}\bar{x}_i$, e_{ij} being the exponent of x_i in r_j. Thus $H_2(K)$ may be computed. Now elements $\xi \in \pi_2(K)$ correspond to relations (in F) of the form $\prod_{i=1}^k u_i^{-1} r_{j_i}^{\varepsilon_i} u_i = 1$, $(\varepsilon_i = \pm 1, u_i \in F)$, and the corresponding element $h(\xi) \in H_2(K)$ is then $\sum_{i=1}^k \varepsilon_i \bar{r}_{j_i}$. It follows that recursively enumerating all relations in F of the above kind, one will eventually obtain a complete set of relations for $H_2(G)$. (I am grateful to Andrew Casson for this observation.) However, Theorem 4 shows that in general there is no way of knowing when a complete set of relations has been obtained.

Proof of Theorem 4. Let Π be a finite presentation of a group G with unsolvable word problem, and, for w a word in the generators of Π, let Π_w, G_w

be as in the proof of Theorem 3. To prove Theorem 4, we could work directly
with (a slight modification of) G_w, but with a view to Theorem 6, we perform
an additional construction.

First note that $H_1(G_w) = 0$, since, for example, setting $[w, a] = [w, \alpha] = 1$
kills G_w. If $w \neq 1$ in G, then a and b have infinite order in G_w. Let H_w be the
HNN extension of G_w with presentation Σ_w obtained from Π_w by adjoining
an additional generator t and the relation $t^{-1}at = b$. If $w = 1$ in G, then
$H_w \cong \mathbb{Z}$, so that $H_2(H_w) = 0$. If $w \neq 1$ in G, then the Mayer-Vietoris
sequence for HNN extensions gives an exact sequence

$$H_2(H_w) \to H_1(\mathbb{Z}) \to H_1(G_w) = 0 \,,$$

so that $H_2(H_w)$ maps onto $\mathbb{Z}$. Thus $H_2(H_w) = 0$ if and only if $w = 1$ in G, and
since G has unsolvable word problem, the question of whether $H_2(H_w) = 0$
is undecidable for the set of presentations $\{\Sigma_w\}$.

Note that in the above proof, the group H_w is generated by the two elements
a and t.

Proof of Theorem 5.   Recall that the *deficiency* def H of a finitely presented
group H is $\sup\{$def $\Pi : \Pi$ a finite presentation of $H\}$, where def Π = number
of generators − number of relators. Also, one has the inequality (see [2])

$$\text{def } H \leq \dim H_1(H; \mathbb{Q}) - \mu H_2(H) \,,$$

where μ means minimum number of generators.

Let H_w be as in the proof of Theorem 4. Then $w = 1$ in G implies $H_w \cong \mathbb{Z}$,
so that def $H_w = 1$, whereas $w \neq 1$ in G implies $\mu H_2(H_w) \geq 1$, giving
def $H_w \leq 1 - 1 = 0$.

Proof of Theorem 6.   Recall [6] that a finitely presented group H is a higher-
dimensional knot group (that is, the fundamental group of the complement
of a smooth n-sphere in the $(n + 2)$-sphere, $n \geq 3$) if and only if $H_1(H) \cong \mathbb{Z}$,
$H_2(H) = 0$, and H has weight 1, that is, is generated by a single conjugacy
class.

Let H_w be as above. Then $H_1(H_w) \cong \mathbb{Z}$. Also, setting $t = 1$ implies $a = b$,
and hence, since G_w is generated by a and b, and $H_1(G_w) = 0$, kills H_w. Thus
H_w has weight 1. Therefore H_w is a higher-dimensional knot group if and
only if $H_2(H_w) = 0$.

References

[1] S.I. Adjan, On algorithmic problems in effectively complete classes of
    groups, *Dokl. Akad. Nauk SSSR* **123** (1958), 13–16.

[2] D.B.A. Epstein, Finite presentations of groups and 3-manifolds, *Quart. J. Math.* **12** (1961), 205–212.

[3] A.P. Goryushkin, Imbedding of countable groups in 2-generator groups *Mat. Zametki* **16** (1974), 231–235.

[4] P. Hall, Embedding a group in a join of given groups, *J. Austral. Math. Soc.* **17** (1974), 434–495.

[5] G. Higman, B.H. Neumann, and H. Neumann, Embedding theorems for groups, *J. London Math. Soc.* **24** (1949), 247–254.

[6] M.A. Kervaire, On higher dimensional knots, in: *Differential and Combinatorial Topology,* A Symposium in Honor of Marston Morse (S.S. Cairns, ed.), Princeton University Press, Princeton, New Jersey, 1965, pp. 105–119.

[7] R.C. Lyndon and P.E. Schupp, *Combinatorial Group Theory,* Ergebnisse der Mathematik und ihrer Grenzgebiete **89** Springer-Verlag, Berlin-Heidelberg-New York, 1977.

[8] C.F. Miller, III, *On Group-Theoretic Decision Problems and their Classification,* Ann. of Math. Studies **68**, Princeton University Press, Princeton, New Jersey, 1971.

[9] C.F. Miller, III, Decision problems in algebraic classes of groups, in: *Word Problems,* Studies in Logic **71** (W.W. Boone, F.B. Cannonito and R.C. Lyndon, eds.), North-Holland Publishing Company, Amsterdam, 1973, pp. 507–523.

[10] C.F. Miller, III, Decision problems for groups — survey and reflections, in: *Algorithms and Classification in Combinatorial Group Theory* (G. Baumslag and C.F. Miller, III, eds.), Mathematical Science Research Institute Publications **23**, Springer-Verlag, New York, 1992, pp. 1–59.

[11] B.H. Neumann, An essay on free products of groups with amalgamations, *Philos. Trans. Roy. Soc., London Ser. A* **246** (1954), 503–554.

[12] M.O. Rabin, Recursive unsolvability of group theoretic problems, *Ann. of Math.* **67** (1958), 172–174.

[13] P.E. Schupp, Embeddings into simple groups *J. London Math. Soc.* **13** (1976), 90–94.

Department of Mathematics,
The University of Texas at Austin,
Austin,
TX 78712

Some results on bounded cohomology

R.I. GRIGORCHUK[1]

Abstract

The structure of the second bounded cohomology group is investigated. This group is computed for a free group, a torus knot group and a surface group. The description is based on the notion of a pseudocharacter. A survey of results on bounded cohomology is given.

Introduction

If we use the standard bar resolution then the definition of bounded cohomology of the trivial G-module $\mathbb{R}$ differs from the definition of ordinary cohomology in that instead of arbitrary cochains with values in $\mathbb{R}$ one should consider only the bounded cochains.

Bounded cohomology was first defined for discrete groups by F.Trauber and then for topological spaces by M.Gromov [39]. Moreover, M.Gromov developed the theory of bounded cohomology and applied it to Riemannian geometry, thus demonstrating the importance of this theory. The second bounded cohomology group is related to some topics of the theory of right orderable groups and has applications in the theory of groups acting on a circle [33], [55], [56].

In [9] R.Brooks made a first step in understanding the theory of bounded cohomology from the point of view of relative homological algebra. This approach was developed by N.Ivanov [48], whose paper probably contains the best introduction in the subject.

Actually the theory of bounded cohomology of discrete groups is a part of the theory of cohomology in topological groups [34] and in Banach algebras [49] introduced at the beginning of the sixties if we consider the trivial (that is $gx = x = xg$) $l_1(G)$-module $\mathbb{R}$.

On the other hand the papers of Gromov, Brooks, Ghys, Mitsumatsu, Matsumoto, Morita and others give excellent examples of applications of abstract theory of cohomology in Banach algebras in Riemannian geometry, topology, dynamics and other branches of mathematics.

An important feature of the theory is that the bounded cohomology of a topological space and its fundamental group coincide [39], [9], [47]. That makes it possible to study them simultaneously from two basic view points: group theory and topology.

The bounded cohomology, $H_b^*(G, \mathbb{R})$, of an amenable group G is zero (Trauber's theorem). In [39], [9], [55], [56], [57] some examples are given

---

[1]Part of the results presented here were obtained under the financial support of the Russian Fund for Fundamental Research 93-011-239 and of the International Science Foundation, Grant number MVI000.

showing that for nonamenable groups bounded cohomology may be nonzero and even infinite dimensional.

The first dimension in which bounded cohomology should be investigated is dimension 2 because $H_b^{(0)}(G,\mathbb{R}) = \mathbb{R}$ and $H_b^{(1)}(G,\mathbb{R}) = 0$ for any group G.

In this paper we develop a method based on the use of the space of pseudocharacters and apply it to the calculation of $H_b^{(2)}(G,\mathbb{R})$ in some important cases, first of all for a free group.

The notion of a pseudocharacter comes from functional analysis and is related to some problems of stability [2], [43], [44], [45], [74]. It is a particular case of the notion of a pseudorepresentation [52], [66] that plays some role in the theory of representations.

Our observation is that $H_b^{(2)}(G,\mathbb{R})$ contains a subspace $H_{b,2}^{(2)}(G,\mathbb{R})$ that is naturally isomorphic to the space of pseudocharacters and the factor space $H_b^{(2)}(G,\mathbb{R})/H_{b,2}^{(2)}(G,\mathbb{R})$ is canonically isomorphic to the bounded part of an ordinary cohomology group $H^{(2)}(G,\mathbb{R})$ which has been the object of consideration in many papers of a geometric character.

In the case of a free group, or free semigroup, F_m the space $PX(F_m)$ of pseudocharacters was described by V.A.Faiziev [27], [28], [31]. We clarify the formulation and the proof of Faiziev's result and establish some additional information about the space. This also gives a description of the second bounded cohomology group for a free group. The modification of the method allows us to compute $H_b^{(2)}(G,\mathbb{R})$ in the case when G is a torus knot group or a surface group.

Moreover we establish some results of a general character on bounded cohomology and formulate a number of open problems.

The paper is organized as follows. In Section 1 we define l_1-homology and l_∞-cohomology of a discrete group and establish some kind of duality between these two spaces. In Section 2 we prove the extension of Trauber's theorem. In Section 3 we define the notion of a pseudocharacter and establish its connection with bounded cohomology. Some corollaries of this connection are deduced. Section 4 contains some general results on bounded cohomology and in Section 5 we compute $H_b^{(2)}(G,\mathbb{R})$ for a free group, torus group and surface group.

1 l_1-homology and l_∞-cohomology of discrete groups

For the definition of l_1-homology and l_∞-cohomology we use the standard bar-resolution [11].

The ordinary homology group $H_*(G,\mathbb{R})$ of a group G is given by the homology of the chain complex $C_*(G,\mathbb{R})$ (in future we usually will omit the second argument in the notation for (co)homology groups and the corresponding complexes)

$$\longrightarrow C_n(G) \xrightarrow{\partial_n} C_{n-1}(G) \longrightarrow \cdots \longrightarrow C_2(G) \xrightarrow{\partial_2} C_1(G) \xrightarrow{\partial_1 = 0} \mathbb{R} \xrightarrow{\partial_0 = 0} 0,$$

where the vector space $C_n(G)$ has the $\mathbb{R}$-basis consisting of n-tuples $[g_1|\cdots|g_n]$ and where the differential $\partial = (\partial_n)_{n \geq 1}$,

$$\partial_n : C_n(G) \to C_{n-1}(G)$$

is given by the formula

$$\partial_n[g_1|\ldots|g_n] = [g_2|\ldots|g_n]$$
$$+ \sum_{i=1}^{n-1}(-1)^i[g_1|\ldots|g_ig_{i+1}|\ldots|g_n] + (-1)^n[g_1|\ldots|g_{n-1}], \qquad (1.1)$$

if $n \geq 2$, and ∂_1, ∂_0 are by definition the zero operators.

Then $H_*(G) = (H_n(G))_{n \geq 0}$, where

$$H_n(G) = \ker \partial_n / \operatorname{Im} \partial_{n+1}.$$

Now let $\overline{C}_n(G)$ be the completion of $C_n(G)$ with respect to the norm

$$\left\| \sum_{g_1,\ldots,g_n} \alpha_{[g_1|\ldots|g_n]}[g_1|\ldots|g_n] \right\|_1 = \sum_{g_1,\ldots,g_n} |\alpha_{[g_1|\ldots|g_n]}|.$$

We have the chain complex

$$\longrightarrow \overline{C}_n(G) \xrightarrow{\overline{\partial}_n} \overline{C}_{n-1}(G) \longrightarrow \cdots \longrightarrow \overline{C}_2(G) \xrightarrow{\overline{\partial}_2} \overline{C}_1(G) \xrightarrow{\overline{\partial}_1 = 0} \mathbb{R} \xrightarrow{\overline{\partial}_0 = 0} 0$$

of Banach spaces and bounded operators, where the operators $\overline{\partial}_n$, $n \geq 2$ are determined by formula (1.1), and the operator $\overline{\partial}_n$ is of norm $\leq n+1$ if $n \geq 2$.

The l_1-homology of the group G is the homology $H_*^{l_1}(G) = (H_n^{l_1}(G))_{n \geq 0}$ of this complex, where

$$H_n^{l_1}(G) = \ker \overline{\partial}_n / \operatorname{Im} \overline{\partial}_{n+1}, \quad n \geq 0. \qquad (1.2)$$

We will see later that

$$H_1^{l_1}(G) = 0$$

for every group G.

The vector space (1.2) carries the structure of a seminormed space over $\mathbb{R}$ equipped with the seminorm

$$\|[\omega]\|_1 = \inf_{\omega':[\omega]=[\omega']} \|\omega'\|_1,$$

where by $[\omega]$ we denote the element of $H_n^{l_1}(G)$ corresponding to an element $\omega \in \ker \overline{\partial}_n$. This seminorm is a norm if and only if $\operatorname{Im}\overline{\partial}_{n+1}$ is a closed subspace. As we will see later, $\operatorname{Im}\overline{\partial}_2 = \ker \overline{\partial}_1$ and so $\operatorname{Im}\overline{\partial}_2$ is closed. It is not known if $\operatorname{Im}\overline{\partial}_n$ is closed for every $n \geq 3$.

The inclusion $C_*(G) \subset \overline{C}_*(G)$ induces a homomorphism $H_*(G) \to H_*^{l_1}(G)$, the image of which we call the finitary part of the l_1-homology group and denote $H_{*,1}^{l_1}(G)$.

Let $H_{*,2}(G) \subseteq H_*(G)$ be the subspace

$$H_{*,2}^{(n)}(G) = \ker \partial_n \cap \operatorname{Im} \overline{\partial}_{n+1} / \operatorname{Im} \partial_{n+1}.$$

Proposition 1.1 *A natural isomorphism of vector spaces*

$$H^{l_1}_{*,1}(G) \cong H_*(G)/H_{*,2}(G)$$

holds.

Proof. We have the exact sequence

$$0 \to \ker \partial_n \cap \operatorname{Im} \overline{\partial}_{n+1}/\operatorname{Im} \partial_{n+1} \to \ker \partial_n/\operatorname{Im} \partial_{n+1}$$
$$\to \ker \partial_n/\ker \partial_n \cap \operatorname{Im} \overline{\partial}_{n+1} \to 0$$

and the last factor space is isomorphic to the space

$$\ker \partial_n + \operatorname{Im} \overline{\partial}_{n+1}/\operatorname{Im} \overline{\partial}_{n+1},$$

which is the n-dimensional part of $H^{l_1}_{*,1}(G)$.

Corollary 1.2 *An isomorphism of vector spaces*

$$H_*(G) \cong H^{l_1}_{*,1}(G) \oplus H_{*,2}(G)$$

holds.

The ordinary cohomology group $H^*(G)$ is given by the cohomology of the cochain complex $C^*(G)$:

$$\longleftarrow C^{(n)}(G) \xleftarrow{\delta^{(n-1)}} C^{(n-1)}(G) \longleftarrow \cdots$$
$$\cdots \longleftarrow C^{(2)}(G) \xleftarrow{\delta^{(1)}} C^{(1)}(G) \xleftarrow{\delta^{(0)}=0} \mathbb{R} \xleftarrow{\delta^{(-1)}=0} 0,$$

where $C^{(n)}(G)$, $n \geq 0$ consists of mappings

$$\underbrace{G \times \cdots \times G}_{n} \to \mathbb{R}$$

and where the differential $\delta = (\delta^{(n)})_{n\geq 0}$:

$$\delta^{(n)} : C^{(n)}(G) \to C^{(n+1)}(G)$$

is given by the formula

$$(\delta^{(n)} f)(g_1, \ldots, g_{n+1}) = f(g_2, \ldots, g_{n+1}) +$$
$$+ \sum_{i=1}^{n} (-1)^i f(g_1, \ldots, g_{i-1}, g_i g_{i+1}, \ldots, g_{n+1}) + (-1)^{n+1} f(g_1, \ldots, g_n).$$

Now let us consider bounded cochains $f \in C^{(n)}(G, \mathbb{R})$, that is, cochains for which there exists $M_f > 0$ such that

$$|f(g_1, \ldots, g_n)| \leq M_f$$

for all $g_1, \ldots, g_n \in G$.

We have the cochain complex $C_b^*(G)$:

$$\longleftarrow C_b^{(n)}(G) \overset{\delta^{(n-1)}}{\longleftarrow} C_b^{(n-1)}(G) \longleftarrow \cdots$$

$$\cdots \longleftarrow C_b^{(2)}(G) \overset{\delta_b^{(1)}}{\longleftarrow} C_b^{(1)}(G) \overset{\delta_b^{(0)}=0}{\longleftarrow} \mathbb{R} \overset{\delta_b^{(-1)}=0}{\longleftarrow} 0$$

of bounded cochains with values in $\mathbb{R}$ and can define l_∞ (or bounded)-cohomology

$$H_b^*(G, \mathbb{R}) = H^*(C_b^*(G));$$

that is

$$H_b^{(n)}(G) = \ker \delta_b^{(n)} / \operatorname{Im} \delta_b^{(n-1)}, \quad n \geq 0,$$

where

$$\delta_b^{(n)} = \delta^{(n)}|_{C_b^{(n)}(G)}$$

is the bounded differential operator (the restriction of $\delta^{(n)}$ to the bounded cochain complex).

It is easy to see that $H_b^{(1)}(G) = 0$ for all G. The fact is that there do not exist nontrivial bounded homomorphisms $G \to \mathbb{R}$.

The group $H_b^{(n)}(G)$ carries the structure of vector space over $\mathbb{R}$. Moreover the following seminorm in $H_b^{(n)}(G)$ is important for applications. There is a natural norm $\| \cdot \|_\infty$ in the space $C_b^{(n)}(G)$:

$$\|f\|_\infty = \sup_{g_1, \ldots, g_n \in G} |f(g_1, \ldots, g_n)|$$

which turns it into a Banach space. Thus in $H_b^{(n)}(G)$ there is a seminorm

$$\|g\| = \inf \|f\|_\infty,$$

where $g \in H_b^{(n)}(G)$ and the infimum is taken over all cochains $f \in \ker \delta_b^{(n)}$ lying in the cohomology class corresponding to g. This seminorm is a norm if and only if $\operatorname{Im} \delta_b^{(n-1)}$ is closed subspace of $C_b^{(n)}(G)$. All that is known is that the space $\operatorname{Im} \delta_b^{(1)}$ is closed [56],[48] and so it turns $H_b^{(2)}(G)$ into a Banach space.

Remark. In [48], to prove that $\operatorname{Im} \delta_b^{(1)}$ is closed the left inverse to the bounded operator $\delta_b^{(1)}$ was constructed with the help of the generalized Banach limit. Using the description of the space $\ker \delta_b^{(1)}$ which will be given in the next section one can modify Ivanov's arguments using only the ordinary limit.

On the other hand in [56] it was shown that $\operatorname{Im} \overline{\partial}_2$ is closed with the help of the right inverse to $\overline{\partial}_2$, that is the operator $S : \overline{C}_1(G) \to \overline{C}_2(G)$ given by

$$S(g) = \sum_{k=1}^{\infty} \frac{1}{2^k} [g^{2^k} | g^{2^k}], \quad g \in G,$$

which is of norm 1 (this operator was first constructed in [57]). As is easy to see, the facts that each of the spaces $\operatorname{Im}\overline{\partial}_2$, $\operatorname{Im}\delta_b^{(1)}$ is closed are equivalent.

The space $C_b^{(n)}(G)$ is the dual space of $\overline{C}_n(G)$: if $\alpha \in \overline{C}_n(G)$ and $\beta \in C_b^{(n)}(G)$ with

$$\alpha = \sum \alpha_{[g_1|\ldots|g_n]}[g_1|\ldots|g_n]$$

then

$$\langle \alpha, \beta \rangle = \sum_{g_1,\ldots,g_n} \alpha_{[g_1|\ldots|g_n]}\beta(g_1,\ldots,g_n).$$

The operator $\delta_b^{(n-1)}$ is the adjoint of the operator $\overline{\partial}_n$. Indeed, if $\beta \in C_b^{(n-1)}(G)$, then

$$\langle \alpha, \delta_b^{(n-1)}\beta \rangle = \sum \alpha_{[g_1|\ldots|g_n]}(\delta_b^{(n-1)}\beta)(g_1,\ldots,g_n) =$$

$$= \sum \alpha_{[g_1|\ldots|g_n]}\Big\{\beta(g_2,\ldots,g_n) + \sum_{i=1}^{n-1}(-1)^i\beta(g_1,\ldots,g_ig_{i+1},\ldots,g_n)+$$

$$+(-1)^n\beta(g_1,\ldots,g_{n-1})\Big\}$$

$$= \Big\langle \sum \alpha_{[g_1|\ldots|g_n]}([g_2|\ldots|g_n] + \sum_{i=1}^{n-1}(-1)^i[g_1|\ldots|g_ig_{i+1}|\ldots|g_n]+$$

$$+(-1)^n[g_1,|\ldots|g_{n-1}]), \beta \Big\rangle$$

$$= \langle \overline{\partial}_n\alpha, \beta \rangle.$$

Proposition 1.3 $H_1^{l_1}(G) = 0$ *for any group* G.

Proof. This is a corollary of the triviality of $H_b^1(G)$ because the equality $\operatorname{Im}\overline{\partial}_2 = \overline{C}_1(G)$ holds if and only if the adjoint operator $\delta_b^{(1)}$ is one-to-one and its image is norm closed in $C_b^{(2)}(G)$. But $\delta_b^{(1)}$ has trivial kernel since only the zero character $G \to \mathbb{R}$ belongs to $C_b^{(1)}(G)$ and, as we know, $\operatorname{Im}\delta_b^{(1)}$ is closed.

Actually a more general statement holds.

Proposition 1.4 ([49]) *If* $\operatorname{Im}\overline{\partial}_n$ *is a closed subspace and* $H_{n+1}^{l_1}(G) = 0$, *then* $H_b^{(n+1)}(G) = 0$. *If* $\operatorname{Im}\delta_b^{(n)}$ *is closed and* $H_b^{(n)}(G) = 0$, *then* $H_n^{l_1}(G) = 0$.

In particular we have

Corollary 1.5 (compare [56], Corollary 1.3) *The condition* $H_n^{l_1}(G) = 0$ *for* $n \geq m$ *and* $\operatorname{Im}\overline{\partial}_m$ *is closed is equivalent to the condition* $H_b^{(n)}(G) = 0$ *for* $n \geq m$.

Remark 1.6 The equality $H_1^{l_1}(G) = 0$ is a special case of the statement: $H_1(G, X) = 0$ for any reflexive left Banach G-module X ([49], Theorem 3.4).

We are going to compare l_1-homology and l_∞-cohomology using the duality of the corresponding complexes. Here we continue the investigations of [57], [56]. Let us recall some notions and results of duality theory (see [61], for instance).

If E is a vector space with a seminorm then E^* denotes the normed dual of E (it is always a Banach space [26], section 1.10.6).

The $*$-weak topology of E^* is the weakest topology that makes all functionals $y \to \langle x, y \rangle$ continuous.

If $H \subset E$ is a subspace then $\langle H \rangle_\#$ denotes the norm closure of H and if $F \subset E^*$ then $\langle F \rangle_*$ is the $*$-weak closure of F.

Suppose E is a Banach space, K is a subspace of E and L is a subspace of E^*. Their annihilators $K^\perp$ and $^\perp L$ are defined as follows:

$$
\begin{aligned}
K_{E^*}^\perp &= \{y \in E^* : \langle x, y \rangle = 0 \text{ for all } x \in K\}, \\
^\perp L_E &= \{x \in E : \langle x, y \rangle = 0 \text{ for all } y \in L\}
\end{aligned}
\tag{1.3}
$$

(usually we shall omit the lower subscripts). Then $^\perp(K^\perp)$ coincides with $\langle K \rangle_\#$, and $(^\perp L)^\perp$ coincides with $\langle L \rangle_\#$ ([61], Theorem 4.7).

If $A : H_1 \to H_2$ is a bounded operator, then $A^* : H_2^* \to H_1^*$ is its adjoint operator and if H_1, H_2 are Banach spaces then

$$
\begin{aligned}
\ker A^* &= (\operatorname{Im} A)^\perp \\
\ker A &= {}^\perp(\operatorname{Im} A^*).
\end{aligned}
\tag{1.4}
$$

The following statement is not very original but we do not have a suitable reference. It also clarifies the situation of problem 398 from [51].

Theorem 1.7 *Let*

$$
H_1 \xrightarrow{A} H_2 \xrightarrow{B} H_3,
$$

$$
H_1^* \xleftarrow{A^*} H_2^* \xleftarrow{B^*} H_3^*
$$

be dual sequences of Banach spaces and bounded operators with $\operatorname{Im} A \subseteq \ker B$. *Then the inclusion* $\langle \operatorname{Im} B^* \rangle_* \subseteq \ker A^*$ *and the natural isometric isomorphism*

$$
\rho : \ker A^* / \langle \operatorname{Im} B^* \rangle_* \to (\ker B / \operatorname{Im} A)^*
$$

hold.

The map ρ is also a topological isomorphism of the topological vector space $(\ker B / \operatorname{Im} A)^*$ *under the $*$-weak topology and the factor space* $\ker A^* / \langle \operatorname{Im} B^* \rangle_*$ *of the space* $\ker A^*$ *endowed with the topology induced by the $*$-weak topology on* H_2^*.

Proof. The inclusion $\operatorname{Im} B^* \subseteq \ker A^*$ is obvious because if $x \in H_1$, $y \in H_3^*$ then

$$\langle x, A^* B^* y \rangle = \langle BAx, y \rangle = 0.$$

The space $\ker A^*$ is $*$-weakly closed in H_2^* due to relation (1.3) because an annihilator $K^\perp$ is a $*$-weakly closed subspace of the corresponding dual space. Thus $\langle \operatorname{Im} B^* \rangle_* \subseteq \ker A^*$.

To prove the theorem we need the following well-known statement the first part of which is, for instance, Theorem 4.9 from [61] and the second part of which is a corollary of Proposition 8.1.2 from [26].

Theorem 1.8 *I. Let M be a closed subspace of a Banach space X.*

(a) By the Hahn–Banach theorem every functional $m^* \in M^*$ can be extended to some functional $x^* \in X^*$. Let us put

$$\sigma m^* = x^* + M^\perp.$$

This formula correctly defines a map $\sigma : M^* \to X^*/M^\perp$ which is an isometric isomorphism of M^* onto $X^*/M^\perp$.

(b) Let $\pi : X \to X/M$ be the factor-mapping and let $Y = X/M$. For every $y^* \in Y^*$ let us put

$$\tau y^* = y^* \pi.$$

Then τ is an isometric isomorphism of the space Y^* onto $M^\perp$.

II. a) The topological vector space $(M^*, *)$ ($*$ means $*$-weak topology) is topologically isomorphic to the space $X^*/M^\perp$ endowed with the factor topology induced by the $*$-weak topology on X^*.

b) The topological vector-space $((X/M)^*, *)$ is topologically isomorphic to the space $M^\perp$ with topology η induced by the $*$-weak topology of the space X^*.

Proof of part II of Theorem 1.8. We use the standard notation $\sigma(E, E')$ for the weak topology on E defined by the dual pair (E, E') of vector spaces [26], section 8.1. Thus $\sigma(X^*, X)$ is the $*$-weak topology on X^*.

By Proposition 8.1.2 from [26] and Statement I of Theorem 1.8

$$\sigma(X^*/M^\perp, {}^\perp(M^\perp)) = \sigma(X^*/M^\perp, M) = \sigma(M^*, M)$$

is the factor topology of the $*$-weak topology $\sigma(X^*, X)$ modulo $M^\perp$.

On the other hand

$$\sigma(M^\perp, X/{}^\perp(M^\perp)) = \sigma(M^\perp, X/M)$$

is the topology induced on $M^\perp$ by the topology $\sigma((X/M)^*, X/M)$, which is the $*$-weak topology in $(X/M)^*$. This completes the proof of part II of Theorem 1.8.

Using the first part of Theorem 1.8 we get the following sequence of spaces and isometric isomorphisms

$$(\ker B/\operatorname{Im} A)^* \cong (\ker B/\langle \operatorname{Im} A\rangle_\#)^* \cong$$
$$\cong (H_2/\langle \operatorname{Im} A\rangle_\#)^*/(\ker B/\langle \operatorname{Im} A\rangle_\#)^{\perp}_{(H_2/\langle \operatorname{Im} A\rangle_\#)^*} \cong$$
$$\cong (H_2/\langle \operatorname{Im} A\rangle_\#)^*/(H_2/\langle \operatorname{Im} A\rangle_\#/\ker B/\langle \operatorname{Im} A\rangle_\#)^* \cong$$
$$\cong (H_2/\langle \operatorname{Im} A\rangle_\#)^*/(H_2/\ker B)^* \cong$$
$$\cong (\langle \operatorname{Im} A\rangle_\#)^{\perp}_{H_2^*}/(\ker B)^{\perp}_{H_2^*} \cong$$
$$\cong (\operatorname{Im} A)^{\perp}_{H_2^*}/(\ker B)^{\perp}_{H_2^*} \cong \ker A^*/\langle \operatorname{Im} B^*\rangle_*.$$

because

$$(\ker B)^{\perp}_{H_2^*} \cong ({}^{\perp}(\operatorname{Im} B^*))^{\perp}_{H_2^*} = \langle \operatorname{Im} B^*\rangle_*.$$

Applying the second part of Theorem 1.8 we finish the proof of Theorem 1.7.

Proposition 1.9 *For every group G the isometric isomorphism*

$$H_b^{(2)}(G) \cong (H_2^{l_1}(G))^* \tag{1.5}$$

holds.

Proof. The isomorphism (1.5) is a special case of the statement given in Theorem 1.7 because $\operatorname{Im}\delta_b^{(1)}$ is a norm closed subspace of $C_b^{(2)}(G) = \overline{C}_2^*(G)$ and this is equivalent to $\operatorname{Im}\delta_b^{(1)}$ being $*$-weak closed ([61], Theorem 4.14).

Problem 1.10 Are the invariants $H_b^*(G)$, $H_*^{l_1}(G)$ of group isomorphism equivalent or not?

The inclusion homomorphism $C_b^*(G) \to C^*(G)$ induces a homomorphism $\xi: H_b^*(G) \to H^*(G)$ which in general is neither injective nor surjective. The image of this homomorphism is called the bounded part of $H^*(G)$ and will be denoted by $H_{b,1}^*(G)$.

The seminorm on $H_b^*(G)$ induces a seminorm on the bounded part of $H^*(G)$ by the formula:

$$\|a\| = \inf_{\xi(b)=a} \|b\|.$$

Proposition 1.11 ([9]) *The space $H_{b,1}^{(n)}$ with the induced seminorm is a Banach space.*

Proof. Let $f \in C_b^{(n)}(G)$ and $[f] \in H_{b,1}^{(n)}(G)$ where $[f]$ denotes the class of the element f. Then

$$\|[f]\|_{H_{b,1}^{(n)}} = \inf_g \|[f+g]\|_{H_b^{(n)}},$$

where g passes through the space $\operatorname{Im}\delta^{(n-1)} \cap \ker\delta_b^{(n)}$, and

$$\|[f+g]\|_{H_b^{(n)}} = \inf_h \|f+g+h\|_\infty,$$

where h pass through the space $\operatorname{Im} \delta_b^{n-1}$.

Since $[f]$ is a nonzero element of the group $H^{(n)}(G)$ there is $x \in \ker \partial_n$ (x is a finitary function) such that $\langle x, f \rangle \neq 0$. But if $g = \delta^{n-1} g_1$, $h = \delta_b^{n-1} h_1$ then

$$\langle x, f + g + h \rangle = \langle x, f \rangle + \langle \partial_n x, g_1 \rangle + \langle \partial_n x, h_1 \rangle = \langle x, f \rangle.$$

Thus

$$\|[f]\|_{H_{b,1}^{(n)}} \geq \langle x, f \rangle > 0$$

if $[f] \neq 0$ in $H_{b,1}^{(n)}$, completing the proof of the proposition.

Let

$$H_{b,2}^{(n)}(G) \subseteq H_b^{(n)}(G)$$

be the subspace

$$H_{b,2}^{(n)}(G) = \operatorname{Im} \delta^{(n-1)} \cap \ker \delta_b^{(n)} / \operatorname{Im} \delta_b^{(n-1)}.$$

We call $H_{b,2}^*(G)$ the singular part of the bounded cohomology group.

Corollary 1.12 *All degeneration of the seminorm on $H_b^{(n)}$ (if it exists) belongs to the subspace $H_{b,2}^{(n)}$. In other words*

$$\langle \operatorname{Im} \delta_b^{(n-1)} \rangle_* / \operatorname{Im} \delta^{(n-1)}$$

and we have the exact sequence

$$0 \to \langle \operatorname{Im} \delta_b^{(n-1)} \rangle_* / \operatorname{Im} \delta^{(n-1)} \to H_{b,2}^{(n)} \to$$
$$\operatorname{Im} \delta^{(n-1)} \cap \ker \delta_b^{(n)} / \langle \operatorname{Im} \delta_b^{(n-1)} \rangle_* \to 0.$$

Proposition 1.13 *The natural isometric isomorphism*

$$\operatorname{Im} \delta^{(n-1)} \cap \ker \delta_b^{(n)} / \langle \operatorname{Im} \delta_b^{(n-1)} \rangle_* \cong (\ker \overline{\partial}_n / \langle \ker \partial_n \rangle_\#)^* \tag{1.6}$$

holds.

Proof. We denote by K (L) the left (right) hand side of (1.6).

Every element $[f] \in K$ is determined by the values of a representative f on $\ker \overline{\partial}_n$. Indeed, if $f_1, f_2 \in \operatorname{Im} \delta^{(n-1)} \cap \ker \delta_b^{(n)}$, $f_1 = \delta^{n-1} f_1'$, $f_2 = \delta^{n-1} f_2'$ and

$$f_1 \big|_{\ker \overline{\partial}_n} = f_2 \big|_{\ker \overline{\partial}_n},$$

then

$$f_1 - f_2 \in (\ker \overline{\partial}_n)^\perp = ({}^\perp \operatorname{Im}(\delta_b^{(n-1)}))^\perp = \langle \operatorname{Im} \delta_b^{(n-1)} \rangle_*$$

and so $[f_1] = [f_2]$ in K.

Let us show that if $f \in \operatorname{Im} \delta^{(n-1)} \cap \ker \delta_b^{(n)}$ then

$$f \big|_{\langle \ker \partial_n \rangle_\#} = 0.$$

It is enough to show that
$$f\,|_{\langle\ker\partial_n\rangle}=0.$$
But if $x\in\ker\partial_n$ and $f=\delta^{(n-1)}f_1$ then
$$\langle x,f\rangle=\langle x,\delta^{(n-1)}f_1\rangle=\langle\partial_n x,f_1\rangle=0.$$

Thus every element of K can be viewed as an element of L and this embedding ξ is isometric as is easy to see.

Vice versa, if f is a continuous functional in $\ker\overline\partial_n$ then there is a continuous extension $\tilde f$ on the space $\overline C_n$ that covers the functional f. We have
$$\tilde f\,|_{C_n}\in(\ker\partial_n)^\perp=(^\perp(\operatorname{Im}\delta^{(n-1)}))^\perp=\operatorname{Im}\delta^{(n-1)}$$

(annihilators are considered with respect to the algebraic duality). If $\tilde f\,|_{C_n}=\delta^{(n-1)}g$ then $\delta^{n-1}g\in\operatorname{Im}\delta^{(n-1)}\cap\ker\delta_b^{(n)}$ and thus $\delta_b^{(n-1)}$ determines an element of K. Two different continuations of f give the same element of K and we get the inverse to the isometric mapping ξ.

Proposition 1.14 *A natural isomomorphism of vector spaces with semi-norms*
$$H_{b,1}^*(G)\cong H_b^*(G)/H_{b,2}^*(G) \tag{1.7}$$
holds.

Proof. That (1.7) holds follows immediately from the exact sequence
$$0\to(\operatorname{Im}\delta^{(n-1)}\cap\ker\delta_b^{(n)})/\operatorname{Im}\delta_b^{(n-1)}\to$$
$$\to\ker\delta_b^{(n)}/\operatorname{Im}\delta_b^{(n-1)}\to\ker\delta_b^{(n)}/(\operatorname{Im}\delta^{(n-1)}\cap\ker\delta_b^{(n)})\to0$$
because
$$H_{b,1}^{(n)}(G)=\ker\delta_b^{(n)}/(\operatorname{Im}\delta^{(n-1)}\cap\ker\delta_b^{(n)})$$
and the seminorm was defined as a seminorm on a factor-space.

Corollary 1.15 *An isomorphism of vector spaces*
$$H_b^*(G)\cong H_{b,1}^*(G)\oplus H_{b,2}^*(G)$$
holds.

The definition of bounded cohomology $H_b^{(n)}(G,B)$ can be given, in the same manner as above, for the rings $B=\mathbb Z,\mathbb C$ of integers and complex numbers. It is clear that $H_b^{(n)}(G,\mathbb C)$ is the complexification of the space $H_b^{(n)}(G,\mathbb R)$. On the other hand $H_b^{(n)}(G,\mathbb Z)$ can be included in the Mayer–Vietoris exact sequence [32]
$$\cdots\to H^{(n-1)}(G,\mathbb R)\to H_b^{(n)}(G,\mathbb Z)\to$$
$$H^{(n)}(G,\mathbb Z)\oplus H_b^{(n)}(G,\mathbb R)\to H^{(n)}(G,\mathbb R)\to\cdots$$

which sometimes can be used for calculation of $H_b^{(n)}(G, \mathbb{Z})$. For instance $H_b^{(2)}(\mathbb{Z}, \mathbb{Z}) = \mathbb{R}/\mathbb{Z}$. For applications of $H_b^{(2)}(G, \mathbb{Z})$ see [33], [55], [56].

As was observed in the introduction the bounded cohomology of a discrete group G can be viewed as the cohomology of the trivial module $\mathbb{R}$ of the Banach algebra $l_1(G)$. Here we recall the general definition of cohomology in a Banach algebra A with coefficients in a Banach A-module E, following [49].

Let A be a Banach algebra over $\mathbb{C}$. By definition a Banach space E over $\mathbb{C}$ is a Banach A-module if it is a two-sided A-module and there is a positive real number K such that $\|ax\| \leq K\|a\|\|x\|$ and $\|xa\| \leq K\|x\|\|a\|$ for all $a \in A, x \in E$. For a Banach A-module E one can form the Cartan-Eilenberg complex

$$\cdots \xrightarrow{\partial_3} A \hat{\otimes} A \hat{\otimes} E \xrightarrow{\partial_2} A \hat{\otimes} E \xrightarrow{\partial_1} E \xrightarrow{\partial_0} 0 \qquad (1.8)$$

where

$$\partial_n(a_1 \otimes a_2 \cdots \otimes a_n \otimes x) =$$
$$= a_2 \otimes \cdots \otimes a_n \otimes xa_1 + \sum_{i=1}^{n-1}(-1)^i a_1 \otimes \cdots \otimes a_i a_{i+1} \otimes \cdots \otimes a_n \otimes x +$$
$$+ (-1)^n a_1 \otimes \cdots \otimes a_{n-1} \otimes a_n x$$

and $\hat{\otimes}$ denotes projective tensor product of Banach spaces. Then $H_n(A, E) = \ker \partial_n / \operatorname{Im} \partial_{n+1}$ is the n-component of the homology group $H_*(A, E)$.

The dual E^* of E becomes a Banach A-module if one defines

$$\langle ya, x \rangle = \langle y, ax \rangle$$

$$\langle ay, x \rangle = \langle y, xa \rangle, \ a \in A, \ x \in E, \ y \in E^*.$$

Let $L_n(A, E) = A \hat{\otimes} \cdots \hat{\otimes} A \hat{\otimes} E$. The dual space of $L_n(A, E)$ is the space of bounded multilinear functionals n-linear in A and 1-linear in E. $L_n(A, E)^*$ can be identified with $L^n(A, E^*)$, the space of n-linear transformations of A into E^*.

Writing $F = E^*$, $\delta^{(n)} = \partial_n^*$ the dual of (1.8) is

$$\cdots \xleftarrow{\delta^3} L^3(A, F) \xleftarrow{\delta^2} L^2(A, F) \xleftarrow{\delta^{(1)}} L^1(A, F) \xleftarrow{\delta^0} F \xleftarrow{\delta^{-1}} 0, \qquad (1.9)$$

where for $f \in L^{n-1}(A, F)$

$$(\delta^{n-1} f)(a_1, \ldots, a_n) = a_1 f(a_2, \ldots, a_n) +$$
$$+ \sum_{i=1}^{n-1}(-1)^i f(a_1, \ldots, a_i a_{i+1}, \ldots, a_n) + (-1)^n f(a_1, \ldots, a_{n-1})a_n \qquad (1.10)$$

(of course, the complex (1.9) can be defined for any Banach A-module F). Then $H^{(n)}(A, F) = \ker \delta^n / \operatorname{Im} \delta^{n-1}$ is the n-component of the cohomology group $H^*(A, F)$.

There is, due to Hochschild, the classical procedure of reduction of dimension ([49], 1.a):

$$H_{n+p}(A, X) \cong H_n(A, L_p(A, X)),$$
$$H^{(n+p)}(A, X) \cong H^{(n)}(A, L^p(A, X)).$$

If G is a locally compact (in particular discrete) group and X is a two-sided G-module such that there is a constant $K > 0$ with $\|gx\| \leq K\|x\|$, $\|xg\| \leq K\|x\|$ for all $g \in G, x \in X$ and for any $y \in X$ the maps $G \to X$ given by $g \to gy$ and $g \to yg$ are continuous, then X is called a Banach G-module. There is a one-to-one correspondence between Banach G-modules and Banach $l_1(G)$-modules. Thus one can define $H_n(G, X)$, $H^{(n)}(G, X)$ as

$$H_n(G, X) = H_n(l_1(G), X),$$
$$H^{(n)}(G, X) = H^{(n)}(l_1(G), X).$$

A standard procedure associates to any two-sided Banach G-module X the new Banach G-module X^0 with the same underlying Banach space but with actions $g \circ x = g^{-1}xg$, $x \circ g = x$.

Then the isomorphism

$$H_n(G, X) \cong H_n(G, X^0)$$

holds [49]. Thus in cohomology theory it is enough to consider Banach G-modules with one-side trivial action. The bounded cohomology group in the sense of Gromov and Brooks is the cohomology of the trivial G-module $\mathbb{R}$, where $gx = x = xg$ for any $x \in \mathbb{R}$, $g \in G$.

Remark 1.16 The definition of $H_b^*(G)$ in the same manner as above can be given in the case when G is a semigroup and again we have the equality $H_b^{(n)}(G) = H^{(n)}(l_1(G), \mathbb{R})$. The idea to consider cohomology of semigroups is not new [1], [70]. We hope that bounded cohomology can be useful for investigation of amenability properties of semigroups. On the other hand there is some connection between bounded cohomology of a cancellative semigroup and its group of quotients. Semigroups can provide examples of Banach algebras with new cohomological properties. Thus it makes sense to develop the theory of bounded cohomology for semigroups as well.

2 Bounded Cohomology of amenable semigroups

The first result of the theory of bounded cohomology was Trauber's theorem stating that the bounded cohomology of an amenable group is zero. We extend this result for the case of semigroups.

Recall that a group G is amenable if, in the Banach space $B(G)$ of bounded real valued functions on G with the norm

$$\|f\|_\infty = \sup_{g \in G} |f(g)|,$$

there is a left invariant mean (LIM), that is a linear functional $m \in B^*(G)$ with the following properties:

(i) $m(1_G) = 1$, where 1_G is the constant function on G with value 1.
(ii) $m(f) \geq 0$ if $f \geq 0$.
(iii) $m(L_g f) = m(f)$ for every $g \in G$ and $f \in B(G)$.

Here L_g is the left shift operator:

$$(L_g f)(x) = f(gx).$$

If there is a left invariant mean on G then there is a right invariant mean on G and conversely [35].

The definition of a left (or a right) invariant mean on a semigroup is the same as in the case of a group. However a semigroup can be left (right) amenable but not right (left) amenable [35].

Theorem 2.1 *Let S be left or right amenable semigroup. Then the bounded cohomology $H_b^*(S, \mathbb{R})$ is zero.*

Proof. Suppose S is left amenable and let $m \in B^*(S)$ be a left invariant mean. We use the notation

$$m(f) = \int f(x) dx \qquad (2.1)$$

for $f \in B(S)$.

Suppose that $H_b^{(n)}(S) \neq 0$ for some $n \geq 2$ and let $f \in \ker \delta_b^{(n)}$ but $f \notin \mathrm{Im}\, \delta_b^{(n-1)}$. We have

$$(\delta_b^{(n)} f)(h_1, \ldots, h_{n+1}) = f(h_2, \ldots, h_{n+1}) +$$
$$+ \sum_{i=1}^{n} (-1)^i f(h_1, \ldots, h_i h_{i+1}, \ldots, h_{n+1}) + (-1)^{n+1} f(h_1, \ldots, h_n) = 0.$$

Let us integrate the last relation over the variable h_{n+1} in the sense of (2.1):

$$\int f(h_2, \ldots, h_{n+1}) dh_{n+1} +$$
$$+ \sum_{i=1}^{n-1} (-1)^i \int f(h_1, \ldots, h_i h_{i+1}, \ldots, h_n, h_{n+1}) dh_{n+1} +$$
$$+ (-1)^n \int f(h_1, \ldots, h_{n-1}, h_n h_{n+1}) dh_{n+1} +$$
$$+ (-1)^{n+1} \int f(h_1, \ldots, h_n) dh_{n+1} = 0. \qquad (2.2)$$

By property (i)

$$\int f(h_1, \ldots, h_n) dh_{n+1} = f(h_1, \ldots, h_n)$$

and by (iii)

$$\int f(h_1, \ldots, h_{n-1}, h_n h_{n+1}) dh_{n+1} = \int f(h_1, \ldots, h_{n-1}, h_{n+1}) dh_{n+1}.$$

If we denote

$$\Phi(h_2, \ldots, h_n) = \int f(h_2, \ldots, h_n, h_{n+1}) dh_{n+1}$$

(Φ is bounded) we can rewrite (2.2) in the form

$$\Phi(h_2, \ldots, h_n) + \sum_{i=1}^{n-1} (-1)^i \Phi(h_1, \ldots, h_i h_{i+1}, \ldots, h_n) +$$

$$+ (-1)^n \Phi(h_1, \ldots, h_{n-1}) = (-1)^n f(h_1, \ldots, h_n)$$

and thus $f = \delta_b^{(n-1)}((-1)^n \Phi)$ which leads to a contradiction.

We have proved that the bounded cohomology group of a left amenable semigroup vanishes. A similar proof can be given in the case of a right amenable semigroup; only now we should integrate (2.2) over the variable h_1:

$$\int f(h_2, \ldots, h_{n+1}) dh_1 - \int f(h_1 h_2, h_3, \ldots, h_{n+1}) dh_1 +$$

$$+ \sum_{i=2}^{n} (-1)^i \int f(h_1, \ldots, h_{i-1} h_i, \ldots, h_{n+1}) dh_1 +$$

$$+ (-1)^{n+1} \int f(h_1, \ldots, h_n) dh_1 = 0. \qquad (2.3)$$

Let

$$\Psi(h_2, \ldots, h_n) = \int f(h_1, \ldots, h_n) dh_1.$$

Then (2.3) can be rewritten in the form

$$f(h_2, \ldots, h_{n+1}) =$$

$$= \Psi(h_3, \ldots, h_{n+1}) + \sum_{i=2}^{n} (-1)^{i-1} \Psi(h_2, \ldots, h_{i-1} h_i, \ldots, h_{n+1}) +$$

$$+ (-1)^n \Psi(h_2, \ldots, h_n) = (\delta_b^{(n-1)} \Psi)(h_2, \ldots, h_{n+1})$$

and again we have a contradiction. This completes the proof of the Theorem.

Theorem 2.5 [49] of B.E. Johnson states that $H^{(1)}(G, E^*) = 0$ for all Banach G-modules E if and only if G is an amenable group and this is equivalent to the triviality of $H^{(n)}(G, E^*)$ for $n = 1, 2, \ldots$ and any Banach G-module E. Thus Trauber's theorem is a corollary of Johnson's theorem. The converse of Trauber's theorem is not true as the example given at the end of Section 3 shows.

Johnson [49] calls a Banach algebra A amenable if $H^{(1)}(A, E^*) = 0$ for all Banach A-modules E. Another version of the definition was given by A.Y.Khelemskiĭ (see survey [50]). A locally compact group G is amenable if and only if the algebra $L_1(G)$ is amenable. For semigroups this is not true because the algebra $l_1(G)$ of a discrete amenable semigroup can be not amenable (see [23], for instance). Thus Theorem 2.1 is not a corollary of Johnson's theorem.

As is written in Gromov's paper [39] (page 246) "Bounded cohomology first appeared in the group theoretic context. I learned this notion from Phillip Trauber who explained to me his (unpublished) version of the Theorem of Hirsch–Thurston:

if a group π is amenable then the bounded cohomology vanishes"
and (page 247) "... but the original argument of Trauber is shorter and also yields the following more general fact.

Let $f : Y \to X$ be a regular covering with an amenable Galois group π. Then the induced map $\hat{f}^* : H_b^*(X) \to H_b^*(Y)$ is injective and isometric relative to the norm $\| \cdot \|_\infty$."
(We have adopted Gromov's original notation for the bounded cohomology group).

In group theoretic language this statement can be reformulated as follows. Let $H \trianglelefteq G$ be a normal subgroup with G/H being amenable and let $f : H \to G$ be the inclusion map. Then the induced map $f^* : H_b^*(G) \to H_b^*(H)$ is injective and isometric relative to the norm $\| \cdot \|_\infty$.

This statement can be applied, for instance, to the pair $G, [G, G]$ consisting of a group and its commutator subgroup. In some important cases (surface groups, fundamental groups of the complements of some knots) the commutator $[G, G]$ is a free group. Thus the computation of the bounded cohomology of a free group is of special interest.

3   Quasicharacters and pseudocharacters

In this section G can be a group or a semigroup.

Definition 3.1 A function $f : G \to \mathbb{R}$ is a *quasicharacter* if there is a constant $C \geq 0$ s.t. $\forall x, y \in G$

$$|f(x) + f(y) - f(xy)| \leq C.$$

We use the following notation:
$X(G)$ - the space of additive characters (homomorphisms) $G \to \mathbb{R}$.
$QX(G)$ - the space of quasicharacters.
    A function $f : G \to \mathbb{R}$ belongs to $QX(G)$ if and only if $\delta^{(1)}f$ is bounded.

Proposition 3.2 *An isomorphism of vector spaces*

$$H_{b,2}^{(2)}(G) \cong QX(G)/(X(G) \oplus B(G))$$

holds.

Proof. By definition

$$H_{b,2}^{(2)}(G) = (\mathrm{Im}\,\delta^{(1)} \cap \ker\delta_b^{(2)})/\mathrm{Im}\,\delta_b^{(1)}.$$

We have a sequence of linear mappings

$$QX(G) \xrightarrow{\delta^{(1)}} \mathrm{Im}\,\delta^{(1)} \cap \ker\delta_b^{(2)} \xrightarrow{\lambda} \mathrm{Im}\,\delta^{(1)} \cap \ker\delta_b^{(2)}/\mathrm{Im}\,\delta_b^{(1)},$$

where $\delta^{(1)}$ is onto and λ is the canonical mapping. It is clear that $\lambda(\delta^{(1)}f) = 0$ if and only if there is a bounded function g such that $\delta^{(1)}f = \delta^{(1)}g$ and so $f - g \in X(G)$, as the kernel of $\delta^{(1)}$ coincides with $X(G)$. Therefore

$$\ker\lambda\delta^{(1)} = X(G) \oplus B(G)$$

and this finishes the proof.

Definition 3.3 A function $f : G \to \mathbb{R}$ is a *pseudocharacter* if $f \in QX(G)$ and in addition

$$f(g^n) = nf(g),$$

where $n \in \mathbb{Z}$ if G is a group and $n \in \mathbb{Z}_+ = \{0, 1, \ldots\}$ if G is a semigroup.

We denote by $PX(G)$ the space of pseudocharacters and use the notation $f \sim g$ for a pair of functions that differ by a bounded function.

Lemma 3.4 *Any pseudocharacter $f \in PX(G)$ is constant on the conjugacy classes of a group G.*

Proof. Suppose

$$f(x) - f(y^{-1}xy) = \xi \neq 0$$

for some $x, y \in G$. Then the difference

$$f(x^n) - f(y^{-1}x^ny) = n\xi$$

is unbounded when $n \to \infty$. On the other hand

$$f(y^{-1}x^ny) \sim f(y^{-1}) + f(x^n) + f(y) = f(x^n)$$

and we get a contradiction.

Theorem 3.5 *An isomorphism of vector spaces*

$$H_{b,2}^{(2)}(G) \cong PX(G)/X(G)$$

holds.

Proof. The next lemma, which we are going to use to prove the Theorem, was first proven by D.Hyars [43] in case $G = \mathbb{R}$. P.F. Kurchanov rediscovered it in the more general situation of amenable groups. Here we set forth his proof. A more general form of the statement was given by D.Kazhdan [52], Theorem 1 for ε-representations but with the restriction $\varepsilon < 1/100$.

Lemma 3.6 *Let G be a left or right amenable semigroup and $f \in QX(G)$ with*

$$|f(x) + f(y) - f(xy)| \leq C_f \qquad (3.1)$$

for all $x, y \in G$ and some constant $C_f \geq 0$.

Then there exists an additive character $g \in X(G)$ such that

$$|f(x) - g(x)| \leq C_f$$

with the same constant C_f as in (3.1).

Proof. Let

$$h(x, y) = f(xy) - f(x) - f(y).$$

Then $|h(x, y)| \leq C_f$.

Let us suppose for instance that G is left amenable and put

$$H(x) = \int h(x, t)dt,$$

where integration is in the sense of a left invariant mean on G. Then $|H(x)| \leq C_f$ and

$$-\delta^{(1)}H(x, y) = H(xy) - H(x) - H(y) = -f(xy) + f(x) + f(y) = (\delta^{(1)}f)(x, y). \qquad (3.2)$$

Indeed

$$\int [h(xy, t) - h(x, t) - h(y, t)]dt =$$

$$= \int [f(xyt) - f(xy) - f(t) - f(xt) + f(x) + f(t) - f(yt) + f(y) + f(t)]dt =$$

$$= \int [-f(xy) + f(x) + f(y)]dt + \int [f(xyt) - f(yt)]dt - \int [f(xt) - f(t)]dt =$$

$$= -f(xy) + f(x) + f(y)$$

because the last two integrals are equal to zero.

The function

$$g(x) = f(x) + H(x)$$

is a character by (3.2) and this finishes the proof of the lemma.

To every element $g \in G$ there corresponds the semigroup $\langle g \rangle = \{g, g^2, \ldots\}$ generated by this element. The semigroup $\langle g \rangle$ is commutative and so left and right amenable.

By Lemma 3.6 for every $f \in QX(G)$ for which (3.1) holds and for any $g \in G$ there is a character $h_g \in X(\langle g \rangle)$ such that

$$|f(x) - h_g(x)| \leq C_f \qquad (3.3)$$

if $x \in \langle g \rangle$.

If for given $g_1, g_2 \in G$ the subsemigroups $\langle g_1 \rangle, \langle g_2 \rangle$ have nontrivial intersection Z_{g_1,g_2} then h_{g_1} and h_{g_2} coincide on Z_{g_1,g_2}. Indeed, if this intersection is a finite semigroup then h_{g_1}, h_{g_2} must be zero characters and so coincide, and if Z_{g_1,g_2} is infinite, $Z \in Z_{g_1,g_2}$ and $h_{g_1}(Z) \neq h_{g_2}(Z)$ then

$$\lim_{n \to \infty} |h_{g_1}(Z^n) - h_{g_2}(Z^n)| = \infty$$

in contradiction with (3.3). Thus we can define the pseudocharacter $F(x) = h_g(x)$ if $x \in \langle g \rangle$ with the property that

$$|f(x) - F(x)| \leq C_f, \quad x \in G.$$

After this the isomorphism

$$QX(G)/X(G) \oplus B(G) \cong PX(G)/X(G)$$

is obvious and the theorem follows from Proposition 3.2.

Corollary 3.7 *Let a semigroup G be periodic, i.e., for every $g \in G$ there are two numbers $m(g), n(g) \in \mathbb{N}$, $m(g) \neq n(g)$ such that*

$$g^{m(g)} = g^{n(g)}.$$

Then $H_{b,2}^{(2)}(G) = 0$.

Definition 3.8 A group G has *finite width* (or *bounded generation*) if there are elements $a_1, \ldots, a_m \in G$ and a number $k \in \mathbb{N}$ such that every $g \in G$ can be expressed in the form

$$g = a_{i_1}^{\xi_1} \ldots a_{i_k}^{\xi_k},$$

where $1 \leq i_j \leq m$, $\xi_j \in \mathbb{Z}$, $j = 1, \ldots, k$.

Example 3.9 [Carter and Keller [14]] The groups $SL(n, K)$, $n \geq 3$, where K is a ring of integers of an algebraic number field have finite width (see [1] for elementary proof in case of groups $SL(n, \mathbb{Z})$, $n \geq 3$). This result was essentially extended by O.I.Tavgen [72].

Corollary 3.10 *If G has finite width then $H_{b,2}^{(2)}(G)$ is finite dimensional.*

Proof. Because G has finite width every pseudocharacter $f \in PX(G)$ modulo a bounded function is determined by the values

$$f(a_1), \ldots, f(a_m),$$

where $a_1, \ldots, a_m \in G$ are elements from the definition. But if two pseudocharacters differ by a bounded function then they coincide. Thus

$$\dim H_{b,2}^{(2)}(G) \leq m.$$

Definition 3.11 A group is called *ambivalent* if every element is conjugate to its inverse.

Corollary 3.12 *If G is ambivalent then $H_{b,2}^{(2)}(G) = 0$.*

Example 3.13 As it is easy to see the symmetric group $S(\infty)$ over an infinite set is ambivalent. Another example is the group of invertible automata over a finite alphabet with the operation of superposition of automata [71].

Definition 3.14 A group G is said to be *uniformly perfect* if for some $N > 0$ any $g \in G$ can be represented as a product of at most N commutators.

Corollary 3.15 ([56]) *If G is uniformly perfect then $H_{b,2}^{(2)} = 0$.*

Example 3.16 By R.C. Thompson [73] $SL(n,k)$ is uniformly perfect when k is a field and $n \geq 3$. The same is true when k is a commutative principle ideal domain [58], [21]. Thus $SL(n,\mathbb{Z})$ is uniformly perfect when $n \geq 3$ and this gives the equality

$$H_b^{(2)}(SL(n,\mathbb{Z})) = 0$$

if $n \geq 3$, because as is well known, $H^{(2)}(SL(n,\mathbb{Z}),\mathbb{R}) = 0$. At the same time $H_b^{(2)}(SL(2,\mathbb{Z}))$ is infinite dimensional (see Section 5).

Remark 3.17 The groups with trivial (or infinite dimensional) $H_b^{(2)}$ are of some interest. For instance, if $H_b^{(2)}(G) = 0$ then any two actions of the group G on the circle S^1 are semiconjugate in the sense of E. Ghys [33] if and only if the corresponding rotation numbers of each generator coincide. As we have seen there are many nonamenable groups with this property.

When this paper had been written Ghys informed me that the notion of pseudocharacter, under the name homogeneous quasi-morphism, was used by Besson [7] to describe the kernel of the homomorphism $H_b^{(2)}(G,\mathbb{R}) \to H_b^{(2)}(G,\mathbb{R})$, that is $H_{b,2}^{(2)}(G)$ (see [7], Proposition 3.3.1). On the other hand Bavard [6] gave a characterization of groups with $H_{b,2}^{(2)}(G) = 0$ in terms of so-called "longueur stable" $\|g\|$ of an element $g \in G' = [G,G]$.

Namely for $g \in G'$ let $C(g)$ be the minimal number of commutators needed to represent g as a product. Then

$$\|g\| = \lim_{n \to \infty} \frac{C(g^n)}{n}$$

(the limit exists).

Theorem 3.18 ([7]) $H_{b,2}^{(2)}(G) = 0$ *if and only if* $\|g\| = 0$ *for any* $g \in G'$.

Let $PLF(\mathbb{R})$ be the group of orientation-preserving homeomorphisms of the real line which are piecewise-linear with respect to a finite subdivision of $\mathbb{R}$ (see [8] for information about this group). Actually this group should be called the Chehata group (see [17], [53]).

One interesting and important subgroup of $PLF(\mathbb{R})$ is Thompson's group F, which can be given by two generators and two relations. A central problem in this context is Geoghegan's question about the existence of an invariant mean on F. This group provides the first example of an infinite dimensional torsion-free FP_∞ group [13]. One way to prove that F is not amenable is to show that $H_b^{(n)}(F) \neq 0$ for some $n > 1$.

Our conjecture is that F is not amenable and thus $H^{(1)}(F, X^*) \neq 0$ for some Banach F-module X. On the other hand in [49], page 39 the class of groups with $H^{(2)}(G, X^*) = 0$ for any Banach G-module X was considered. As shown in [49], any group G with $H^{(2)}(G, X^*) = 0$ for all Banach G-modules X does not contain a free subgroup $\mathcal{F}_2$ with two generators and so any such group with $H^{(1)}(G, X^*) \neq 0$ for some X would be a non-amenable group containing no subgroup isomorphic to $\mathcal{F}_2$. Groups with the last property where constructed by A. Yu. Olshanskiĭ and S. I. Adyan (they where unknown when [49] was written, see [37] for the history of the question). At the same time examples of groups with $H^{(1)}(G, X^*) \neq 0$ for some X but $H^{(2)}(G, Y^*) = 0$ for every Y are unknown up to now. Thompson's group F is a good candidate for such an example. In any case it does not contain $\mathcal{F}_2$ [13] and $H_b^{(2)}(F) = 0$, as was recently observed by E. Ghys and V. Sergiescu.

Problem 3.19 Is it true that $H^{(2)}(F, Y^*) = 0$ for any Banach F-module Y?

The same questions may be asked about the group $PLF(\mathbb{R})$ and its subgroups $G(P)$ defined in [8].

Remark 3.20 In [56] using the uniformly perfect property it is shown that

$$H_b^{(2)}(\mathrm{Homeo}_+(S^1)) \cong H_b^{(2)}(SL(2, \mathbb{R})) \cong \mathbb{R}$$

(all groups are considered as discrete). At the same time $H_b^{(n)}(\mathrm{Homeo}_k(\mathbb{R}^n)) = 0$ if $n \geq 1$, where $\mathrm{Homeo}_k(\mathbb{R}^n)$ is the group of all homeomorphisms of $\mathbb{R}^n$ with compact support.

This group is not amenable because it contains a free subgroup with two generators. Thus the example shows that the converse of Trauber's theorem is not true.

The group $\mathcal{G}$ of measure preserving transformations of Lebesgue space $([0, 1], \mu)$ is also uniformly perfect (actually every element $g \in \mathcal{G}$ is a commutator [62]) and so it has trivial $H_{b,2}^{(2)}$.

We see that "large" groups have the tendency to have "small" second bounded cohomology groups.

4 Bounded cohomology of some group constructions

To any homomorphism $\varphi : G \to H$ there corresponds, in a natural way, the induced homomorphism $\varphi^* : H_b^*(H) \to H_b^*(G)$ ([39], Section 3.0). The following statement is an algebraic version of Gromov's Mapping theorem [39], Section 3.1.

Theorem 4.1 (N. Ivanov [47]) *If H is an amenable normal subgroup of G then the map*

$$\varphi^* : H_b^*(G/H) \to H_b^*(G)$$

induced by the canonical homomorphism

$$\varphi : G \to G/H$$

is an isometric isomorphism.

Proposition 4.2 *An isomorphism of vector spaces*

$$H_{b,1}^{(n)}(G_1 * G_2) \cong H_{b,1}^{(n)}(G_1) \oplus H_{b,1}^{(n)}(G_2) \tag{4.1}$$

holds.

Proof. The canonical imbeddings $G_i \to G_1 * G_2$, $i = 1, 2$, induce homomorphisms

$$\begin{aligned} \varphi_i &: H_b^*(G_1 * G_2) \to H_b^*(G_i), \\ \psi_i &: H^*(G_1 * G_2) \to H^*(G_i), \end{aligned}$$

(all cohomology groups are taken for a trivial module $\mathbb{R}$), $i = 1, 2$ which are onto and the diagram

$$
\begin{array}{ccc}
H_b^*(G_1 * G_2) & \xrightarrow{\;\varphi = \varphi_1 \times \varphi_2\;} & H_b^*(G_1) \oplus H_b^*(G_2) \\
{\scriptstyle \varepsilon}\big\downarrow & & \big\downarrow{\scriptstyle \delta = \varepsilon_1 \times \varepsilon_2} \\
H^*(G_1 * G_2) & \xrightarrow{\;\psi = \psi_1 \times \psi_2\;} & H^*(G_1) \oplus H^*(G_2)
\end{array}
$$

is commutative where ε, ε_i, $i = 1, 2$ are the homomorphisms the images of which are the bounded parts of the usual cohomology groups.

Since, as is well known, ψ is an isomorphism and φ is surjective, we have

$$\psi_i(\varepsilon(H_b^*(G_1 * G_2))) = \varepsilon_i(H_b^*(G_i)), \quad i = 1, 2,$$

and thus

$$\psi(\varepsilon(H_b^*(G_1 * G_2)) = \varepsilon_1(H_b^*(G_1)) \oplus \varepsilon_2(H_b^*(G_2)),$$

which is equivalent to (4.1).

We are going to describe the second cohomology group of a free product. For this purpose we establish the formula which expresses the singular part of the second bounded cohomology group of a free product $G = C * D$ in terms of corresponding singular parts of the groups C, D.

Let $G = C * D$, $F = [C, D]$ be the normal subgroup in G generated by commutators $[c, d]$, $c \in C$, $d \in D$. It is known that F is a free group with the basis

$$E = \{[c, d], \ c \in C, \ d \in D, \ c, d \neq 1\}.$$

Let $\Gamma \leq \operatorname{Aut} F$ be the subgroup generated by the inner automorphisms $\hat{c}, \hat{d}$, $c \in C, d \in D$, defined by

$$e^{\hat{c}} = c^{-1}ec, \quad e^{\hat{d}} = d^{-1}ed, \quad e \in F$$

(we write automorphisms as powers) and let denote by $PX_\Gamma(F, E)$ the space of pseudocharacters on F that are bounded on the set E of generators and are Γ-invariant:

$$PX_\Gamma(F, E) = \{f \in PX(F), \ f|_E \in l_\infty(E), \ f(e) = f(e^\gamma), \ e \in F, \ \gamma \in \Gamma\}.$$

Proposition 4.3 *An isomorphism of vector spaces*

$$H_{b,2}^{(2)}(C * D) \cong H_{b,2}^{(2)}(C) \oplus H_{b,2}^{(2)}(D) \oplus PX_\Gamma(F, E)$$

holds.

Proof. We denote by $QX_0(G)$ the subspace of $QX(G)$ consisting of the quasicharacters equal to zero at the identity element.

The restriction $f_F = f|_F$ of a function $f \in QX(G)$ is bounded on the set E of generators because

$$f([c, d]) = f(c^{-1}d^{-1}cd) \sim -f(c) - f(d) + f(c) + f(d) = 0.$$

Furthermore f_F is almost invariant under the action of the group Γ in the sense that there is a constant $C = C(f)$ such that $|f_F(e) - f_F(e^\gamma)| \leq C$ for all $e \in F$ and $\gamma \in \Gamma$.

We denote by $QX_0^\Gamma(F, E)$ the space of the quasicharacters f on F with these two properties satisfying the additional property that $f(1) = 0$. If $f \in PX(G)$ then obviously f_F is Γ-invariant.

The mapping

$$f \rightarrow f_C \oplus f_D \oplus f_F,$$

where

$$f_C = f|_C, \ f_D = f|_D$$

defines the homomorphism

$$\xi : QX_0(G) \rightarrow QX_0(C) \oplus QX_0(D) \oplus QX_0^\Gamma(F, E),$$

which is surjective. Indeed, for any triple f_C, f_D, f_F, where $f_C \in QX_0(C)$, $f_D \in QX_0(D)$ and $f_F \in QX_0^\Gamma(F, E)$ we can define $f \in QX_0(G)$ by the rule:

$$f(cde) = f_C(c) + f_D(d) + f_F(e),$$

where $c \in C$, $d \in D$, $e \in F$. It is clear that $f(1) = 0$ and $f \in QX(G)$ because

$$f(c_1 d_1 e_1 c_2 d_2 e_2) = f(c_1 d_1 c_2 d_2 e_1^{\hat{c}_2 \hat{d}_2} e_2) = f(c_1 c_2 d_1 d_2 [d_1, c_2]^{\hat{d}_2} e_1^{\hat{c}_2 \hat{d}_2} e_2) =$$
$$= f_C(c_1 c_2) + f_D(d_1 d_2) + f_F([d_1, c_2]^{\hat{d}_2} e_1^{\hat{c}_2 \hat{d}_2} e_2) \sim$$
$$\sim f_C(c_1) + f_C(c_2) + f_D(d_1) + f_D(d_2) + f_F(e_1) + f_F(e_2) =$$
$$= f(c_1 d_1 e_1) + f(c_2 d_2 e_2).$$

We have the inclusion

$$\ker \xi \subseteq B_0(G) = \{f \in B(G) : f(1) = 0\},$$

which follows from the equivalence

$$f(cde) \sim f(c) + f(d) + f(e)$$

and thus

$$\ker \xi = \{f \in B_0(G) : f_C = f_D = f_F = 0\}.$$

For any group K the isomorphism

$$H_{b,2}^{(2)}(K) \cong QX_0(K)/(B_0(K) \oplus X(K)),$$

holds and moreover we have the isomorphisms

$$\begin{aligned} X(G) &\cong X(C) \oplus X(D), \\ QX_0^\Gamma(F, E) &\cong B_0(F) \oplus PX_\Gamma(F, E), \qquad (4.2) \\ B_0(G) &\cong \ker \xi \oplus B_0(C) \oplus B_0(D) \oplus B_0(F). \end{aligned}$$

Only (4.2) need be explained but it is a direct corollary of the Lemma 3.6. Thus

$$H_{b,2}^{(2)}(G) = QX_0(G)/(B_0(G) \oplus (X(G)) \cong$$
$$\cong (QX_0(C)/B_0(C) \oplus X(C)) \oplus (QX_0(D)/B_0(D) \oplus X(D)) \oplus PX_\Gamma(F, E) \cong$$
$$\cong H_{b,2}^{(2)}(C) \oplus H_{b,2}^{(2)}(D) \oplus PX_\Gamma(F, E)$$

which finishes the proof.

Remark 4.4 The statement about the space of pseudocharacters of a free product given in [27] (Theorem 4) is not correct. To get the correct statement one should change the space $BPX(F, C \cup D)$ (we use the notation of [27]) in the corresponding formula to the space $PX_\Gamma(F, E)$ (our notation).

Let $C \rtimes D$ be a semidirect product, where C is the passive group, D is the active group and $\Gamma \leq \operatorname{Aut} C$ is the group generated by automorphisms $c \to c^d$, where $c^d = d^{-1}cd$, $c \in C, d \in D$. We denote by $PX_\Gamma(C)$ the space of pseudocharacters on C that are Γ-invariant and by $X_\Gamma(C)$ the space of characters on C that are Γ-invariant.

Theorem 4.5

$$H_{b,2}^{(2)}(C \rtimes D) = H_{b,2}^{(2)}(D) \oplus PX_\Gamma(C)/X_\Gamma(C). \tag{4.3}$$

Proof. The space of pseudocharacters of a semidirect product was described in [30] and the description is as follows:

$$PX(C \rtimes D) = PX(D) \oplus PX_\Gamma(C).$$

But

$$X(C \rtimes D) = X_\Gamma(C) \oplus X(D).$$

An application of Theorem 3.5 gives (4.3).

Corollary 4.6 *The singular parts of the second bounded cohomology groups of the braid group $B(n)$ and the pure braid group $K(n)$ with $n \geq 3$ strings are infinite dimensional spaces.*

Proof. The pure braid group $K(n)$ with n strings is isomorphic to a semidirect product of a free group of rank $n-1$ (passive group) and a pure braid group $K(n-1)$ (active group). Thus

$$K_n \cong F_{n-1} \rtimes (F_{n-2} \rtimes \dots (F_2 \rtimes \mathbb{Z}) \dots)$$

(recall that $K(2) \cong \mathbb{Z}$).

The group $K(3) = F_2 \rtimes \mathbb{Z}$ is a group generated by the elements $x_1 = \sigma_1^2$, $x_2 = \sigma_2 \sigma_1^2 \sigma_2^{-1}$, $y = \sigma_1^2$ where σ_1, σ_2 are standard generators of the group $B(3)$. The elements x_1, x_2 generate a free group F_2 of rank 2 and y is the generator of an infinite cyclic group which acts on F_2. One can easily verify that y acts on F_2 as conjugation by the element $x_1^{-1} x_2^{-1}$, thus

$$H_{b,2}^{(2)}(K(3)) = H_{b,2}^{(2)}(F_2).$$

But $H_{b,2}^{(2)}(F_2)$ is a direct summand of $H_{b,2}^{(2)}(K(n))$, if $n \geq 3$, and $\dim H_{b,2}^{(2)}(F_2) = \infty$ (see section 5 below).

To prove that $H_{b,2}^{(2)}(B(n))$ is infinite dimensional, we need the following statement.

Proposition 4.7 *Let H be a subgroup of finite index of a finitely generated group G. Then $\dim PX(G) < \infty$ if and only if $\dim PX(H) < \infty$.*

Proof. If $g \in G$ then $g^n \in H$ for some $n \in N$ and so every pseudocharacter on G is determined by its values on H which gives an injection

$$\tau : PX(G) \to PX(H), \qquad (4.4)$$

defined by restriction to H of a psuedocharacter on G. Let $E = \{g_1, \ldots, g_k\}$ be a system of representatives of cosets gH, $g \in G$. If two pseudocharacters f_1, f_2 coincide on E and on H, then they differ by a bounded function and thus coincide. This shows that the factor space $PX(H)/\tau(PX(G))$ is finite dimensional, as required.

We have the exact sequence

$$1 \to K(n) \to B(n) \to S(n) \to 1,$$

where $S(n)$ is a symmetric group on n elements. Using the last proposition and information about $H_{b,2}^{(2)}(K(n))$ we finish the proof of Corollary 4.6.

Remark 4.8 The braid group B_3, modulo its (amenable) centre, is isomorphic to the free product $\mathbb{Z}_2 * \mathbb{Z}_3$ and we can apply Theorem 4.1, Proposition 4.2 and Propsition 4.3 to show that $H_b^{(2)}(\mathbb{Z}_2 * \mathbb{Z}_3) = H_b^{(2)}(F_2)$, where F_2 is a free group of rank 2. On the other hand the second bounded cohomology group of a free product of two cyclic groups is described in Section 5.

Remark 4.9 The bounded parts $H_{b,1}^{(n)}(B_n)$ and $H_{b,1}^{(n)}(K_n)$ are finite dimensional spaces in every dimension n, because the usual cohomology groups $H^{(n)}(B_n, \mathbb{R})$ and $H^{(n)}(K_n, \mathbb{R})$, with $n \geq 1$, are finite dimensional.

Problem 4.10 To construct a basis for $H_b^{(2)}(B_n)$.

Let $H \trianglelefteq G$ be a normal subgroup of finite index and let

$$PX_G(H) = \{f \in PX(H): \; f(x) = f(x^g), \; g \in G\},$$

$$X_G(H) = \{f \in X(H): \; f(x) = f(x^g), \; g \in G\},$$

where G acts on H by conjugation.

Proposition 4.11 *The natural isomorphism of vector spaces*

$$H_{b,2}^{(2)}(G) \cong PX_G(H)/X_G(H)$$

holds.

Proof. It is clear that $\tau(PX(G)) \subset PX_G(H)$ where τ is the homomorphism 4.4. On the other hand every $f \in PX_G(H)$ can be extended to the pseudocharacter $\hat{f}$ on G: $\hat{f}(g) = \frac{1}{N}f(g^N)$ where $N = n!$, $n = |G : H|$.

To see this, let $g_1, \ldots, g_n$ be a system of representatives of cosets gH and for $g \in G$ let $\bar{g}$ denote the chosen representative of gH. Using the equivalence $\sim$ on the space of functions on H we get:

$$\hat{f}(g_i h) = \frac{1}{N} f((g_i h)^N) = \frac{1}{N} f(g_i h g_i^{-1} g_i^2 h g_i^{-2} \ldots g_i^N h g_i^{-N} g_i^N) \sim$$

$$\sim \frac{1}{N} \left(\sum_{k=1}^{N} f(g_i^k h g_i^{-k}) + f(g_i^N) \right) \sim f(h),$$

$$\hat{f}(g_i h_1 g_j h_2) = \hat{f}\left(\overline{g_i g_j} (\overline{g_i g_j})^{-1} g_i g_j g_j^{-1} h_1 g_j h_2 \right) \sim$$

$$\sim f\left((\overline{g_i g_j})^{-1} g_i g_j g_j^{-1} h_1 g_j h_2 \right) \sim f\left((\overline{g_i g_j})^{-1} g_i g_j \right) + f(g_j^{-1} h_1 g_j) +$$

$$+ f(h_2) \sim f(h_1) + f(h_2) \sim \hat{f}(g_i h_1) + \hat{f}(g_j h_2).$$

This shows that $\hat{f} \in PX(G)$. The same arguments can be applied to show that $\hat{f} \in X(G)$ if $f \in X_G(H)$ and this finishes the proof.

Theorem 4.12 *Let $G = S^{-1}S$ $(G = SS^{-1})$ be the group of left (right) quotients for a semigroup S. Then the isomorphism of vector spaces*

$$H_{b,2}^{(2)}(G) \cong H_{b,2}^{(2)}(S) \tag{4.5}$$

holds.

Proof. We consider the case when $G = S^{-1}S$. First we shall establish the isomorphism $X(G) \cong X(S)$. For any $f \in X(G)$ the restriction $\hat{f} = f|_S$ is a character on S and the mapping $\xi : f \to \hat{f}$ is injective, because if $\hat{f} = 0$ then $f(a^{-1}b) = -\hat{f}(a) + \hat{f}(b) = 0$.

We are going to prove that ξ is onto. Let $\hat{f} \in X(S)$ and $f : G \to \mathbb{R}$ be a function defined by the rule:

$$f(a^{-1}b) = -\hat{f}(a) + \hat{f}(b). \tag{4.6}$$

To prove that this is well defined function we must show that

$$-\hat{f}(a) + \hat{f}(b) = -\hat{f}(c) + \hat{f}(d) \tag{4.7}$$

if $a^{-1}b = c^{-1}d$.

By the Ore-Dubreil theorem [19] a semigroup S has a left quotient group if and only if it satisfies the right Ore condition: $\forall a, b \in S \; \exists x, y \in S$ such that $xa = yb$.

Suppose $a^{-1}bd^{-1}c = 1$ and $xb = yd$ for some $x, y \in S$, that gives

$$\hat{f}(b) - \hat{f}(d) = \hat{f}(y) - \hat{f}(x). \tag{4.8}$$

But $a^{-1}bd^{-1}c = a^{-1}x^{-1}yc = 1$, thus $yc = xa$ and

$$\hat{f}(c) - \hat{f}(a) = \hat{f}(x) - \hat{f}(y). \tag{4.9}$$

Using (4.8), (4.9) we get (4.7).

Let us show now that the function f is a character, i.e.,

$$f(a^{-1}bc^{-1}d) = f(a^{-1}b) + f(c^{-1}d)$$

for any $a, b, c, d \in S$. Suppose $bc^{-1} = x^{-1}y$, $x, y \in S$. Then

$$f(a^{-1}bc^{-1}d) = f((xa)^{-1}yd) = -\hat{f}(xa) + \hat{f}(yd) =$$
$$= -\hat{f}(x) - \hat{f}(a) + \hat{f}(y) + \hat{f}(d) = \hat{f}(b) - \hat{f}(c) - \hat{f}(a) + \hat{f}(d) =$$
$$= f(a^{-1}b) + f(c^{-1}d)$$

because $\hat{f}(b) - \hat{f}(c) = \hat{f}(y) - \hat{f}(x)$.

We have constructed the canonical isomorphism $\xi : X(G) \to X(S)$. Now we define an isomorphism $\eta : PX(G) \to PX(S)$ which is an extension of ξ, i.e. ηf is the restriction $f|_S = \hat{f}$, for $f \in PX(G)$.

If $\hat{f} = 0$ then f is bounded because for some constant $C \geq 0$

$$|f(a^{-1}b) + f(a) - f(b)| \leq C,$$

$a, b \in S$, and so f is the zero pseudocharacter.

Let us prove that η is onto. Having $\hat{f} \in PX(S)$ we again define f by formula (4.6). If, in all the relations used when proving that f is a character under the condition $\hat{f} \in X(G)$, we change the sign of equality to the sign "$\sim$" of equivalence ($f_1 \sim f_2$ means that the difference $f_1 - f_2$ is a bounded function) we will get that $f \in PX(G)$. This proves that η is an isomorphism and thus (4.5) holds.

Example 4.13 Let

$$G = \langle a, b| \ ab = b^2 a \rangle \tag{4.10}$$

$$S = \langle a, b| \ ab = b^2 a \rangle_\# \tag{4.11}$$

be the group and the semigroup given by the same presentation (the subscript # means that we consider the corresponding presentation as a presentation of a semigroup given by generators and relators). Then S is a subsemigroup of G and $G = S^{-1}S$ but $S^{-1}S$ is a proper subset of G.

The group G is solvable and so amenable. Thus S is right amenable but not left amenable [36]. Using Theorem 4.12 and Trauber's theorem we get that $H_{b,2}^{(2)}(S) = 0$ (the same follows from Theorem 2.1).

Example 4.14 Let F again be Thompson's group. It can be defined by the presentation

$$F = \langle \ x_0, x_1, \ldots \ | \ x_n x_i = x_i x_{n+1}, \ i < n \ \rangle.$$

Let S be the corresponding semigroup

$$S = \langle\ x_0, x_1, \ldots\ |\ x_n x_i = x_i x_{n+1},\ i < n\ \rangle_{\#}.$$

Then S is a subsemigroup of F and $F = SS^{-1}$ but $F \neq S^{-1}S$ [13]. This shows that S is not right amenable [36]. Using Theorem 4.12 and the claim of Ghys and Sergiescu, mentioned near the end of Section 3, we get $H_{b,2}^{(2)}(F) = H_{b,2}^{(2)}(S)$.

5   Some examples

A free group

Let F_m be a free group of finite or countable rank m, $2 \leq m \leq \infty$ with a basis $A = \{a_1, \ldots, a_m\}$. It is clear that the bounded part $H_{b,1}^{(n)}(F_m)$ is zero for every $n = 2, 3, \ldots$.

In [49], [9], [39] it is pointed out that $H_b^{(2)}(F_m)$ is nonzero when $m \geq 2$. Moreover in [9], page 54, is written: "In §3(a) we construct explicitly an infinite set of linearly independent generators of $H_b^{(2)}(\mathbb{Z} * \mathbb{Z})$". Here $\mathbb{Z} * \mathbb{Z}$ is the free product of two copies of an infinite cyclic group; that is a free group with two generators. However in [9] there is no discussion about the generation of $H_b^{(2)}(\mathbb{Z} * \mathbb{Z})$ but only the statement that the dimension of $H_b^{(2)}(\mathbb{Z} * \mathbb{Z})$ is infinite. Let us analyze this statement.

For every reduced word of length ≥ 2 over an alphabet of free generators and their inverses, Brooks defines a quasicharacter f_W:

$\qquad f_W(g)\quad =$ number of times W occurs in g
$\qquad\qquad\qquad\ -$ number of times W^{-1} occurs in g,

where the phrase "W occurs in g" means that W is a subword of the reduced word representing g (similar quasicharacters were introduced by B.E. Johnson in [49], Proposition 2.8).

It is easy to verify that

$$|\delta^{(1)} f_W(g, h)| \leq |W|,$$

where $|W|$ is the length of a word W. In [9] the following assertion appears as a proposition (page 58)

Assertion 5.1 *The bounded 2-cocycles $\delta^{(1)} f_W$ are all non-trivial cocycles, and are all linearly independent.*

But the last statement of the assertion is not correct as the following example shows. Let a, b be generators of $\mathbb{Z} * \mathbb{Z}$,

$$f = f_{ab} + f_{a^{-1}b} + f_{ab^{-1}} + f_{a^{-1}b^{-1}}.$$

It is easy to check that

$$f(g) = \begin{cases} 1 & \text{if } g \text{ begins with } a^{\pm 1} \text{ and ends with } b^{\pm 1}, \\ -1 & \text{if } g \text{ begins with } b^{\pm 1} \text{ and ends with } a^{\pm 1}, \\ 0 & \text{otherwise.} \end{cases}$$

Thus f is bounded and so

$$\delta^{(1)} f = \delta^{(1)} f_{ab} + \delta^{(1)} f_{a^{-1}b} + \delta^{(1)} f_{ab^{-1}} + \delta^{(1)} f_{a^{-1}b^{-1}}$$

represents the zero element in $H_b^{(2)}(\mathbb{Z} * \mathbb{Z})$.

In [57], Proposition 4.5, a rigorous proof was given that the space $H_b^{(2)}(\mathbb{Z} * \mathbb{Z})$ is infinite dimensional. At the same time in [57] Proposition 5.1 Mitsumatsu extracted from $\{\delta^{(1)} f_w\}_{|w| \geq 2}$ an infinite independent subsystem. In fact, in the notation of [57], he showed that the cocycles $F_n = \delta^{(1)} f_{[a^n, b^n]}$, $n = 1, 2, \dots$ are independent because $\langle F_n, E_k \rangle = -\delta_{n,k}$, for any n, k (where E_k is a sequence of l_1-cycles constructed in [57]).

Now we give a description of $H_b^{(2)}(F_m)$ using the results of V. A. Faiziev [27], [31] and of this paper.

As usual we identify elements of a free group with reduced words representing them.

Let W be a reduced word over the alphabet $A \cup A^{-1}$ and let $\overline{W}$ be the cyclic word corresponding to W. We call $\overline{W}$ the *cyclic closure* of W. If $W = T^{-1} W_1 T$, where W_1 is cyclically reduced then $\overline{W}_1$ is the cyclic closure of W. We agree that the positive rotation when reading a cyclic word is the anticlockwise rotation.

An element $W \in F_m$ is *without self-overlapping* if the intersection $B(W) \cap E(W)$ is empty, where $B(W)$ is the set of beginnings of W and $E(W)$ is the set of ends of W:
if $W = a_{i_1}^{\xi_1} a_{i_2}^{\xi_2} \dots a_{i_k}^{\xi_k}$, then

$$B(W) = \{a_{i_1}^{\xi_1}, a_{i_1}^{\xi_1} a_{i_2}^{\xi_2}, \dots, a_{i_1}^{\xi_1} a_{i_2}^{\xi_2} \dots a_{i_{k-1}}^{\xi_{k-1}}\},$$

and

$$E(W) = \{a_{i_2}^{\xi_2} \dots a_{i_{k-1}}^{\xi_{k-1}} a_{i_k}^{\xi_k}, \dots, a_{i_{k-1}}^{\xi_{k-1}} a_{i_k}^{\xi_k}, a_{i_k}^{\xi_k}\}.$$

Now for every $W \in F_m$ we define the function on F_m:
$$e_W(V) = \text{number of times } W \text{ occurs in } \overline{V}$$
$$\quad\quad - \text{ number of times } W^{-1} \text{ occurs in } \overline{V},$$
where $V \in F_m$.

Lemma 5.2 *For every $W \in F_m$ the function e_W is a quasicharacter. Moreover if W is without self-overlapping then e_W is a pseudocharacter.*

A proof of the lemma is easy and is omitted .

Lemma 5.3 ([31]) *Let $W \in F_m$ be without self-overlapping. Then*

$$|e_W(x) + e_W(y) - e_W(xy)| \leq 6, \tag{5.1}$$

if $x, y \in F_m$.

Remark 5.4 In [31] this inequality is given with the constant 15 on the right hand side. (5.1) can be rewritten in the form $\|\delta^{(1)}e_W\|_\infty \leq 6$, where $\|\cdot\|_\infty$ is the sup norm and thus

$$\|\delta^{(1)}e_W\| \leq 6.$$

The exact value of $\|\delta^{(1)}e_W\|_\infty$ is of some interest because it gives an upper bound for the norm of elements $\delta^{(1)}e_W \in H_b^2(F_m)$ which constitute a basis of $H_b^2(F_m)$ under some additional conditions on W (see below). One can verify that the number 6 is best possible in (5.1).

Sketch proof. Our arguments are different from those given in [31]. Let $x, y \in F_m$ be two elements, $x = t^{-1}x_1 t$, $y = u^{-1}y_1 u$, where x_1, y_1 are cyclically reduced. This can be demonstrated by the diagrams of Figure 1.

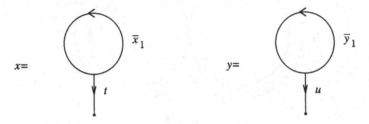

Figure 1

Starting from the diagram of the product xy, shown in Figure 2, and per-

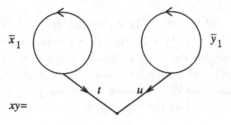

Figure 2

forming, in a geometric way, cancellations in this product we will get a diagram of the type of one of Cases 1–4 below, with the boundary word representing the cyclic word $\overline{xy}$ (the punctured line on each of these diagrams shows the way of reading $\overline{xy}$). Cases 1, 2 and 3 are shown in Figures 3, 4 and 5 respectively.

Case 4. In this case $\overline{xy}$ is a cyclic word corresponding to a subword of one of the cyclic words $\overline{x}_1$ or $\overline{y}_1$ (this case may happen when all letters of y_1 or x_1 cancel).

  Suppose for instance that $\overline{xy}$ is a cyclic word corresponding to a subword $Q = T^{-1}Q_1 T$ of $\overline{y}_1$. Then $\overline{x}_1$ is the cyclic closure of some word E^{-1} and $\overline{y}_1$ is the cyclic closure of the word EQ, as illustrated in Figure 6.

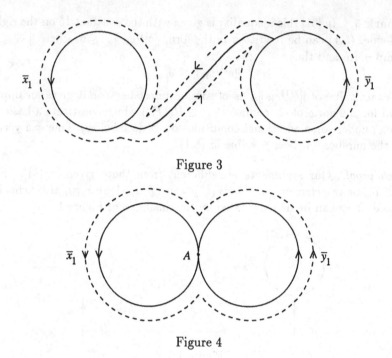

Figure 3

Figure 4

The proof of the proposition involves consideration of all possible cases. We point out only the typical cases, giving the maximal value for $\delta^{(1)}e_W(x,y)$ in each of Cases 1–4.

Every occurrence of W (W^{-1}) in $\bar{x}$ $(\bar{y})$ that does not take place in $\overline{xy}$ or every occurrence of W (W^{-1}) in $\overline{xy}$ that does not take place in $\bar{x}$ or $\bar{y}$ can be located only in a neighborhood of a singular point of a diagram. (We don't give a formal definition of a singular point. On a diagram singular points are marked A or B.) Because W is without self-overlapping, the number of such occurrences is determined by the topology of a diagram and does not exceed 6.

Diagrams for Cases 1–4 are shown in Figures 7–10, respectively.

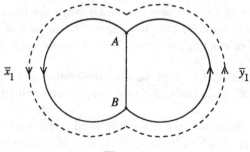

Figure 5

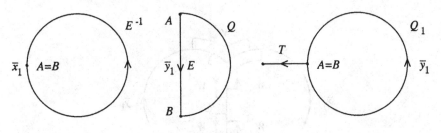

Figure 6

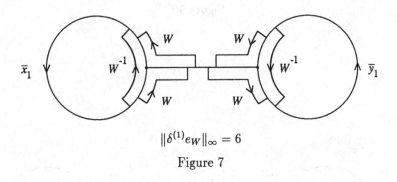

$$\|\delta^{(1)}e_W\|_\infty = 6$$

Figure 7

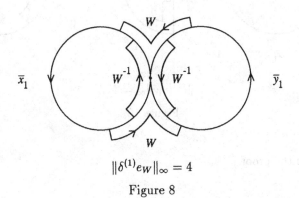

$$\|\delta^{(1)}e_W\|_\infty = 4$$

Figure 8

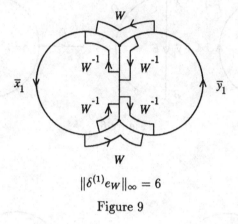

$$\|\delta^{(1)} e_W\|_\infty = 6$$

Figure 9

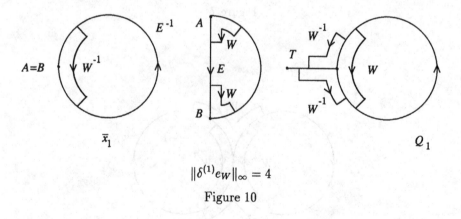

$$\|\delta^{(1)} e_W\|_\infty = 4$$

Figure 10

This finishes the proof.

An element of a free group is *simple* if it is not a nontrivial power of another element. The *length* of a conjugacy class by definition is the length of the shortest representative of a class.

Lemma 5.5 ([31]) *There is a system P of representatives of the length ≥ 2 conjugacy classes of simple elements in the free group F_m which satisfies the following conditions:*

a) if $W \in P$ then W is a shortest element in its conjugacy class (equivalently W is cyclically reduced).

b) if $W \in P$ then $W^{-1} \in P$.

c) if $W \in P$ then $B(W) \cap E(W) = \emptyset$.

Proof. (due to P.F.Kurchanov). We shall prove first that every simple cyclically reduced element of length ≥ 2 in a free group is conjugate to an element satisfying condition c).

For this purpose we introduce the ordering

$$a_1 \prec a_2 \prec \ldots \prec a_m$$

on the set of generators and extend it to the lexicographical ordering $\preceq$ on F_m.

Let $V \in F_m$ be any simple cyclically reduced element of length ≥ 2 and let W be a maximal (with respect to the ordering) element among the elements which are conjugate to V. We claim that

$$B(W) \cap E(W) = \emptyset.$$

Firstly we observe that any beginning U of W is greater than or equal to any subword D of W with length $|U|$. This remark will be used in the proof a number of times.

Suppose $B(W) \cap E(W) \neq \emptyset$, $D \in B(W) \cap E(W)$ and D is a shortest word with this property.

If $2|D| > |W|$ then we can find a shorter subword of W with the property of being a beginning and an end of W. Thus we can suppose that $2|D| \leq |W|$ and $W = DBD$ for some word B.

Let us show that in this case W has the form

$$W = D \ldots DCD, \tag{5.2}$$

where $|C| < |D|$. Indeed, if this is not the case and $C = C_1 C_2$ with $|C_1| = |D|$ then $C_1 \preceq D$ and if the last inequality is strict, then

$$W \prec D \ldots DC \tag{5.3}$$

in contradiction with our assumption.

Thus $C_1 = D$ and (5.2) can be transformed to

$$W = DD \ldots DDC_2 D,$$

so we can consider the case when $|C| < |D|$ in (5.2). Let $D = D_1 D_2$ with $|D_1| = |C|$. If $C \prec D_1$ then we have again (5.3) which leads to a contradiction. Thus $C = D_1$.

Now it is easy to se that if some power C^k of the word C occurs in W then the beginning of W of length $k|C|$ must coincide with C^k.

We may apply the last remark a number of times to the word

$$\cdot W = D \ldots DCC D_2$$

and to get that either D is a power of C or some beginning of D coincides with the corresponding end of D. This leads to a contradiction with the assumption that D is the shortest word from the set $B(W) \cap E(W)$.

It is well known that in a free group the elements g and g^{-1} lie in different conjugacy classes, if $g \neq 1$. Thus we can fix some system $P = P^- \cup P^+$, $P^- \cap P^+ = \emptyset$ of the length ≥ 2 representatives of a conjugacy classes, satisfying condition a), b), c) of the proposition. This finishes the proof.

Let P_n^+, $n = 2, 3, \ldots$ be the set of words of length n in P^+.

Before formulating the next result we recall some notions from functional analysis (see for instance [65]).

Definition 5.6 A sequence $\{x_n\}$ in an infinite dimensional Banach space E is called a *basis* of E if for every $x \in E$ there exists a unique sequence of scalars $\{\alpha_n\}$ such that

$$x = \sum_{i=1}^{\infty} \alpha_i x_i \qquad (5.4)$$

(i.e. such that

$$\lim_{n \to \infty} \| x - \sum_{i=1}^{n} \alpha_i x_i \| = 0 \quad).$$

More generally a sequence $\{x_n\}$ in a topological linear space F is called a *basis* of F if for every $x \in F$ there exists a unique sequence of scalars $\{\alpha_n\}$ such that (5.4) holds in the sense of convergence in the topology of F.

A sequence $\{e_n\}$ in the dual space E^* (of a Banach space E) is called a *-weak basis* of E^* if $\{e_n\}$ is a basis of the topological linear space F obtained by endowing E^* with the *-weak topology $\sigma(E^*, E)$.

Let $\{x_n\}$ be a basis of a topological linear space F. The sequence of linear functionals $\{f_n\}$ defined by

$$f_j(x) = \alpha_j \quad (x = \sum_{i=1}^{\infty} \alpha_i x_i \in F, \ j = 1, 2, \ldots)$$

is called the sequence of *coefficient functionals* associated to the basis $\{x_n\}$.

A basis $\{x_n\}$ of a topological linear space F is said to be a *Schauder basis* of F, if all the coefficient functionals f_n $(n = 1, 2, \ldots)$ are continuous on F (i.e. if $f_n \in F^*$ for $n = 1, 2, \ldots$).

We would like to state that $E = \{\delta_b^{(1)} e_W\}_{W \in P^+}$ is a Schauder basis of the space $H_b^{(2)}(F_m)$ endowed with the $*$-weak topology $\sigma = \sigma(H_b^{(2)}(F_m), H_2^{l_1}(F_m))$. If $H_2^{l_1}(F_m)$ were a Banach space this would give the existence of a basis of $H_2^{l_1}(F_m)$, [65], Theorem 14.1. Unfortunately we are not able to prove that E is a basis with respect to the topology σ but we can state that E is a basis of the space $H_b^{(2)}(F_m)$ endowed with some topology ξ that is weaker than σ.

Namely the topology $\tilde{\xi}$ of pointwise convergence in the space $l_\infty(G) = C_b^{(1)}(G)$ is weaker than the $*$-weak topology $\sigma(l_\infty(G), l_1(G))$. Actually it is the weakest topology that coincides with the $*$-weak topology on balls

$$\{x \in l_\infty(G) : \|x\|_\infty \leq R\},$$

$0 < R < \infty$. The topology $\tilde{\xi}$ induces the topology ξ on the space $H_b^{(2)}(F_m) = \ker \delta_b^{(2)} / \operatorname{Im} \delta_b^{(1)}$. As is easy to see $\tilde{\xi}$ is the topology of pointwise convergence on the space $PX_0(F)$.

Theorem 5.7 *The second bounded cohomology group $H_b^{(2)}(F_m)$ of the free group F_m of rank m, $2 \leq m \leq \infty$ is isomorphic to the space $PX_0(F_m)$ of pseudocharacters vanishing on the set of free generators and is an infinite dimensional space. This space being endowed with the topology of pointwise convergence has the Schauder basis $\{e_W\}_{W \in P^+}$ and thus every element $f \in PX_0(F_m)$ can be uniquely expressed in the form*

$$f = \sum_{W \in P^+} \alpha_W e_W \tag{5.5}$$

with some coefficients $\alpha_W \in \mathbb{R}$, $W \in P^+$.

Proof. The space $PX(F_m)$ of pseudocharacters can be expressed in the form

$$PX(F_m) = X(F_m) \oplus PX_0(F_m)$$

where $PX_0(F_m)$ is the space of pseudocharacters vanishing on the set of generators A. By Theorem 3.5

$$H_b^{(2)}(F_m) \cong PX_0(F_m).$$

The elements $e_W \in PX_0(F_m)$, $W \in P^+$ are linearly independent. Indeed let us suppose that

$$e = \sum_{i=1}^{n} \alpha_{W_i} e_{W_i} = 0$$

for some different words $W_i \in P^+$, with $i = 1, \ldots, n$, and for some coefficients $\alpha_{W_1}, \ldots, \alpha_{W_n}$, not equal to zero. If

$$|W_1| \leq |W_2| \leq \ldots \leq |W_n|$$

then $e_{W_n}(W_n) = 1$ while $e_{W_i}(W_n) = 0$, $i = 1, \ldots, n-1$ and thus $e(W_n) = \alpha_{W_n} \neq 0$, which leads to a contradiction.

We are going to show that $\{e_W\}_{W \in P^+}$ is a basis. If $f \in PX_0(F_m)$, then $f(a_i^\xi) = 0$, $\xi = \pm 1$, $i = 1, \ldots, m$. Recall that f is constant on conjugacy classes. If $V \in F_m$ is any element of length 2 and $W \in P^+$ is in the same conjugacy class as V, then the pseudocharacter $f - \alpha_W e_W$, where $\alpha_W = f(V)$ is zero on the conjugacy class of the element V.

Taking the other representatives of the conjugacy classes of length 2 we will find a linear combination

$$f_2 = \sum_{W \in P_2^+} \alpha_W e_W$$

such that $f - f_2$ is zero on elements of length ≤ 2.

Repeating this argument we get for every $n = 3, 4, \ldots$ a linear combination

$$f_n = f_{n-1} + \sum_{W \in P_n^+} \alpha_W e_W$$

such that $f - f_n$ is zero on the set of elements of length $\leq n$. This gives (5.5) because there are only a finite number of nonzero terms in the sum $\sum_{W \in P^+} \alpha_W e_W(f)$ for any $g \in F_m$. It is obvious that the expression (5.5) is unique, and that the coefficient functionals α_W are continuous with respect to the topology of pointwise convergence; thus E is a Schauder basis.

Remark 5.8 It is a question of some interest under what conditions on the coefficients α_W, $W \in P^+$ the sum (5.5) represents an element of $PX(F_m)$. The next two statements give some information about this.

Proposition 5.9 *Let* $\{\alpha_W\}_{W \in P^+} \in l_1(P^+)$ *that is*

$$\sum_{W \in P^+} |\alpha_W| < \infty.$$

Then

$$f = \sum_{W \in P^+} \alpha_W e_W$$

is the element of $PX_0(F_m)$.

Proof. Let $x, y \in F_m$. Then

$$|f(x) + f(y) - f(xy)| = |\sum_{W \in P^+} \alpha_W(e_W(x) + e_W(y) - e_W(xy))| \leq$$

$$\leq \sum_{W \in P^+} |\alpha_W| \cdot |(e_W(x) + e_W(y) - e_W(xy))| \leq 6 \sum_{W \in P^+} |\alpha_W|,$$

by the lemma, and thus f is a pseudocharacter and moreover an element of $PX_0(F_m)$.

A naïve conjecture would be that f, given by (5.5), is a pseudocharacter if and only if the coefficients $\{\alpha_W\}_{W\in P^+}$ satisfy the condition $\{\alpha_W\}_{W\in P^+} \in l_1(P^+)$. This conjecture is not correct as the next example shows.

Let $V_n = a_1^n(a_1a_2)^n a_2$, $n = 1, 2, \ldots$ be a sequence of elements of F_m. It is easy to see that for any m, n, $(m \neq n)$ the word V_m is not a subword of the word V_n and V_m, V_n do not overlap. Moreover V_n does not overlap with itself.

We can extend the set $\{V_n\}_{n\in N}$ to the system P^+ described in Lemma 5.5.

Proposition 5.10 *For every bounded sequence $\{\alpha_n\}_{n=1}^\infty$ of coefficients the function*

$$f = \sum_{n=1}^\infty \alpha_n e_{V_n}$$

is a pseudocharacter.

Proof. For any $x, y \in F_m$ we have

$$|f(x) + f(y) - f(xy)| \leq \sum_{n=1}^\infty |\alpha_n||e_{V_n}(x) + e_{V_n}(y) - e_{V_n}(xy)|. \qquad (5.6)$$

But for a given pair x, y of elements there are at most 6 nonzero terms in (5.5).

Indeed, let us suppose that for a product xy after cancellation we have one of the diagrams presented in Figures 7–10. Let it be the diagram given in Figure 7 for instance.

As follows from the proof of Lemma 5.3 the number

$$e_{V_n}(x) + e_{V_n}(y) - e_{V_n}(xy)$$

is nonzero only if the word V_n occurs in a neighborhood of one of the singular points of the diagram. If V_n occurs in a neighbourhood of a singular point of a diagram, (for instance as shown in Figure 11), then because different words

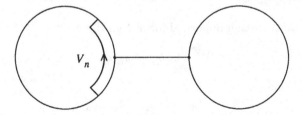

Figure 11

do not overlap the corresponding place will be forbidden for occurrences of words V_m, $m \in N$ in the same area (see Figure 12).

We have at most 6 places for occurrences related to singular points and thus at most 6 nonzero terms in (5.6). This shows that $f \in PX_0(F_m)$.

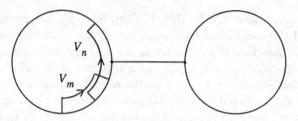

Figure 12

Torus knot groups

Let $G = \langle a, b \mid a^m = b^n \rangle$ be the fundamental group of the complement of a torus knot. The subgroup $Z = \langle a^m \rangle$ generated by the element a^m is central and we have the short exact sequence

$$1 \to Z \to G \to N \to 1$$

where

$$N = \langle a, b \mid a^m = b^n = 1 \rangle \cong \mathbb{Z}_m * \mathbb{Z}_n.$$

By Theorem 4.1 we have $H_b^{(n)}(G) = H_b^{(n)}(N)$, $n \geq 1$, and $H_{b,1}^{(n)}(N) = 0$, $n \geq 1$, by Proposition 4.2.

We can describe $H_{b,2}^{(2)}(N)$ in a much the same way as we described $H_{b,2}^{(2)}$ in the case of the free group. Let $A = \{a_1, \ldots, a_{m-1}, b_1, \ldots, b_{n-1}\}$ be the system of generators of N where $a_i = a^i$, $1 \leq i \leq m - 1$, $b_j = b^j$, $1 \leq j \leq n - 1$.

Every element $g \in N$ can be uniquely written as a word W over the alphabet A satisfying the following condition: after every symbol a_i in W follows some symbol b_j and after b_j follows some a_k etc. Such a form is called a normal form of an element.

Thus $a_i a_j$, $b_k b_l$ are forbidden subwords in normal forms. We call a cyclic word cyclically reduced if it does not contain a forbidden word of this form and consider the length of elements $g \in N$ with respect to the system of generators A.

If W is a word written in normal form, e.g., the word

$$W = a_{i_1} b_{j_1} \ldots a_{i_k} b_{j_k},$$

then by definition

$$W^- = b_{n-j_k} a_{m-i_k} \ldots b_{n-j_1} a_{n-i_1}.$$

It is clear that $WW^- = 1$ in the group N.

Repeating the arguments used when proving the Lemma 5.5 we show that there is a system $P = P^- \cup P^+$ of representatives of the length ≥ 2 conjugacy classes of N given by words in normal form satisfying the conditions a), c) of Lemma 5.5 and the condition: b') if $W \in P^+$ then $W^- \in P^-$.

For every $W \in P$ we define the pseudocharacter

$e_W(g)$ = number of times W occurs in $\bar{g}$ − number of times W^- occurs in $\bar{g}$,

where $\bar{g}$ is the reduced cyclic word corresponding to the element $g \in N$.

Repeating the arguments used when proving Theorem 5.7 we get

Theorem 5.11 *The space $H_{b,2}^{(2)}(N)$ is infinite dimensional. It is isomorphic to the space $PX(N)$ and every element $f \in PX(N)$ can be uniquely written in the form*

$$f = \sum_{W \in P^+} \alpha_W e_W.$$

for some coefficients $\alpha_W \in \mathbb{R}$.

Remark 5.12 The commutator subgroup $[\mathbb{Z}_m * \mathbb{Z}_n, \mathbb{Z}_m * \mathbb{Z}_n]$ of $\mathbb{Z}_m * \mathbb{Z}_n$ is free and has finite index. Thus for computation of $H_b^2(\mathbb{Z}_m * \mathbb{Z}_n)$ we can apply Proposition 4.11 as well.

In particular the commutator subgroup of $\mathbb{Z}_2 * \mathbb{Z}_3$ is a free group $\mathcal{F}_2$ of rank 2 generated by the elements $\xi = abab^{-1}$, $\eta = ab^{-1}ab$. The automorphisms $\hat{a}$ and $\hat{b}$ act on these generators by

$$\begin{cases} \xi^{\hat{a}} = \xi^{-1} \\ \eta^{\hat{a}} = \eta^{-1} \end{cases} \qquad \begin{cases} \xi^{\hat{b}} = \xi^{-1}\eta \\ \eta^{\hat{b}} = \xi^{-1}. \end{cases}$$

Thus $H_b^{(2)}(\mathbb{Z}_2 * \mathbb{Z}_3)$ is isomorphic to the subspace of $PX(F)$ consisting of those pseudocharacters that are $\hat{a}$ and $\hat{b}$ invariant and are zero on the set $\{\xi, \eta\}$. It is clear that this space is a proper subspace of the space $PX_0(\mathcal{F}_2)$ consisting of pseudocharacters vanishing on the basis $\{\xi, \eta\}$.

Surface groups

Let Γ_g be the fundamental group of a closed oriented surface S_g of genus $g \geq 2$. The standard presentation of Γ_g is

$$\Gamma_g = \langle a_1, \ldots, a_g, b_1, \ldots, b_g | \prod_{i=1}^{g} [a_i, b_i] = 1 \rangle \qquad (5.7)$$

where $[a, b] = a^{-1}b^{-1}ab$.

In [10], [57] it is shown that $H_b^{(2)}(\Gamma_g)$ is infinite dimensional. Additional information about $H_b^{(2)}(\Gamma_g)$ can be found in [4], [5]. See also [52], Theorem 2.

We are going to construct a basis for this space. To do this we need to have a normal form for elements of Γ_g satisfying some special properties. The standard form constructed in [10] will be suitable for our purposes. We recall some definitions taken from [10].

Let $CG(\Gamma_g)$ be the Cayley graph of Γ_g corresponding to the system of generators given by (5.7). It consists of infinitely many copies of the basic polygon Q, which corresponds to the defining relation from the presentation (5.6).

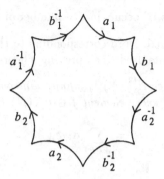

Figure 13

An R-cycle (respectively L-cycle) of length l in a word V is a sequence of letters in V which occurs as a sequence of l consecutive edges in the boundary ∂Q traversed clockwise (respectively anticlockwise).

Suppose Q_1, Q_2 are polygons in $CG(\Gamma_g)$ with a common edge and that $E \subset \partial Q_1$, $F \subset \partial Q_2$ are both R-cycles or both L-cycles in $\partial Q_1, \partial Q_2$ with one common vertex (the end of E and the beginning of F). Then E and F are called consecutive right or left cycles. An R-chain of type $p_1, p_2, \ldots, p_n$, $p_i \in \mathbb{N}$, is a sequence of $n \geq 2$ consecutive R-cycles of lengths $p_1, p_2, \ldots, p_n$. Such a chain constitutes part of the clockwise boundary of a chain of polygons $Q_1 \cup \ldots \cup Q_n$ in $CG(\Gamma_g)$, where Q_i and Q_{i+1} intersect along one edge, $i = 1, \ldots, n-1$, and $Q_i \cap Q_j = \emptyset$ otherwise. Replacing the R-chain by the sequence of L-cycles which represent the complementary part of $\partial(Q_1 \cup \ldots \cup Q_n)$ traversed anticlockwise, we obtain the complementary L-chain which is of type $4g - p_1 - 1, 4g - p_2 - 2, \ldots, 4g - p_{n-1} - 2, 4g - p_n - 1$. By a chain in $CG(\Gamma_g)$ we mean a sequence of polygons Q_j, $1 \leq j \leq n$, such that $|\partial Q_j \cap \partial Q_{j+1}| = 1$, $1 \leq j \leq n$, and such that $Q_j \cap Q_k = \emptyset$, $k - j \geq 2$.

Theorem 5.13 ([10], Theorem 2.1) *Any element $x \in \Gamma_g$ can be uniquely expressed as a word V in the generators $a \in A \cup A^{-1}$, where $A = \{a_1, \ldots, a_g, b_1, \ldots, b_g\}$, satisfying the following rules:*

Rule 1. There are no occurrences $c^{-1}c$, $c \in A \cup A^{-1}$; in other words V is reduced.

Rule 2. There are no cycles of length $> 2g$ in V.

Rule 3. There are no L-cycles of length $2g$ in V.

Rule 4. There are no R-chains of type $2g, 2g - 1, \ldots, 2g - 1, 2g$ in V.

Moreover any word satisfying the rules 1–4 is represented by a shortest path in the graph $CG(\Gamma_g)$.

A sequence of letters satisfying these rules is called *admissible*, and if $x \in \Gamma_g$ is expressed in this way we say it is in *standard form*.

Throughout this section, we assume w is a reduced word and following [10] define $h_w(x)$, $x \in \Gamma_g$, to be the number of times that w occurs as a subword in the admissible word representing x. If U is any word in the generators, not necessarily admissible, then $H_w(U)$ denotes the number of times that w occurs in U.

The function $\Gamma_g \to \mathbb{Z}$

$$f_w(x) = h_w(x) - h_{w^{-1}}(x) - h_w(x^{-1}) + h_{w^{-1}}(x^{-1})$$

satisfies the relation $f_w(x^{-1}) = -f_w(x)$. In [10] it is proven that if w contains no cycles of lengths greater than 1 then $f_w \in QX(\Gamma_g)$. We state that the same holds for any w.

Proposition 5.14 *For any freely reduced w the function f_w is a quasicharacter.*

Proof. We must prove that

$$f_w(x_1 x_2) \sim f_w(x_1) + f_w(x_2) \tag{5.8}$$

(recall that the equivalence $f_1 \sim f_2$ means that $f_1 - f_2$ is a bounded function).

We shall use the statement of the main lemma from [64] a number of times. In our situation this can be reformulated as

Lemma 5.15 *Let L be a simply connected region in $CG(\Gamma_g)$ so that ∂L has no self-intersections. Then either*
(i) L consists of one polygon; or
(ii) L is a chain; or
(iii) ∂L contains at least three disjoint cycles of length $4g - 2$.

The next lemma is an easy corollary of the previous one.

Lemma 5.16 *Let $x = a_1 \ldots a_n$, $a_i \in A \cup A^{-1}$ be in standard form. Then either $x^{-1} = a_n^{-1} \ldots a_1^{-1}$ is in standard form or the standard form y of x^{-1} is a word of the same length as x and such that the word xy constitutes the boundary label reading from the point A of the diagram consisting of chains*

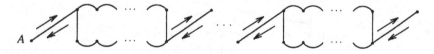

Figure 14

of polygons joined by segments in the way described in Figure 14.

The following lemma is almost trivial.

Lemma 5.17 *If $x_1, x_2, x_1 x_2, x_1^{-1}, x_2^{-1}, x_2^{-1} x_1^{-1}$ are in standard form then*

$$|f_w(x_1 x_2) - f_w(x_1) - f_w(x_2)| \leq 2|w|$$

($|w|$ is the length of the word w).

As shown in [10], Proposition 3.2, to prove (5.8) we need only consider the following situations.

Case 1. x_1, x_2 and x_1x_2 are all admissible.

Case 2. x_1, x_2 are admissible and $l(x_1x_2) \leq 1$ $\big(l(x)$ is the length of the admissible word representing the element $x \in \Gamma_g$ with respect to the generating system $A\big)$.

Case 3. When $x_1, x_2, x_1^{-1}, x_2^{-1}$ are written in standard form there are violations of rules 2, 3, 4 at the joins of x_1x_2 and $x_2^{-1}x_1^{-1}$. These violations contain cycles of lengths at most $4g - 2$.

Let us consider these cases separately.

Case 1. Let U, V, W be the standard forms of the elements $x_2^{-1}, x_1^{-1}, x_2^{-1}x_1^{-1}$. If the word W coincides with the word $x_2^{-1}x_1^{-1}$ then we can apply Lemma 5.17. Now let us suppose that these words are different. Then we have the diagram of Figure 15, where the label of the path ABC is x_1 (we shall denote by $\varphi(\rho)$ the label of a path ρ) and similarly

$$\varphi(CDEF) = x_2,$$

$$\varphi(FEKDC) = U,$$

$$\varphi(CBA) = V,$$

$$\varphi(FEKLBA) = W.$$

We have

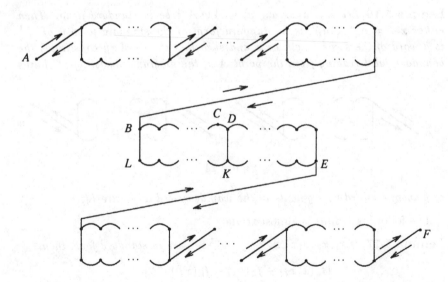

Figure 15

$$f_w(x_1x_2) = H_w(x_1x_2) - H_{w^{-1}}(x_1x_2) - H_w(W) + H_{w^{-1}}(W),$$

$$f_w(x_1) = H_w(x_1) - H_{w^{-1}}(x_1) - H_w(V) + H_{w^{-1}}(V),$$

$$f_w(x_2) = H_w(x_2) - H_{w^{-1}}(x_2) - H_w(U) + H_{w^{-1}}(U).$$

But

$$H_w(x_1x_2) \sim H_w(x_1) + H_w(x_2),$$

$$H_{w^{-1}}(x_1x_2) \sim H_{w^{-1}}(x_1) + H_{w^{-1}}(x_2)$$

and if we denote $Z = \varphi(KLBA)$ then

$$H_w(W) \sim H_w(U) + H_w(Z),$$

$$H_{w^{-1}}(W) \sim H_{w^{-1}}(U) + H_{w^{-1}}(Z).$$

Comparing all these equalities and equivalences we see that to prove (5.8) we should prove that

$$- H_w(Z^{-1}) + H_{w^{-1}}(Z^{-1}) - H_w(V) + H_{w^{-1}}(V) \sim 0. \qquad (5.9)$$

But $Z^{-1}\varphi(KDC)V$ represents the identity element and the length of the word $\varphi(KDC)$ is not greater than $2g$. This gives (5.9).

Case 2. If $x_1x_2 = 1$ then $f_w(x_1x_2) = 0$ and

$$f_w(x_1) + f_w(x_2) = f_w(x_1) + f_w(x_1^{-1}) = 0.$$

If $x_1x_2 = a$ where $a \in A \cup A^{-1}$ then one can easily check that

$$|\delta^{(1)}f_w(x_1, x_2)| \le 16g.$$

Case 3. Suppose that $x_1 = x_1'e_1 \ldots e_p$, $x_2 = e_{p+1} \ldots e_q V_2'$ are admissible where $e_1 \ldots e_q$ is the maximum cycle across the join of x_1 and x_2. Let $f_1 \ldots f_r$ be the complementary cycle. Then $2 \le r \le 2g$ and $r = 2g$ only if $e_1 \ldots e_q$ is an L-cycle. In [10], page 35, when considering Case 3 it is shown that $x_1'f_1 \ldots f_r x_2'$ is admissible and so

$$|h_w(x_1x_2) - H_w(x_1x_2)| \le |w| + 2g,$$

$$|h_{w^{-1}}(x_1x_2) - H_{w^{-1}}(x_1x_2)| \le |w| + 2g.$$

Similar reasoning holds for the inverse elements x_1^{-1}, x_2^{-1}, written in standard form. Using the fact that the function $H_w(g)$ is a quasicharacter on a free group we finish the proof.

We call a word V *cyclically admissible* if the cyclic word $\overline{V}$ (which is the cyclic closure of V) satisfies the rules 1–4 (if we replace V by $\overline{V}$ in these rules). In this case $\overline{V}$ is called *admissible*.

The following property for the group Γ_g holds: if an element $x \in \Gamma_g$ is conjugate to its inverse x^{-1} then $x = 1$ (this follows for instance from the fact that Γ_g is residually free and the above property holds for free groups).

Let $\Gamma_g = \Gamma_g^+ \cup \Gamma_g^- \cup \{1\}$ be any division of the group Γ_g into three non-intersecting parts with the property that if $x \in \Gamma_g^+$ then $x^{-1} \in \Gamma_g^-$. Let P^+ be a system of representatives of conjugacy classes of elements from Γ_g^+ having shortest length in the corresponding conjugacy class.

We identify the elements of P^+ with the standard forms of these elements. The system P^+ can be chosen in such a way that if $W \in P^+$ then $\overline{W}$ is admissible. Indeed let us suppose that for some $W \in P^+$ the cyclic word $\overline{W}$ is admissible. Then some L-cycle of length $2g$ occurs in $\overline{W}$ (an occurrence of another forbidden word in $\overline{W}$ is impossible, because in this case W is not the shortest word in its conjugacy class). We can change in $\overline{W}$ all L-cycles to complementary R-cycles (of the same length $2g$). After doing this we will get a reduced cyclic word $\overline{W}'$ corresponding to a word W' written in standard form, that belongs to the same conjugacy class as W and has the same length as W. (We will not get a new L-cycle of length $2g$ after this change because different L-cycles of length $2g$ do not overlap in W).

Now we suppose that for every $W \in P^+$ the cyclic word $\overline{W}$ is admissible.

If $W \in P^+$ is self-overlapping then using arguments given in the proof of Lemma 5.5 we can change W to a word W' which is a cyclic shift of the word W and which has no self-overlapping. Thus we can suppose that all elements of P^+ satisfy this property.

For every $W \in P^+$ and $x \in \Gamma_g$ we define

$$e_W(x) = h_W(\overline{x}) - h_W(\overline{x^{-1}}) - h_{W^{-1}}(\overline{x}) + h_{W^{-1}}(\overline{x^{-1}})$$

where $h_W(\overline{x})$ is the number of times that W occurs as a subword of the admissible cyclic word $\overline{x}$ corresponding to an element x.

The space of pseudocharacters $PX(\Gamma_g)$ can be expressed as the sum $X(\Gamma_g) \oplus PX_0(\Gamma_g)$ where $X(\Gamma_g)$ is $2g$-dimensional space of (additive) characters and $PX_0(\Gamma_g)$ is the space of pesudocharacters equal to zero on the set of generators.

Lemma 5.18 *If $W \in P^+$ then $e_W \in PX_0(\Gamma_g)$.*

Proof. It is clear that $e_W(x^{-1}) = -e_W(x)$, $x \in \Gamma_g$. On the other hand, if V is an admissible word then V^n is an admissible word for every $n \in \mathbb{N}$. This shows that

$$e_W(x^n) = n e_W(x),$$

for $x \in \Gamma_g$, $n \in \mathbb{Z}$.

The proof that e_W is a quasicharacter can be given in a similar way to the proof of Proposition 5.14 and is omitted.

The pseudocharacters e_W have the following important property: if $x \in \Gamma_g$ then

$$e_W(x) = \begin{cases} 0 \text{ if } l(x) < l(W) \text{ or if } l(x) = l(W) \text{ but } x \text{ is not conjugate to } W \\ 2 \text{ if } x \text{ is conjugate to } W. \end{cases}$$

Theorem 5.19 *The bounded part $H_{b,1}^{(2)}(\Gamma_g)$ of the second cohomology group $H_b^{(2)}(\Gamma_g)$ has dimension 1 and is generated by the dual of the fundamental class of the closed oriented surface S_g of genus g, if we identify $H_b^{(2)}(\Gamma_g)$ and $H_b^{(2)}(S_g)$ via the canonical isomorphism.*

The singular part $H_{b,2}^{(2)}(\Gamma_g)$ has infinite dimension. It is isomorphic (as a vector space) to the space $PX_0(\Gamma_g)$ and every element $f \in PX_0(\Gamma_g)$ can be uniquely expressed in the form

$$f = \sum_{W \in P^+} \alpha_W e_W,$$

where $\alpha_W \in \mathbb{R}$.

Proof. The first part of the theorem is a corollary of Thurston's theorem [39]: if M is a compact Riemannian manifold of negative curvature then the image of the canonical homomorphism $H_b^*(M) \rightarrow H^*(M)$ contains the dual of the fundamental class of the manifold, and the well known fact that $H^{(2)}(S_g, \mathbb{R}) \cong \mathbb{R}$.

The proof of the second part of the theorem is a repetition of the arguments used in the proof of Theorem 5.7.

Remark 5.20 Let G be a free product of cyclic groups or a surface group. Then the commutator group $G' = [G, G]$ is a free group.

Now let $\Gamma \leq \text{Aut } G'$ be the subgroup generated by the automorphisms $\hat{g} : G' \rightarrow G'$, $\hat{g}(h) = g^{-1}hg$, $g \in G$, $h \in G'$. Then the space of pseudocharacters on the group G is canonically isomorphic to the space of Γ-invariant pseudocharacters on the commutator subgroup G'.

This example raises the following question.

Problem 5.21 Given a subgroup $\Gamma \leq \text{Aut } F$, with F free, describe the space of pseudocharacters on F that are Γ-invariant. An important subcase of this problem arises from the pair F_m, $B(m)$, where $B(m)$ is the braid group acting on a free group F_m by automorphisms (Artin's representation) (compare with Problem 4.10).

Remark 5.22 For a torsion-free small cancellation group $\Gamma = \langle A; R \rangle$, satisfying the condition $C'(\frac{1}{8})$, E.Rips and Z.Sela proved [60] the existence of canonical representatives $\Theta : \Gamma \rightarrow (A \cup A^{-1})^*$ where $(A \cup A^{-1})^*$ is the set of words over the alphabet $A \cup A^{-1}$ satisfying the following conditions:

α) Θ is injective.

β) There is a constant M such that if $x, y, z \in \Gamma$; $xyz = 1$ then there exist $u_i, v^{ij}, t_0^i, t_1^i \in (A \cup A^{-1})^*$, $i = 1, 2, 3$ so that:

$$\Theta(x) = t_0^1 v^{11} t_1^1 u_1 (t_1^2)^{-1} v^{21} (t_0^2)^{-1},$$

$$\Theta(y) = t_0^2 v^{22} t_1^2 u_2 (t_1^3)^{-1} v^{32} (t_0^3)^{-1},$$

$$\Theta(z) = t_0^3 v^{33} t_1^3 u_3 (t_1^1)^{-1} v^{13} (t_0^1)^{-1}.$$

γ) $u_1 u_2 u_3, v^{11} v^{13}, v^{21} v^{22}, v^{32} v^{33}$ are elements of the normal subgroup generated by the set R of defining relators in a the free group with basis A.

δ) $l(u_i) \leq M$, $l(v^{ij}) \leq M$, $1 \leq i, j \leq 3$.

It is clear that if a group Γ has a normal form satisfying the conditions α)–δ) then the functions f_w, e_w defined above are quasicharacters.

It is an open question if for every hyperbolic group (even without torsion) there is an analogous normal form.

To get a wide class of quasicharacters on a group Γ it is enough to have a normal form $\Theta : \Gamma \to (A \cup A^{-1})^*$ (Θ being injective) satisfying the weaker condition:

There is some constant $c > 0$ such that

$$d_H(\Theta(xy), \overline{\Theta(x)\Theta(y)}) \leq C, \tag{5.10}$$

where $d_H(U, V)$ is the Hamming distance between a words U, V, and the bar means reduction.

Problem 5.23 Does any hyperbolic group have a normal form satisfying (5.10)?

6   Acknowledgement

I owe my gratitude to P.Kurchanov, E.Ghys and A.Machi for their encouragement and helpful conversations. I would like to thank A.Duncan, N.Gilbert and J.Howie for organization on a high level the Workshop on Geometric and Combinatorial Methods in Group Theory, March 28 — April 8, 1993, Edinburgh where the results were announced. The author wishes to thank S.Shahshahani for his hospitality and help during the author's stay in the Institute for Studies in Theoretical Physics and Mathematics in Tehran where the paper was written. I owe my thanks to A.Khelemskiĭ and A.M.Sinclair for discussions on the connection between bounded cohomology and cohomology in Banach algebras. Also I thank Miss Sami' and Miss O.Dorofeeva for their help in preparing this manuscript.

7 Guide to the References

Trying to make the bibliography more complete we have included in the list of references all papers known to us having even a little connection with bounded cohomology and related topics. We hope that this will make our paper more informative. We do not refer in the text to every one of the papers from the list. We hope that this will not confuse the reader.

References

[1] S.I.Adian, J.Mennicke: *On bounded generation of* $SL_n(\mathbb{Z})$. International Journal of Algebra and Computation, 2, (1992), 357–365.

[2] J.Baker, J.Lawrence, Z.Zorzitto: *The stability of the equation* $f(xy) = f(x)f(y)$. Proc. Amer. Math. Soc., 74, (1979), 242–246.

[3] J.Barge, E.Ghys: *Cocycles bornés et actions de groupes sur les arbres réels.* In: Group Theory from a Geometric Viewpoint (Trieste 1990), World Sci Publ., New Jersey, 1991, 617–622.

[4] J.Barge, E.Ghys: *Cocycles d'Euler et de Maslov.* Math. Ann., 1992, 294, 235–265.

[5] J.Barge, E.Ghys: *Surfaces et cohomologie bornée.* Invent. Math., 92, 1988, 509–526.

[6] C.Bavard: *Longueur stable des commutateurs.* L'Enseignement Math., t.37, 1991, 109–150.

[7] G.Besson: *Sem. de Cohomologie bornée.* ENS Lyon (Fév. 1988).

[8] M.G.Brin, C.C.Squier: *Groups of piecewise linear homeomorphisms of the real line.* Invent. Math., 79, 1985, 485–498.

[9] R.Brooks: *Some remarks on bounded cohomology.* In: Riemann surfaces and related topics, Ann. of Math. Studies, 97 (1981), 53–63.

[10] R.Brooks, C.Series: *Bounded cohomology for surface groups.* Topology, 23, (1984), 29–36.

[11] K.S.Brown: *Cohomology of Groups.* Springer–Verlag, New-York, 1982.

[12] K.S.Brown: *The Geometry of Rewriting Systems: A Proof of the Anick–Groves–Squier Theorem.* In: Algorithms and classification in Combinatorial Group Theory (G. Baumslag, C.F. Miller III eds.) MSRI publ. 23, Springer-Verlag New York, 1992, 137–163.

[13] K.S.Brown, R.Geoghegan: *An infinite-dimensional torsion-free* FP_∞ *group.* Invent. Math., 77 (1984), 367–381.

[14] D.Carter, G.Keller: *Bounded Elementary generation on* $SL_n(\mathcal{O})$. Amer. J. of Math., 105 (1983), 673–687.

[15] D.Cenzer: *The stability problem for transformations of the circle.* Proc. Roy. Soc. Edinburgh, N84 (1979), 279–281.

[16] J.Cheeger, M.Gromov: L_2-*cohomology and group cohomology.* Topology, 25, (1986), 189–215.

[17] C.G.Chehata: *An algebraically simple ordered group.* Proc. Lond. Math. Soc., 2, 1952, 183–197.

[18] I.M.Chiswell, P.H.Kropholler: *Soluble right orderable groups are locally indicable.* Canad. Math. Bull. 36 (1993) 22–29.

[19] A.H.Clifford, G.B.Preston: *The algebraic theory of semigroups.* AMS, 1964.

[20] A.Connes, H.Moscovici: *Cyclic cohomology, the Novikov Conjecture and hyperbolic groups.* Topology, 29, (1990), 345–388.

[21] R. Dennis, L. Vaserstein: *Commutators in linear groups.* K-Theory, 2 (1989), 761–767.

[22] A.Domic, D.Toledo: *The Gromov norm of the Kaehler class of symmetric domains.* Math. Ann., 276 (1987), 425–432.

[23] J.Duncan, A.L.T. Paterson: *Amenability for discrete convolution semigroup algebras.* Math. Scand., 66, 1990, 141–146.

[24] B.Eckmann: *Amenable groups and Euler characteristic.* Comment. Math. Helveltici, 67 (1992), 383–393.

[25] B.Eckmann: *Manifolds of even Dimension with Amenable fundamental Group.* Preprint, Zurich, 1993.

[26] R.E.Edwards: *Functional Analysis.* Holt, Rinehart and Winston, 1965.

[27] V.A.Faiziev: *Pseudocharacters on free groups and on some group constructions.* Russian Math. Surveys (UMN), 43, N5 (1988), 225–226.

[28] V.A.Faiziev: *Pseudocharacters on free products of semigroups.* Matemat. Zametki, 52, N6 (1992), 119–130.

[29] V.A.Faiziev: *Pseudocharacters on the group* $SL(2,\mathbb{Z})$. Func. Analysis and its Applications, 26, (1992), 293–295.

[30] V.A.Faiziev: *Pseudocharacters on semidirect products of semigroups.* Matemat. Zametki, 53, N3 (1993), 132–139.

[31] V.A.Faiziev: *Pseudocharacters on free groups.* Izv. Russ. Akad. Nauk., 1993.

[32] S.M.Gersten: *Bounded cocycles and combings of groups.* International Journal of Algebra and Computation, 2, (1992), 307-326.

[33] E.Ghys: *Groupes d'homeomorphisms du cercle et cohomologie bornée.* In: The Lefschetz centennial conference, part III (Mexico City 1984), Contemporary Math., 58, III, Amer. Math. Soc. (1987), 81-105.

[34] A.Guichardet: *Cohomologie des groupes topologiques et des algèbres de Lie.* Cedic / Fernand Nathan, Paris, 1980.

[35] F.P.Greenleaf: *Invariant Means on Topological Groups.* Van Nostrand, Math. Stud. 16 (1969).

[36] R.I.Grigorchuk: *Growth and amenability of a semigroup and its group of quotients.* In: Proceedings of the International Symposium on Semigroup theory, Kyoto, 1990, 103-108.

[37] R.I.Grigorchuk, P.F.Kurchanov: *Some Questions of Group Theory Related to Geometry.* In: Encycl. Math. Sci.,58, Part II, Springer-Verlag Berlin, 1993.

[38] M.Gromov: *Asymptotic invariants of infinite groups.* Geometric Group Theory Vol. 2 (G.A. Niblo, M.A. Roller eds.) LMS Lect. Notes 182, Cambridge Univ. Press 1993, 1993.

[39] M.Gromov: *Volume and bounded cohomology.* Publ. Math. IHES, 56 (1982), 5-100.

[40] K.Grove, H.Karcher, E.A.Roh: *Jacobi fields and Finsler metrics on compact Lie groups with a application to differentiable pinching problems.* Math. Ann., 211, (1974), 7-21.

[41] P. de la Harpe, M.Karoubi: *Représentations approchés d'un groupe dans une algèbre de Banach.* Manuscripta Math., 22 (1977), 293-310.

[42] M.W.Hirsch, W.P.Thurston: *Foliated bundles, invariant measures and flat manifolds.* Math. Ann., 101, (1975), 777-781.

[43] D.H.Hyers: *On the stability of the linear functional equations.* Proc. Nat. Acad. Sciences USA, 27, (1941), 222-224.

[44] D.H.Hyers, S.M.Ulam: *On approximate isometry.* Bull. Amer. Math. Soc., N51 (1945), 288-292.

[45] D.H.Hyers, S.M.Ulam: *Approximate isometry on the space of continuous functions.* Ann. of Math., 48, (1947), 285-289.

[46] H.Inoue, K.Yano: *The Gromov invariant of negatively curved manifolds.* Topology, 21, (1981), 83–89.

[47] N.Ivanov: *Foundation of the theory of bounded cohomology.* Zapiski Nauchnykh Seminarov Leningradskogo otdeleniya Matematicheskogo Instituta im. V. A. Steklova AN SSSR, 143 (1985), 69–109.

[48] N.Ivanov: *The second bounded cohomology group.* Zapiski Nauchnykh Seminarov Leningradskogo otdeleniya Matematicheskogo Instituta im. V. A. Steklova AN SSSR,167 (1988), 117–120.

[49] B.E.Johnson: *Cohomology in Banach Algebras.* Mem. AMS, 127 (1972).

[50] A.Y.Khelemskiĭ: *Flat Banach modules and amenable algebras.* Trans. Moscow Math. Soc. (AMS Translation 1985), 47 (1984), 199-244.

[51] A.A. Kirillov, A.D.Gvishiani: *Theorems and problems in functional analysis.* Nauka, Moscow,1979.

[52] D.Kazhdan: *On ε-representations.* Israel Jour. Math., 43, (1982), 315–323.

[53] A.I.Kokorin, V.M.Kopytov: *Linear ordered groups.* Nauka, Moscow, 1972.

[54] J. Lawrence: *The stability of multiplicative semigroup homomorphisms to real normed algebras.* I. Aequat. Math., 28, (1985), 94–101.

[55] S. Matsumoto: *Numerical invariants for semiconjugacy of homeomorphisms of the circle.* Proc. of the AMS, 98, (1986), 163–168.

[56] S.Matsumoto, S.Morita: *Bounded cohomology of certain groups of homeomorphisms.* Proc. of the AMS, 94, (1985) 539–544.

[57] Y.Mitsumatsu: *Bounded cohomology and l^1-homology of surfaces.* Topology, 23, N4 (1984), 465–471.

[58] M.Newman: *Unimodular commutators.* Proc. AMS, 101 (1987) 605–609.

[59] G.A.Noskov: *Algebras of rapidly decreasing functions on groups and cocycles of polynomial growth.* Siberian Math. Journal, 33, N4 (1992), 634–640.

[60] E.Rips, Z.Sela: *Canonical representatives and equations in hyperbolic groups.* IHES / M / 91 / 84. Preprint.

[61] W.Rudin: *Functional Analysis.* McGraw-Hill, New-York, 1973.

[62] V.V.Ryzhikov: *Representation of transformations preserving the Lebesgue measure, in the form of a product of periodic transformations.* Matem. Zametki, 38, N6, 1985, 860–865.

[63] Z.Sela: *Acylindrical accessibility for groups.* Max-Plank-Institute für Mathematik, 1992, Preprint.

[64] C.Series: *The infinite word problem and limit sets in Fuchsian groups.* Ergod. Th. & Dynam. Sys. (1981), 1, 337–360.

[65] I.Singer: *Bases in Banach Spaces I.* Springer-Verlag, New-York—Heidelberg—Berlin, 1970.

[66] A.I.Shtern: *Quasirepresentations and pseudorepresentations.* Functional Analysis and its Applications, 25, (1991), 70–73.

[67] A.I.Shtern: *Stability of homomorphisms to the group* $\mathbf{R}^*$. Vestnik Moskov. Univ. Ser. I Mat.-Mekh., N3 (1982), 29–32.

[68] T.Soma: *The Gromov invariant for links.* Invent. Math., 64, (1981), 445–454.

[69] C.Soulé: *The cohomology of $SL_3(\mathbb{Z})$.* Topology, 17 (1978), 1–22.

[70] C.C.Squier: *Word problems and homological finiteness conditions for monoids.* J. Pure Appl. Algebra, 49 (1987), 201–217.

[71] V.I.Sushchanskii: *Ambivalence of the automorphism group of a regular tree.* Dokl. Akad. Nauk. Ukraiy, N11, (1992), 6–9.

[72] O.I.Tavgen: *Finite width of arithmetic subgroups of Chevalley groups of rank ≥ 2.* Soviet Math. Dokl., 41, N1 (1990), 136–140.

[73] R.C.Thompson: *Commutators in the special and general linear groups.* Trans. Amer. Math. Soc., 101, (1961), 16–33.

[74] S.M.Ulam: *A collection of Mathematical Problems.* Los Alamos Scientific Laboratories, New Mexico, 1961.

[75] K.Yano: *Gromov invariant and S^1-actions.* J. Fac. Sci. U. Tokyo, Sec. 1A Math., 29, N3 (1982), 493–501.

R. I. Grigorchuk,
Moscow Institute of Railway Transportation,
ul. Obraztsova 15,
101475 Moscow, Russia.

On perfect subgroups of one-relator groups

JENS HARLANDER

If P is a group of operators on a group A, then we denote by $d_P(A)$ the minimal number of generators of A as a P-group.

Suppose G is a group and F/N is a presentation of G. Then F acts on N by conjugation and induces an action of G on N_{ab}. This $\mathbb{Z}G$-module is called the relation module of the presentation F/N.

Definition. A presentation F/N of a group G is said to have a *relation gap* if $d_G(N_{ab})$ is strictly less than $d_F(N)$.

It is an open problem whether there exists a presentation that has a relation gap (see Harlander [H1], [H2] and Baik, Pride [B-P]). Such a presentation would be interesting not only to group theorists. In [D] Dyer shows that a presentation with a relation gap could be used to settle an open question concerning complexes dominated by a 2-complex (see also Wall [W] and Ratcliffe [R]).

In [H1] and [H2] the author studies groups that have cyclic relation modules. Such groups are quotients G/P, with G a one-relator group, say presented by $\langle X \mid r \rangle$, and P a perfect normal subgroup G. Now if P is not of the form $\langle w \rangle^F / \langle r \rangle^F$, where $\langle w \rangle^F$ denotes the normal closure of the element w of the free group F on X, then G/P has a presentation with a relation gap.

The main point of this article is to present a method for constructing one-relator groups containing at most one perfect normal subgroup of the form $\langle w \rangle^F / \langle r \rangle^F$. We also study the perfect subgroup structure of one-relator groups arising from free groups via standard group constructions.

Suppose G is a one-relator group with presentation

$$(*) \qquad\qquad \langle t, b, c, \ldots \mid r \rangle$$

where r is cyclically reduced and the exponent-sum of t in r is zero. We assume furthermore that r involves all the generators $b, c, \ldots$. The relator r can be rewritten in terms of

$$b_i = t^i b t^{-i}, c_i = t^i c t^{-i}, \ldots, i \in \mathbb{Z}$$

as described in Lyndon-Schupp [L-S], page 198, yielding a cyclically reduced word s of shorter length than r in case t occurs in r.

For each generator $x \in \{b, c, ...\}$, let $m_x(r)$ and $M_x(r)$ be the minimum and maximum subscript on x actually occuring in s. Let H be the subgroup of G generated by

$$t^{m_b(r)}bt^{-m_b(r)}, ..., t^{M_b(r)}bt^{-M_b(r)}, t^{m_c(r)}ct^{-m_c(r)}, ..., t^{M_c(r)}ct^{-M_c(r)}, ...$$

It is well known (see [L-S], page 198, for example) that H has a presentation

$$\langle b_{m_b(r)}, ..., b_{M_b(r)}, c_{m_c(r)}, ...c_{M_c(r)}, ... \mid s \rangle.$$

and that it is the vertex group of an HNN-decomposition of G with free edge group. This method of decomposing a one-relator group is implicit in Magnus' original paper [M]. It has been the basis of a number of results about one-relator groups. We refer to H as the Magnus base of the presentation (*).

Before we can state our first result we need more notation. For $x \in \{b, c, ...\}$ let

$$\gamma_x(r) = M_x(r) - m_x(r)$$

and

$$\gamma(r) = max\{\gamma_x(r), x \in \{b, c, ...\}\}.$$

We call $\gamma(r)$ the *complexity* of r.

Theorem 1. *Let G be a one-relator group with a presentation as in (*) with Magnus base H. Suppose P is a non-trivial perfect normal subgroup of G of the form*

$$\langle w \rangle^F / \langle r \rangle^F$$

where w is a cyclically reduced word of F, the free group on $\{t, b, c, ...\}$. If $x \in \{b, c, ...\}$ is a generator actually occuring in w then

$$\gamma_x(w) \le \gamma_x(r).$$

When equality holds $H \cap P$ is a non-trivial perfect subgroup of H.

Proof. For $\gamma_x(w)$ to be defined, we have to check that the exponent-sum of t in w is zero. Since $\langle w \rangle^F / \langle r \rangle^F$ is perfect, we have

$$\langle w \rangle^F = [\langle w \rangle^F, \langle w \rangle^F]\langle r \rangle^F$$

and since r contains t with exponent-sum zero, w does too.

Without loss of generality we may assume that $x = b$. Cyclically permuting r and w, we can further assume that $m_b(r) = m_b(w) = 0$. Let $M = M_b(r)$ and $M' = M_b(w)$. Now r and w can be rewritten in the letters $\{b_0, b_1, ..., c_i, ..., i \in \mathbb{Z}\}$ to obtain cyclically reduced elements s and v of the free group on $\{b_0, b_1, ..., c_i, ..., i \in \mathbb{Z}\}$.

The group $G = \langle t, b, c, \ldots \mid r \rangle$ is an HNN-extension

$$\langle t, b_0, \ldots, b_M, c_i, \ldots \mid s, tb_jt^{-1} = b_{j+1}, j = 0, \ldots, M-1, tc_it^{-1} = c_{i+1}, \ldots, i \in \mathbb{Z} \rangle,$$

whereas the quotient $G/P = \langle t, b, c, \ldots \mid w \rangle$ is an HNN-extension

$$\langle t, b_0, \ldots, b_{M'}, c_i, \ldots \mid v, tb_jt^{-1} = b_{j+1}, j = 0, \ldots, M'-1, tc_it^{-1} = c_{i+1}, \ldots, i \in \mathbb{Z} \rangle,$$

the isomorphisms sending t to t, b_i to t^ibt^{-i}, c_i to t^ict^{-i},...etc (see [L-S], page 199). The edge group of the first HNN-extensions is a free group on basis $\{b_0, \ldots, b_{M-1}, c_i, \ldots, i \in \mathbb{Z}\}$ by the Freiheitssatz for one-relator groups (see [L-S], page 198), while the edge group of the second HNN-extension is free on $\{b_0, \ldots, b_{M'-1}, c_i, \ldots, i \in \mathbb{Z}\}$. Suppose $M \leq M'$. Then the quotient map from G to G/P sends b_i to b_i, c_i to c_i, ...etc. Strict inequality would imply that $b_0, \ldots, b_M, c_i, \ldots, i \in \mathbb{Z}$, is a basis of a free subgroup of G since $b_0, \ldots, b_M, \ldots, b_{M'-1}, c_i, \ldots, i \in \mathbb{Z}$, is the basis of a free subgroup of G/P. But s is a non-trivial relation in $b_0, \ldots, b_M, c_i, \ldots, i \in \mathbb{Z}$. So the assumption $M \leq M'$ implies $M = M'$. In that case the quotient map from G to G/P restricts to the identity map on the edge groups of the HNN-extensions. Thus the perfect subgroup P intersects the edge group of the HNN-decomposition of G trivially. Hence, by the subgroup theorem for HNN-extensions (see [L-S], page 212), P is a free product of conjugates of the intersection of the vertex group with P and a free group. Notice that the vertex group of the HNN-extension of G is just a free product of the Magnus base of the presentation $\langle t, b, c, \ldots | r \rangle$ and a free group. It follows that P is a free product of conjugates of $H \cap P$ and a free group. Since P is non-trivial perfect $H \cap P$ has to be non-trivial perfect and the free factor in the free product has to be trivial.

Let us say a few words about the connection between the complexity $\gamma(r)$ of the relator r and the perfect subgroup structure of the group G defined by $\langle t, b, c, \ldots \mid r \rangle$ as in (*).

When $\gamma(r) = 0$, the group G is a free product $H * C$ with C infinite cyclic. So a non-trivial normal subgroup P of G is perfect only if the intersection $P \cap H$ is non-trivial perfect.

When $\gamma(r) = 1$, the perfect subgroup structure can already be complicated, even in case the Magnus base does not contain non-trivial perfect subgroups. To see this let $r_0 = x_0$ and define recursively $r_i = r_{i-1}[r_{i-1}, x_ir_{i-1}x_i^{-1}]$. Set $r = r_n[r_n, tr_nt^{-1}]$ and let G be the group defined by the presentation

$$\langle t, x_0, \ldots, x_n \mid r \rangle.$$

Let P_i be the quotient $\langle r_i \rangle^F / \langle r \rangle^F$. Note that $\bar{r}_n$, the image of r_n in G, is equal to $[\bar{r}_n, t\bar{r}_n\bar{t}^{-1}]^{-1}$. Thus $P_n = \langle \bar{r}_n \rangle^G$ is contained in its commutator subgroup and hence must be perfect. In a similar way one checks that all the other subgroups P_i are perfect as well and that $P_n \langle \ldots \langle P_0$ is a chain of distinct non-trivial perfect normal subgroups of G. Notice that the relator r

has complexity equal to one. We will next show that the Magnus base H of the given presentation of G does not contain non-trivial perfect subgroups. We will need the following

Lemma 1. *Let G be the group defined by the presentation*

$$\langle x_1, ..., x_n \mid r \rangle$$

where r is not a proper power. Let $P(G)$ be the maximal perfect subgroup of G. If G_{ab} is free abelian of rank $n-1$ then the quotient $G/P(G)$ is torsion-free.

Proof. Using the given presentation of the torsion-free one-relator group G we can build a resolution

$$0 \to \mathbb{Z}G \xrightarrow{\partial_2} \mathbb{Z}G^n \xrightarrow{\partial_1} \mathbb{Z}G \to \mathbb{Z} \to 0$$

for G as described in Lyndon [L]. If we apply $\mathbb{Z} \otimes_{\mathbb{Z}G} -$ to it we get

$$\mathbb{Z} \xrightarrow{\bar{\partial}_2} \mathbb{Z}^n \xrightarrow{0} \mathbb{Z}.$$

Since G_{ab} is free abelian of rank $n - 1$, the map $\bar{\partial}_2$ is injective. Hence G is an E-group (see Strebel [S] for the definition of an E-group) and therefore $P(G)$ is an E-group as well (see [S]). In particular $P(G)$ is superperfect, i.e. $H_1(P(G)) = H_2(P(G)) = 0$. So applying $\mathbb{Z} \otimes_{\mathbb{Z}P(G)} -$ to the above resolution for G yields a resolution

$$0 \to \mathbb{Z}(G/P(G)) \to \mathbb{Z}(G/P(G))^n \to \mathbb{Z}(G/P(G)) \to \mathbb{Z} \to 0$$

for $G/P(G)$. In particular $G/P(G)$ has finite cohomological dimension and therefore is torsion-free.

Let us now return to our example. The Magnus base H of the presentation $\langle t, x_0, ..., x_n \mid r \rangle$ has a presentation

$$\langle y_0, ..., y_n, z_0, ..., z_n \mid u[u, v] \rangle$$

where $u = r_n(y_0, ..., y_n)$ and $v = r_n(z_0, ..., z_n)$. Writing this presentation as

$$\langle a, Y, Z \mid aua^{-1} = u^2, a = v \rangle$$

we see that H has a decomposition

$$(F(Y) *_{C_1} a) *_{C_2} F(Z),$$

the infinite cyclic groups C_1 and C_2 being generated by $\bar{u}$ and $\bar{v}$, the images of u and v in H. Let $P(H)$ be the maximal perfect subgroup of H. Suppose $P(H)$ intersects C_1 non-trivially, say $\bar{u}^k \in P(H)$, k a positive integer bigger than zero. Since by Lemma 1 the quotient $H/P(H)$ is torsion-free, we see that then $\bar{u} \in P(H)$.

So $H/P(H)$ has a presentation $\langle Y, Z \mid u, ... \rangle$. From the way u was constructed, we see that the normal closure of the image of y_0 in $H/P(H)$ is perfect. Since the only perfect subgroup of $H/P(H)$ is the trivial group, we conclude that $P(H)$ contains $\bar{y}_0$, the image of y_0 in H. To see that this is impossible consider the normal closure N of the elements $\bar{y}_1, ..., \bar{y}_n, \bar{z}_1, ... \bar{z}_n$ in H. The quotient H/N is presented by

$$\langle y_0, z_0 \mid z_0 y_0 z_0^{-1} = y_0^2 \rangle$$

and hence is metabelian. Thus N contains $P(H)$ but certainly not $\bar{y}_0$.

Assuming that $P(H)$ intersects C_2 non-trivially leads to a contradiction in just the same way. So the perfect normal subgroup $P(H)$ does not intersect C_1 or C_2 and therefore, by the subgroup theorem for amalgamations and HNN-extensions (see [L-S], page 212) $P(H)$ is free and hence trivial.

When in addition G is generated by two elements, one can say more about perfect subgroups.

Theorem 2. *Let G be the group defined by a presentation*

$$\langle t, b \mid r \rangle$$

as in (*) *and suppose the Magnus base of this presentation does not contain non-trivial perfect subgroups. If $\gamma(r) = 1$ then*

1. *G contains at most one non-trivial perfect normal subgroup of the form $\langle w \rangle^F / \langle r \rangle^F$, where w is a cyclically reduced element of the free group F on t and b;*

2. *If the abelianization of $\langle \bar{b} \rangle^G$ is not torsion, then G does not contain non-trivial perfect subgroups ($\bar{b}$ denotes the image of b in G).*

Proof. Let P be a non-trivial perfect normal subgroup as in 1. It follows from Theorem 1 that $\gamma(w)\langle \gamma(r) = 1$. Hence $w = b^k$ for some positive integer k. The quotient G/P is therefore presented by $\langle t, b \mid b^k \rangle$ and thus does not contain non-trivial perfect subgroups. This shows that P is the maximal perfect subgroup of G.

To prove 2., observe first that since $\gamma(r) = 1$, G decomposes as $H *_C t$, the infinite cyclic edge group C being generated by a conjugate of $\bar{b}$. If P is a non-trivial perfect normal subgroup of G, then P has to intersect the edge group in this HNN-extension non-trivially. Otherwise, by the subgroup theorem for HNN-extensions ([L-S], page 212), $H \cap P$ is a non-trivial perfect subgroup of H. But we assumed H not to contain non-trivial perfect subgroups. Thus

P contains $\bar{b}^k$ for some poitive integer k. Since $N = \langle\bar{b}\rangle^G$ contains P and $N_{ab} = (N/P)_{ab}$, we conclude that N_{ab} is torsion.

Groups as in Theorem 2 can be constructed in the following way. Let H be a one-relator group with presentation

$$\langle b_0, b_1 \mid r(b_0, b_1)\rangle$$

which does not contain non-trivial perfect subgroups. Let G be the one-relator group defined by

$$\langle t, b \mid r(b, tbt^{-1})\rangle.$$

The relator in this presentation has complexity one and the Magnus base is just H.

To illustrate the use of Theorem 2 we will now look at some examples.

Example 1. Let G be the group defined by

$$\langle t, b \mid b[b, tbt^{-1}]\rangle.$$

The relator in this presentation has complexity one and its Magnus base is presented by $\langle b_0, b_1 \mid b_0[b_0, b_1]\rangle$ and thus is metabelian. So Theorem 2 applies. Note that $[G, G]$ is the non-trivial maximal perfect subgroup and that it is of the form $\langle b\rangle^F/N$, where F is the free group on t and b, and N is the normal closure in F of the relator in the presentation of G. It now follows from Theorem 2 that if J/N is another non-trivial perfect normal subgroup P of G different from the maximal one, then J can not be the normal closure of a single element. But the relation G/P-module J_{ab} of the presentation F/J is generated by the single element $b[b, tbt^{-1}][J, J]$. So F/J has a relation gap. Unfortunately, we do not know if G actually contains such a subgroup P.

Example 2 . Let G be given by the presentation

$$\langle t, b \mid tb^i t^{-1} b^j\rangle,$$

where i and j are nonzero integers. Then G does not contain non-trivial perfect subgroups.

To see this, note that the relator in the above presentation has complexity one and the Magnus base is presented by

$$\langle b_0, b_1 \mid b_1^i b_0^j\rangle$$

and thus does not contain non-trivial perfect subgroups (see Theorem 3 below). So Theorem 2 applies to G. Using the Reidemeister-Schreier rewriting process (see [L-S], page 102), we see that

$$\langle \{b_k\}_{k\in\mathbb{Z}} \mid \{b_{k+1}^i b_k^j\}_{k\in\mathbb{Z}}\rangle$$

is a presentation for $\langle \bar{b} \rangle^G$ from which one can read off that the abelianization of $\langle \bar{b} \rangle^G$ is not torsion. Indeed, $\langle \bar{b} \rangle^G_{ab}$ contains $\langle b_0, b_1 \mid b_1^i b_0^j, [b_0, b_1] \rangle$ as a subgroup.

We have seen that in order to construct groups as in Theorem 2 one needs to start out with a one-relator group H that does not contain non-trivial perfect subgroups. There are known classes of one-relator groups with that property. For example a positive one-relator group does not contain non-trivial perfect subgroups (see [B1]). We will next study the perfect subgroup structure for one-relator groups that arise from free groups via standard group constructions. The following result is due to Baumslag [B1].

Theorem 3 *(Baumslag). A generalized free product*

$$F(X) *_C F(Y)$$

with free vertex groups and infinite cyclic edge group does not contain non-trivial perfect subgroups.

In fact Baumslag showed in the proof of Theorem 3 in [B1] that a generalized free product $G = A *_C B$, with A and B residually torsion-free nilpotent and C infinite cyclic has a normal residually torsion-free nilpotent subgroup T such that the quotient G/T is torsion-free nilpotent. This implies that G is residually solvable and hence can not contain non-trivial perfect subgroups.

Theorem 4. *An HNN-extension where both vertex and edge group are infinite cyclic does not contain non-trivial perfect subgroups.*

Proof. The groups considered in our second example are precisely the HNN-extensions with infinite cyclic edge and vertex groups.

One relator groups that are HNN-extensions $F(X) *_C t$ with free vertex group and infinite cyclic edge group can contain non-trivial perfect subgroups. For instance, the group

$$G = \langle t, b, c \mid tbt^{-1} = [b, cbc^{-1}] \rangle$$

contains a non-trivial perfect normal subgroup. Indeed, since $\bar{b}$, the image of b in G, is equal to $\bar{t}^{-1}[\bar{b}, \bar{c}\bar{b}\bar{c}^{-1}]\bar{t}$ in G, we see that $\langle \bar{b} \rangle^G = [\langle \bar{b} \rangle^G, \langle \bar{b} \rangle^G]$. Hence $\langle \bar{b} \rangle^G$ is perfect.

Theorem 5. *Let G be the group defined by the presentation*

$$\langle t, b, c, \dots \mid tbt^{-1} = v(b, c, \dots) \rangle$$

Then the maximal perfect subgroup of G is either trivial or the normal closure of $\bar{b}$ in G.

The proof of this theorem requires the following

Lemma 2. *Let G be a group as in Theorem 5. Then the normal closure N of $\bar{b}, \bar{c}, \ldots$ in G is an E-group.*

For the definition of an E-group we refer to [S].
Proof. Using the Reidemeister-Schreier rewriting process (see [L-S], page 102), one sees that N has a presentation

$$\langle b_i, c_i, \ldots i \in \mathbb{Z} \mid r_i, i \in \mathbb{Z} \rangle$$

with $r_i = b_i^{-1} v(b_{i-1}, c_{i-1}, \ldots)$. We use this presentation of N to build a partial $\mathbb{Z}N$-resolution

$$\bigoplus_R \mathbb{Z}N \xrightarrow{\partial_2} \bigoplus_X \mathbb{Z}N \xrightarrow{\partial_1} \mathbb{Z}N \to \mathbb{Z} \to 0,$$

where R denotes the set of relators and X the set of generators in the above presentation of N as described in [L-S], page 100. Applying $\mathbb{Z} \otimes_{\mathbb{Z}N} -$ to this partial resulotion yields

$$\bigoplus_R \mathbb{Z} \xrightarrow{\bar{\partial}_2} \bigoplus_X \mathbb{Z} \xrightarrow{\bar{\partial}_1} \mathbb{Z}$$

with $\bar{\partial}_1 = 0$ and $\bar{\partial}_2(e_i) = (exp_x(r_i))_{x \in X}$, where e_i is a standard generator of $\bigoplus_R \mathbb{Z}$ and $exp_x(r_i)$ is the exponent-sum of the generator x in the relator r_i. Using the fact that $exp_{b_j}(r_i) = 0$ for $i \langle j$ and $= -1$ for $i = j$, it is not difficult to show that $\bar{\partial}_2$ is injective and that $N_{ab} = (\bigoplus_X \mathbb{Z})/im(\bar{\partial}_2)$ is torsion-free. But that says that N is an E-group.

Proof of Theorem 5. Let F be the free group on $t, b, c, \ldots$ and denote by N the normal closure of the single element $tbt^{-1}v^{-1}$ in F. Let J be a normal subgroup of F such that J/N presents the maximal perfect subgroup $P(G)$ of G. If $J \cap \langle b \rangle = 1$, then $P(G) \cap \langle \bar{b} \rangle = 1$ and this implies, since G is an HNN-extension with edge group $\langle \bar{b} \rangle$ and free vertex group, that $P(G)$ itself is trivial (see [L-S], page 212).

So suppose now that $J \cap \langle b \rangle = \langle b^k \rangle$ for some integer $k \geq 1$. By Lemma 2 the normal closure N of $\{\bar{b}, \bar{c}, \ldots\}$ is an E-group with maximal perfect subgroup $P(N) = P(G)$ (the quotient G/N is abelian) and therefore $P(G)$ is superperfect (see [S]). The same arguments as in the proof of Lemma 1 now show that $G/P(G)$ is torsion-free. Thus the integer k has to be equal to one and hence $\langle b \rangle^F$ is contained in J. Notice that to show that J equals $\langle b \rangle^F$, is suffices to show that v is contained in $\langle b \rangle^F$. To see this, note first that since J/N is perfect, $J = NJ^{(n)}$ for any positive integer n ($J^{(n)}$ denotes the n-th term of the commutator series of J). Assuming $v \in \langle b \rangle^F$ implies that N is contained in $\langle b \rangle^F$ and therefore J is contained in $\langle b \rangle^F F^{(n)}$ for any n. But the intersection of all the $\langle b \rangle^F F^{(n)}$ is $\langle b \rangle^F$ because $F/\langle b \rangle^F$ is free and hence residually-solvable (see [L-S], page 14).

Let us now show that v is indeed contained in $\langle b \rangle^F$. In the following $F_{(j)}$ will denote the j-th term in the lower central series of F. By definition, the first term in the lower central series of a group is always the group itself. Suppose that v is contained in $\langle b \rangle^F F_{(j-1)}$ for j an integer greater or equal than two. Then $v = wq$, where w is an element of $\langle b \rangle^F$ and q is an element of $F_{(j-1)}$. Since b is contained in J and J is contained in $NF_{(j)}$, there are elements $f_1, ..., f_n$ of F and $\epsilon_1, ..., \epsilon_n$, $\epsilon_i \in \{\pm 1\}$, such that

$$bF_{(j)} = \prod_{i=1}^{n} f_i (tbt^{-1}v^{-1})^{\epsilon_i} f_i^{-1} F_{(j)}$$

$$= \prod_{i=1}^{n} f_i (tbt^{-1}q^{-1}w^{-1})^{\epsilon_i} f_i^{-1} F_{(j)}$$

$$= \prod_{i=1}^{n} f_i (tbt^{-1}w^{-1})^{\epsilon_i} f_i^{-1} q^{-l} F_{(j)},$$

where $l = \sum_{i=1}^{n} \epsilon_i$. So

$$q^l F_{(j)} = b^{-1} \prod_{i=1}^{n} f_i (tbt^{-1}w^{-1})^{\epsilon_i} f_i^{-1} F_{(j)}$$

and therefore $q^l \in \langle b \rangle^F F_{(j)}$. Note that l can not be zero since that would imply by the last equation that b is contained in $F_{(2)}$, which is of course not the case. Since q is contained in $\langle b \rangle^F F_{(j-1)}$ and $\langle b \rangle^F F_{(j-1)}/\langle b \rangle^F F_{(j)}$ is torsion-free we see that $l = \pm 1$. Thus q is contined in $\langle b \rangle^F F_{(j)}$ and, because $v = wq$, this shows that v is also contained in $\langle b \rangle^F F_{(j)}$.

Therefore v is contained in the intersection of all the $\langle b \rangle^F F_{(i)}$ which is equal to $\langle b \rangle^F$ since $F/\langle b \rangle^F$ is free and hence residually nilpotent (see [L-S], page 14).

If G has a presentation $\langle t, x_2, ..., x_n \mid tut^{-1} = v \rangle$, G_{ab} is free abelian of rank $n-1$ (in that case G is an E-group) and $\langle x_2, ..., x_n \mid u \rangle$ is residually nilpotent, then the same proof shows that the maximal perfect subgroup of G is either trivial or the normal closure of $\bar{u}$. This is not true in general. For example, if G is

$$\langle t, b, c, d \mid tut^{-1} = [u, dud^{-1}] \rangle$$

with $u = b[b, cbc^{-1}]$, then the maximal perfect subgroup of G is not the normal closure of the image of u, but the normal closure of the image of b in G. We do believe however, that if G is an HNN-extension of the form $F(X) *_C t$, where $F(X)$ is free and C is infinite cyclic and the maximal perfect subgroup of G is not trivial, then the normal closure of the generator of C in G is perfect.

Using similar arguments as in the proof Theorem 2 one can also show that if G is a group as in Theorem 5 then any perfect subgroup of the form $\langle w \rangle^F / \langle tbt^{-1}v^{-1} \rangle^F$ is the maximal perfect subgroup of G.

Finally, we would like to remark that whether $P(G)$ is trivial or not for a group G as in Theorem 5 can only depend on the word v. If v is contained in $[\langle b \rangle^F, \langle b \rangle^F]$ then $P(G)$ is certainly non-trivial. We do not know at this time if the converse also holds, that is, whether $P(G)$ non-trivial implies that v is contained in $[\langle b\rangle^F, \langle b\rangle^F]$.

References

[B1] G. Baumslag, Positive one-relator groups, Trans. Amer. Math. Soc. 156 (1971), 165-183.

[B2] G. Baumslag, A survey of groups with a single defining relator, Proc. of Groups-St. Andrews 1985, London Math. Soc. Lecture Notes 121

[B-P] Y. Baik and S. Pride, Generators of the second homotopy module of presentations arising from group constructions, preprint, Glasgow University, 1992.

[Bi] R. Bieri, Homological dimension of discrete groups, Queen Mary College Lecture Notes, London, 1976.

[D] M. Dyer, On constructing complexes dominated by 2-complexes, preprint, University of Oregon

[H1] J. Harlander, Groups with cyclic relation module, Dissertation, University of Oregon, 1992.

[H2] J. Harlander, Solvable groups with cyclic relation module, to J. Pure Appl. Algebra 90 (1993), 189-198.

[L] R. Lyndon, Cohomology theory of groups with a single defining relation, Ann. of Math. 52, (1950), 650-665.

[L-S] R. Lyndon and P. Schupp, Combinatorial group theory, Springer Verlag, Berlin-Heidelberg-New York, 1977.

[M] W. Magnus, Über diskontinuierliche Gruppen mit einer definierenden Relation (Der Freiheitssatz), J. Reine Angew. Math. 163, (1930), 52-74.

[R] J. Ratcliffe, On complexes dominated by a 2-complex, in Combinatorial Group Theory and Topology, Annals of Mathematical Studies 111, Princeton University Press 1987.

[S] R. Strebel, Homological methods applied to the derived series of groups, Comment. Math. Helv. 56, (1974), 302-332.

[W] C. T. C. Wall, Finiteness conditions for CW-complexes, Ann. Math. 81, (1965), 56-69.

Fachbereich Mathematik der Universität Frankfurt,
Robert-Mayer-Strasse 6-10,
60054 Frankfurt am Main.
E-mail: harlander@math.uni-frankfurt.dbp.de

Weight tests and hyperbolic groups

GÜNTHER HUCK AND STEPHAN ROSEBROCK

Abstract

The notion of reduced diagram plays a fundamental role in small
cancellation theory and in tests for detecting the asphericity of 2-
complexes. By introducing *vertex reduced* as a stricter form of reduced-
ness in diagrams we obtain a new combinatorial notion of asphericity
for 2-complexes, called *vertex asphericity*, which generalizes diagrammatic reducibility and implies diagrammatic asphericity. This leads
to a generalization and simplification in applying the weight test [2]
and the cycle test [6] [7] to detect asphericity of 2-complexes and (for
the hyperbolic versions of these tests) to detect hyperbolic group presentations. In the end, we present an application to labeled oriented
graphs. We would like to thank the referee for his helpful suggestions.

1 Basic Definitions

A p.l. map between 2-complexes is called *combinatorial*, if each open cell
is mapped homeomorphically onto its image. A 2-dimensional finite CW-
complex is called *combinatorial*, if the attaching maps of the 2-cells are combinatorial relative to a suitable polygonal subdivision of their boundary.

Let K_P be the standard 2-complex of the presentation P (we assume all presentations to be finite). A *diagram* is a combinatorial map $f: M \to K_P$, where
M is a combinatorial subcomplex of an orientable 2-manifold. A *spherical diagram* is a diagram $f: S \to K_P$, where S is the 2-sphere. These definitions
may be found for example in [1], [2], [6], [7] or [8].

The *Whitehead graph* W_P of K_P is the boundary of a regular neighborhood
of the only vertex of K_P (see [6] or [7]). It consists of two vertices $+x_i$ and
$-x_i$ for each generator x_i of P which correspond to the beginning and the
end of the edge labeled x_i in K_P. The edges of W_P are the *corners* of the
2-cells of the 2-complex. The *star graph* S_P is the same as the Whitehead
graph if no relator of P is a proper power. Let F denote the free group on the
generators of P. If a relator R_i of P has the form $w_i^{k_i}$ with w_i not a proper
power, then the star graph S_P is the Whitehead graph of the presentation
$\langle x_1, \ldots, x_n \mid w_1, \ldots, w_m \rangle$. We denote by $d(R_i)$ the length of R_i, that is the
sum of the absolute values of the exponents of R_i. For a 2-cell D of a com-
binatorial 2-complex, let $d(D)$ be the number of corners of D. By abuse of
notation, from now on, we will often call an edge in the Whitehead graph (or
star graph) a corner.

A *cycle* is a non-constant closed path in a graph that is *cyclically reduced*,
i.e. no oriented edge in the cyclic sequence is followed immediately by its

inverse. Similarly we define non-closed reduced paths. A cycle with no self-crossings is called a *simple cycle*.

Assume we have a diagram $f: M \to K_P$ into a standard 2-complex K_P. Let $L \subset M$ be the 1-manifold arising as the preimage of the Whitehead graph $W_P \subset K_P$. The diagram M is called *vertex reduced*, if each component $w \subset L$ maps onto a path $f(w) \subset W_P$, where no edge of W_P is passed twice in different directions by $f(w)$. A standard 2-complex K_P is called *vertex aspherical* (VA), if there is no nonempty spherical diagram $f: S \to K_P$ which is vertex reduced. These notions may easily be generalized to all combinatorial 2-complexes.

2 Asphericity

If a diagram M is not vertex reduced then a reduction move as in fig. 1 can be performed which decreases the number of 2-cells of M by two. So M may be transformed into a vertex reduced diagram. In this figure we assume D_1

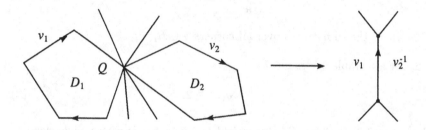

Figure 1: Reduction along a vertex

and D_2 are mapped to the same relator in K_P, but with inverse orientation and in such a way that their corners at the vertex Q map to the same edge in W_P inverse to each other.

There is a slight problem if D_1 and D_2 intersect in more than one component. The result of the reduction may not be a surface in this case (for example it could consist of a one-point union of two spheres). This is discussed in detail in [1].

Certainly every 2-complex that is diagrammatically reducible (see [2] or [9]) is VA. The reduction of fig. 1 could also be achieved by a diamond move and then a standard reduction along an edge. Hence a VA 2-complex is diagrammatically aspherical (see [9]).

Lemma 2.1 *If K_P is vertex aspherical then K_P is aspherical.*

Proof. Observe that the move of fig. 1 is a homotopy. So any given spherical diagram can be reduced by a sequence of these moves until it is empty. The claim follows from the fact that the set of all spherical diagrams is a $\pi_1(K_P)$-module generating set of $\pi_2(K_P)$.

Gersten [2] and Pride [8] invented the weight test. We want to present a changed version of it which is a lot simpler in practice than previous versions:

A *weight* on a standard 2-complex K_P is a real valued function on the corners of W_P or S_P. If g is a weight function and w any path we denote by $g(w)$ the sum of the weights that occur in the path where the weight of an edge that occurs several times in the path is counted with multiplicity. By $d(w)$ we denote the number of edges that are traversed by w, again counting with multiplicity.

We say a presentation P satisfies the *weight test*, if there exists a weight function g for the Whitehead graph W_P that satisfies the following two conditions:

1. for all relators R of P:

$$\sum_{\gamma \in R} g(\gamma) \leq d(R) - 2, \tag{1}$$

   where the sum ranges over all corners $\gamma \in R$,

2. for all simple cycles $z \in S_P$

$$g(z) \geq 2. \tag{2}$$

Theorem 2.2 *If P satisfies the weight test then K_P is vertex aspherical.*

Proof. Assume there is a nonempty vertex reduced spherical diagram $f: S \to K_P$. Pull back the weights onto $L = f^{-1}(W_P)$ via the map f. If D is a 2-cell of S, then $f(D)$ is a 2-cell of K_P which satisfies (1). This implies

$$\sum_{\gamma \in D} g(\gamma) \leq d(D) - 2,$$

Every component $w \subset L$ is a polygonal circle. We will show that $g(w) \geq 2$ and then Proposition 4.4 of [2] gives the desired contradiction.

If $f(w)$ is a simple cycle in W_P then (2) implies $g(w) \geq 2$, so assume $f(w)$ has double points. If $f(w)$ has a double edge, then this edge is passed twice in the same direction since S is vertex reduced. Then we can split this path according to fig. 2, where each component has fewer double edges. Continue this process until each component has only double points. Split at those double points again to end up with a *sum* of simple cycles $z_1, \ldots, z_n$. Their weights sum up to the weight of w. Now (2) implies $g(z_i) \geq 2$ for all of these and $g(w) = \sum g(z_i)$ gives $g(w) \geq 2$.

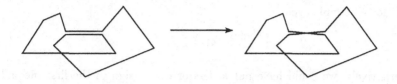

Figure 2: Splitting a path

Example 2.3 *Let K_P be the dunce cap: $P = \langle a \mid a^2 a^{-1} \rangle$. W_P consists of two simple loops of length 1 a_1, a_2 which are connected by an arc b. Define weights $g(a_i) = 2$, $g(b) = -3$. Check that K_P satisfies the weight test. K_P is not diagrammatically reducible.*

There is a simple procedure to determine whether a presentation P satisfies the weight test. In the finite graph W_P there are only finitely many simple cycles which are easy to determine. They give rise to a finite system of inequalities of type (2) where the unknowns are the weights. The finite set of relators of P give rise to another set of inequalities by (1). These two finite sets of inequalities may be tested whether they have a solution, for example with the simplex algorithm.

3   The Hyperbolic Weight Test

When solving the word problem with diagrams, it is more suitable to consider the star graph instead of the Whitehead graph: Consider a relator R of the form $R = u^k$ for some word u in the free group which is not a proper power and $k > 1$. Assume $f : M \to K_P$ is a diagram containing the left part of figure 1, where D_1 and D_2 map with inverse orientation to the same 2-cell corresponding to R in K_P. Assume also that the corners of D_1 and D_2 at the vertex Q correspond to the same edge in S_P but not in W_P. Then the move in figure 1 may still be performed, although it is no longer a homotopy. This gives rise to the following definition:

Let $L \subset M$ be the 1-manifold arising as the preimage of the star graph S_P. The diagram M is called *vertex reduced with respect to* S_P, if each component $w \subset L$ maps onto a path $f(w) \subset S_P$, where no edge of S_P is passed twice in different directions by $f(w)$.

We say a presentation P satisfies the *hyperbolic weight test*, if there exists a weight function g for the star graph S_P that satisfies the following two conditions:

1. for all relators R of P:

$$\sum_{\gamma \in R} g(\gamma) < d(R) - 2, \qquad (3)$$

where the sum ranges over all corners $\gamma \in R$,

2. for all simple cycles $z \in S_P$

$$g(z) \geq 2. \tag{4}$$

Alternatively, we could have put a 'less or equal' sign in the first inequality and a 'strictly bigger' sign in the second one. This would lead to the same statements and to very similar proofs in what follows. We will show that any presentation which satisfies the hyperbolic weight test has a linear isoperimetric inequality, i.e. is word-hyperbolic in the sense of Gromov [3], and therefore has solvable word and conjugacy problems. The proof follows partly the proof of Theorem 2 of [8], but the conditions (3) and (4) are much less restrictive then the corresponding conditions in [8].

Assume $P = \langle x_1, \ldots, x_n \mid R_1, \ldots, R_m \rangle$ satisfies the hyperbolic weight test with weight function g. Let ϵ be the smaller of $\min_i \{ d(R_i) - \sum_{\gamma \in R_i} g(\gamma) - 2 \}$ and 1. Then (3) implies $\epsilon > 0$. There are only finitely many reduced paths in S_P without double edges. Let μ be the smallest weight among these paths, i.e. sum of the weights of the edges of the path. Define $N = \max\{-\mu, 1\}$. Let $f : M \to K_P$ be a diagram where $M = S^2 - D^2$ is a cell decomposition of the 2-sphere minus an open 2-cell. If this diagram is not vertex reduced with respect to S_P then M may be replaced by a new diagram M' with fewer 2-cells but also of the form $S^2 - D^2$ where we read the same word in the boundary. This may be seen by the methods described in [1]. So we assume M to be vertex reduced with respect to S_P. Let $l(\delta M)$ be the sum of the lengths of each boundary component of M, i.e. the sum of the number of edges in each boundary path $\delta(D_i^2)$ (counting edges with multiplicity). Let F be the number of 2-cells of M.

Theorem 3.1
$$F \leq \frac{N+1}{\epsilon} l(\delta M).$$

Proof. We assume $F > 2$, otherwise the result is trivial. Pull back the weights onto $L = f^{-1}(S_P)$ via the map f. Every $w \subset L$ is a polygonal circle or interval. The proof of theorem 2.2 shows that $g(w) \geq 2$ if w is a circle.

We will show that $g(w) \geq -N$ if w is an interval, i.e. if w is the link of a boundary vertex: Let μ be defined as above. We have to show that every reduced path has weight $\geq \mu$. If $f(w)$ has a double edge, then this edge is passed twice in the same direction since M is vertex reduced with respect to S_P. Then we can split this path as in the proof of theorem 2.2. We end up with two components where one is a cycle (with weight ≥ 2) and the other an interval and each component has fewer double edges. Continue this process until each component has only double points and sum up their weights.

The rest of the proof follows closely the proof of Theorem 2 of [8]: Let

$$s = \sum_{D^2 \in M} \sum_{\gamma \in D^2} g(\gamma).$$

Let c be the number of interior vertices of M, e the number of boundary vertices and V the total number of vertices of M. Then $c \geq V - l(\delta M)$ and $e \leq l(\delta M)$. Since $g(w) \geq 2$ at an interior vertex and $g(w) \geq -N$ at a boundary vertex we get $s \geq 2c - Ne$. Altogether we get

$$s \geq 2V - 2l(\delta M) - Nl(\delta M). \tag{5}$$

On the other hand $\sum_{\gamma \in D^2} g(\gamma) \leq d(D^2) - (2 + \epsilon)$ and the definition of s gives: $s \leq \sum_{D^2 \in M}(d(D^2) - (2 + \epsilon))$. If we denote by E the number of edges of M, the last inequality implies: $s \leq 2E - l(\delta M) - (2 + \epsilon)F$. Together with (5) we get $2V - 2l(\delta M) - Nl(\delta M) \leq 2E - l(\delta M) - (2 + \epsilon)F$. This is equivalent to

$$2V - 2E + 2F \leq l(\delta M)(N + 1) - \epsilon F.$$

The Euler characteristic of M gives $2 \leq l(\delta M)(N + 1) - \epsilon F$, which implies the theorem.

4 The Hyperbolic Cycle Test

The cycle test is a generalization of the weight test. It was first defined in [6] as a test for asphericity. The following version which is also presented in [7] implies that the presentation (respectively the group) satisfies a linear isoperimetric inequality.

Let P be a finite presentation. In order to define the hyperbolic cycle test, we need to consider sequences of cycles. A sequence of cycles describes the local incidence configuration of a 2-cell in a diagram: Let $f \colon M \to K_P$ be a diagram which is vertex reduced with respect to S_P. Consider a 2-cell $D \in M$ labeled by the relator R_i of P. Each vertex of D has either a neighborhood in M, described by a cycle in S_P, or a half-disk neighborhood (if the vertex belongs to the boundary of the diagram), described by a reduced path in S_P. When we list these cycles or reduced paths, in order, according to an orientation of the boundary of the 2-cell that follows the word R_i, we obtain what we call a "sequence of cycles for R_i" (which actually consists of cycles and reduced paths, unless the 2-cell is in the interior of the diagram). In addition, each cycle or reduced path in such a sequence of cycles has a preferred corner, namely the "inside corner" of D. This provides the geometric idea of the following definition.

Let R_i be a relator of P and D_i the corresponding 2-cell of K_P. For each R_i there is a (in general infinite) set of *sequences of cycles* $\{Z_i^1, Z_i^2, \ldots\}$. Each *sequence of cyles* Z_i^j is an ordered set of $m = d(R_i)$ cycles or reduced paths $(z_1, \ldots, z_m)$ in S_P, together with a preferred edge β_t in each z_t (called the inside edge), satisfying the following three conditions:

1. No cycle or reduced path z_t of some Z_i^j passes an edge twice in different directions in S_P.

2. If we give the edges β_t the orientation induced by the paths z_t then the sequence of oriented edges $(\beta_1, \ldots, \beta_m)$ is the sequence of oriented corners read along the boundary of D_i in the direction of the word R_i.

3. Suppose $z_t = \gamma_1 \ldots \gamma_\nu$ and $z_{t+1} = \alpha_1 \ldots \alpha_\mu$ ($t + 1 = 1$ if $t = m$) are consecutive cycles or reduced paths of the sequence of cycles Z_i^j and suppose $\beta_t = \gamma_l$ and $\beta_{t+1} = \alpha_k$ are the inside edges of z_t and z_{t+1} in Z_i^j. Then either a) γ_{l+1} and α_{k-1} are adjacent corners of the same relator R_s or b) γ_{l+1} and α_{k-1} do not exist (i.e. γ and α are reduced paths with γ_l the last edge of γ and α_k ($k = 1$) the first edge of α). If, in either case, γ_l ends in a vertex $+a$ ($-a$) of S_P then α_k starts in $-a$ ($+a$) respectively, and, furthermore, in case a), the relators R_s and R_i have an edge a in common. This condition gives intuitively a local diagram (see fig. 3).

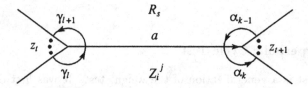

Figure 3: part of the relator R_i in M

We say P satisfies the *hyperbolic cycle test*, if for every sequence of cycles Z_i^j for every relator R_i there exists a weight function $g_i^j \colon \{$corners of $R_i\} \to \mathbb{R}$, that satisfies the following two inequalities:

1. for every sequence of cycles Z_i^j:

$$\sum_{\gamma \in R_i} g_i^j(\gamma) < d(R_i) - 2 \qquad (6)$$

2. Let $z = \gamma_1 \ldots \gamma_\nu$ be a cycle with $\gamma_s \in R_{i_s}$, and for every corner γ_s let $Z_{i_s}^{j_s}$ be a sequence of cycles for R_{i_s} containing z such that the weight $g_{i_s}^{j_s}(\gamma_s)$ is defined. Then:

$$\sum_s g_{i_s}^{j_s}(\gamma_s) \geq 2 \qquad (7)$$

If P satisfies the hyperbolic weight test then it certainly satisfies the hyperbolic cycle test just by giving the corners of S_P fixed weights.

Let P satisfy the cycle test with weight functions g_i^j. As in the proof of Theorem 2.2 and 3.1 we look at cycles in S_P and split them. For a fixed relator R_i one cycle or reduced path z of a sequence of cycles Z_i^j must be compatible with its neighbors z', z'', in the sense that the two corners that are adjacent to the inside corner of z are compatible with the corresponding corners of z' and z'' as expressed in point 2. of the above definition (see fig. 3). So any splitting of cycles has to happen in such a way, that triples of adjacent corners of a cycle or reduced path are not separated. A cycle or reduced path z is called *splittable*, if z traverses a subpath of length two in S_P consisting of edges a and b twice in the same direction. If one "switches tracks" at the vertex between the edges a and b in z (analogously to fig. 2), then one gets two reduced paths z_1 and z_2 that have z as their sum. It is easy to verify, that any triple of adjacent corners in z appears either in z_1 or in z_2. Iterating this process as often as possible produces a set of unsplittable paths (all of which, except possibly one, are cycles) whose sum is z.

For a given presentation P it suffices to find weight functions for sequences of cycles consisting of unsplittable cycles or reduced paths satisfying (6) and (7), in order to prove that P satistisfies the hyperbolic cycle test (see also [6]). Since there are only finitely many of these we can assume w.l.o.g. that for every R_i there are only finitely many different weight functions g_i^j. Let ϵ be the smaller of $\min_{i,j}\{d(R_i) - \sum_{\gamma \in R_i} g_i^j(\gamma) - 2\}$ and 1. Then (6) implies $\epsilon > 0$. Let μ be the smallest weight which occurs among the finite set of unsplittable reduced paths in S_P and define $N = \max\{-\mu, 1\}$. One needs to choose for every cycle or reduced path a fixed decomposition into unsplittable components (this can be done by an algorithm). Let $f: M \to K_P$ be a diagram which is vertex reduced with respect to S_P and $M = S^2 - D^2$ as in the last section. With respect to the fixed decomposition of cycles and reduced paths, the weight functions g_i^j induce unique weights on the corners of M. Let $l(\delta M)$ be the sum of the lengths of each boundary component of M, i.e. the sum of the number of edges in each boundary path $\delta(D_i^2)$ (counting edges with multiplicity). Let F be the number of 2-cells of M.

Theorem 4.1

$$F \leq \frac{N+1}{\epsilon} l(\delta M).$$

The proof is an obvious generalization of the proof of theorem 3.1. There are only finitely many weight functions to consider so it is possible to decide whether a presentation satisfies the hyperbolic cycle test (see also [7]).

5   Reversing Edges in LOGs

Consider a presentation: $P = \langle x_1, \ldots, x_n \mid R_1, \ldots, R_m \rangle$, where each relator is of the form $x_i x_j = x_j x_k$. We can assign a *labeled oriented graph* (LOG) to P by defining a vertex i for every generator x_i and an oriented edge from i to k labeled by j for the relator $x_i x_j = x_j x_k$. LOGs appear in the context of the Whitehead conjecture or in knot theory (see [4] or [5]). An LOG is called *injective*, if every generator occurs at most once as a label of an edge. In our context an LOG is called *reduced*, if each relator consists of three different generators.

Theorem 5.1 *Let P be a presentation of a reduced injective LOG G, which satisfies the weight test (or the hyperbolic weight test). If you change the orientation of any edge of G, then the resulting LOG still satisfies the weight test (or the hyperbolic weight test respectively).*

At first we need:

Lemma 5.2 *Let $P=\langle x_1,\ldots,x_n \mid R_1,\ldots,R_m\rangle$ be a reduced finite presentation which satisfies the weight test or the hyperbolic weight test. If you perform any of the operations below, then the result still satisfies the same test.*

S_1 : *cyclic conjugation or inversion of any relator R_i.*

S_2 : *Inversion of a generator, i.e. $x_i \to x_i^{-1}$.*

S_3 : *Let u,v be nontrivial words in the free group on the $\{x_i\}$. If $R_j = ux_i^{\epsilon}vx_i^{-\epsilon}$, then replace R_j by $R_j' = ux_i^{\epsilon}v^{-1}x_i^{-\epsilon}$ $(\epsilon \in \{\pm1\})$.*

Proof. It is easy to see the assertion for the operation S_1, since the edges, which each relator contributes to the Whitehead graph (or star graph) remain unchanged. Therefore we may give the same weights to its edges. The same argument works with S_3. The operation S_2 leads to an interchange of the labels of $+x_i$ and $-x_i$, but the weights may be kept as before.

Proof of Theorem 5.1. Let $R_i = a^{-1}bcb^{-1}$ be the relator of P, where you want to change the orientation of the corresponding edge in G. Perform an operation S_2 on b, which gives you R_i in the desired form. Now look at all other relators which contain b. Since G is injective, they all have the form $R_j = b^{-1}d^{\epsilon}ed^{-\epsilon}$, $(\epsilon \in \{\pm1\})$. The operation S_2 we just performed changes R_j to $R_j' = bd^{\epsilon}ed^{-\epsilon}$. Now apply S_3 to R_j' with $v = b$ (maybe some of the S_1-operations are necessary before). This changes R_j' back to R_j as desired. The lemma now shows the conclusion.

If P is the presentation of a non-injective reduced LOG G, then the same proof shows, that we may turn around all edges of G with the same label simultaneously, without changing the validity of the (hyperbolic) weight test.

References

[1] D.J. Collins and J. Huebschmann. Spherical diagrams and identities among relations. *Math. Ann.* 261 (1982), 155 – 183.

[2] S. Gersten. Reducible diagrams and equations over groups, In *Essays in group theory*, Math. Sci. Res. Inst. Publ. 8, 15-73, Springer-Verlag (1987).

[3] M. Gromov. *Hyperbolic groups*, In *Essays in group theory*, Math. Sci. Res. Inst. Publ. 8, 75-263, Springer-Verlag (1987).

[4] J. Howie. Some remarks on a problem of J.H.C Whitehead. *Topology* 22 (1983), 475–485.

[5] J. Howie. On the asphericity of ribbon disk complements. *Trans. Amer. Math. Soc.* 289 (1985), 281-302.

[6] G. Huck and S. Rosebrock. Ein verallgemeinerter Gewichtstest mit Anwendungen auf Baumpraesentationen. *Math. Z.* 211 (3) (1992), 351-367.

[7] G. Huck and S. Rosebrock. Applications of diagrams to decision problems. In *Two-dimensional homotopy and combinatiorial group theory*, (A. Sieradski *et. al.* Eds.), LMS Lecture Notes in Math. 197, 189-218, Cambridge University Press (1993).

[8] S. Pride. Star-complexes, and the dependence problems for hyperbolic complexes. *Glasgow Math. J.* 30 (1988), 155–170.

[9] S. Rosebrock. A reduced spherical diagram into a ribbon-disk complement and related examples. In *Topology and Combinatorial Group Theory* (M.P.L. Latiolais ed.) Lect. Notes in Math. 1440, 175-185, Springer-Verlag (1990).

Stephan Rosebrock
Institut f. Didaktik der Mathematik
J.- W.- Goethe Universität
Senckenberganlage 9
60054 Frankfurt/M.
Germany

Günther Huck
Dept. of Math.
Northern Arizona University
Flagstaff AZ 86011
USA

A non-residually finite, relatively finitely presented group in the variety $\mathcal{N}_2\mathcal{A}$

O.G. KHARLAMPOVICH[1] AND M.V. SAPIR[2]

Residually finite varieties of groups were completely described, in [1], by Ol'shanskii. He proved that a group variety is residually finite if and only if it is generated by a finite group with abelian Sylow subgroups.

The next question is: "Which varieties are locally residually finite?" Hall [9] proved that all finitely generated abelian-by-nilpotent groups are residually finite. Hall formulated a conjecture that his result can be extended to the class of abelian-by-polycyclic groups. Jategaonkar [2] proved that finitely generated abelian-by-polycyclic groups are residually finite.

The following result was obtained by Groves [8]. Let $\mathcal{T}_p$ be the variety generated in the variety $\mathcal{B}_p\mathcal{A}$ by all 2-generated groups belonging to $Z\mathcal{A}^2$, (p an odd prime), and let $\mathcal{T}_2$ be the variety generated in the variety $\mathcal{A}_2^2\mathcal{A}$ by all 2-generated groups belonging to $Z\mathcal{A}_2\mathcal{A}$.

Theorem 1 *(Groves) If W is a variety of metanilpotent groups then the following conditions are equivalent.*

1. *W does not contain any $\mathcal{T}_p$.*

2. *W is locally residually finite.*

3. *All finitely generated groups in W satisfy the maximal condition for normal subgroups.*

In [4] it was proved that for odd primes p the variety $\mathcal{T}_p$ coincides with $Z\mathcal{A}^2 \cap \mathcal{B}_p\mathcal{A}$, and $\mathcal{T}_2$ was also described in the language of identities.

Conjecture 1 *The only minimal, non-locally residually finite, varieties of solvable groups are the varieties from the previous theorem and the varieties A_pA_qA (p, q are distinct primes). A variety of solvable groups is locally residually finite if and only if it does not contain any of them.*

Bieri and Strebel proved in [5] that absolutely finitely presented groups belonging to the variety $\mathcal{N}_2\mathcal{A}$ are residually finite. The analog of this result is not true for relatively finitely presented groups belonging to the variety $\mathcal{N}_2\mathcal{A}$.

[1]Research supported by NSERC and FCAR grants.
[2]Research supported in part by an NSF grant and by the Center for Communication and Information Science of the University of Nebraska at Lincoln.

Theorem 2 *In the variety $\mathcal{N}_2\mathcal{A}$ there exists a group which is relatively finitely presented and which is not residually finite. For every prime $p \geq 3$ such a group exists in the variety $\mathcal{N}_2\mathcal{A} \cap \mathcal{B}_p\mathcal{A}$.*

Proof. Let $W = \mathcal{N}_2\mathcal{A} \cap \mathcal{B}_p\mathcal{A}$. Let G be the group defined in the variety W by the set of generators $\{x, B, a, b, a_1, b_1\}$ and defining relations

1. if $y, z \in \{a, b, a_1, b_1\}$, then $[y, z] = 1$;

2. $[a, B] = [a_1, B] = 1$;

3. $B^p = x^p = 1$;

4. if we denote $x_B = [x, B]$, then $[x_B, B] = 1$;

5. $x^{-1}x^a = x^{a_1}$, $x^{-1}x^b = x^{b_1}$, $x_B^{-1}x_B^a = x_B^{a_1}$;

6. $[x^{a^\alpha b^\beta}, x] = 1$, $[x_B^{a^\alpha}, x] = 1$, when $\alpha, \beta \in \{0, 1, -1\}$;

7. if $z \in \{a, b\}$ and $f \in G$, then set $f * z = f^{-1}f^z f^{-z^{-1}} f_1^{z_1^{-1}}$ and $f * B = [f, B]$; define $f * z_1 * z_2 = (f * z_1) * z_2$, $f * z^{(0)} = f$, $f * z^{(m+1)} = f * z^{(m)} * z$. Let $x * a * b = x$ and $x * a * B = 1$.

To complete the proof of Theorem 2 we require Lemmas 1 to 5 and Proposition 1.

Lemma 1 *([6], [7].) Suppose that a group is generated by three sets X, $K = \{a_i | i = 1, \ldots, m\}$ and $K' = \{a_i' | i = 1, \ldots, m\}$ such that*

1. *the subgroup generated by $K \cup K'$ is abelian;*

2. *for every $a \in K$ and every $x \in X$ we have $xx^a = x^{a'}$;*

3. *we have $[x_1^{a_1^{\alpha_1} \ldots a_m^{\alpha_m}}, x_2] = 1$, for every $x_1, x_2 \in X$, and every $\alpha_1, \ldots, \alpha_m \in \{0, 1, -1\}$.*

Then the normal subgroup generated by X in the subgroup $< X \cup K \cup K' >$ is abelian and this subgroup is metabelian.

Lemma 2 *The subgroup of the group G generated by the elements of the form $x^{a^\alpha b^\beta a_1^\gamma b_1^\delta}$ (with $\alpha, \beta, \gamma, \delta \in Z$), is abelian of exponent p. The subgroup generated by the elements of the form $x_B^{a^\alpha a_1^\gamma}$, $x^{a^\alpha a_1^\gamma}$ is abelian of exponent p.*

Proof. This follows from Lemma 1 with use of relations 5 and 6.

Lemma 3 *The following relations hold in the group G*

1. $x * a^{(n)} * b^{(m)} * a * b = x * a^{(n)} * b^{(m)} * b * a$ *and*

2. $x * a^{(n)} * B * a = x * a^{(n+1)} * B$.

Proof. Let us denote $f = x * a^{(n)} * b^{(m)}$ then by Lemma 2 we can use module notation

$$
\begin{aligned}
f * a * b &= f^{(-1+a-a^{-1}+a_1^{-1})(-1+b-b^{-1}+b_1^{-1})} \\
&= f^{(-1+b-b^{-1}+b_1^{-1})(-1+a-a^{-1}+a_1^{-1})} \\
&= f * b * a.
\end{aligned}
$$

Also by Lemma 2 the second relation holds.

Proposition 1 *In G the element $x * B \neq 1$ and $x * a^{(n)} \neq x * a^{(m)}$, for any $m \neq n$.*

Proof. We construct a homomorphic image of the group G such that the images of the elements $x * a^{(n)}$ and $x * a^{(m)}$ are distinct and the image of the element $x * B$ is nontrivial. We denote this group by H.

First consider a semigroup S with 0 generated by the elements q, a, b, B with defining relations

$$ ab = ba, aB = Ba, qab = q, qaB = 0, qq = aq = bq = Bq = Bb = B^2 = 0. $$

Notice that all the nonzero words in S are subwords of the words $qa^n b^m B$ and words obtained from them using commutativity relations. It is clear that $qB \neq 0$ (we can only apply the relation $q = qab$ and the commutativity relations to the word qB, and in this way we never get a subword of the form qaB) and $qa^n \neq qa^m$ in S for $n \neq m$.

For each element $u \in S$ containing q assign symbols $x_{i,u}$, $i \in 1, \ldots, 9$. Let $x_{i,u} = 1$ for $u = 0$. Let T be the abelian group of exponent p with generating set

$$ x_{i,u}; i = 1, \ldots, 9, u \in S, u \neq 0. $$

To each element in the set $a, b, a_1, b_1, B \in G$ assign a map defined on the generators of the group T as follows. Define a and a_1 by

$$
x_{j,u}^a = \begin{cases}
x_{j,u} x_{j+1,u} x_{j+2,u} x_{j,ua} & \text{if } j = 1, 4, 7 \\
x_{j,u} x_{j-1,u}^{-1} & \text{if } j = 2, 5, 8 \\
x_{j-2,u} & \text{if } j = 3, 6, 9
\end{cases}
$$

and $x_{j,u}^{a_1} = x_{j,u}^{-1} x_{j,u}^a$. To define b and b_1 first suppose that u does not contain B. In this case let

$$
x_{j,u}^b = \begin{cases}
x_{j,u} x_{j+3,u} x_{j+6,u} x_{j,ub}, & \text{if } j = 1, 2, 3 \\
x_{j,u} x_{j-3,u}^{-1}, & \text{if } j = 4, 5, 6 \\
x_{j-6,u} & \text{if } j = 7, 8, 9
\end{cases}
$$

and $x_{j,u}^{b_1} = x_{j,u}^{-1} x_{j,u}^b$. If u contains B, then set $x_{j,u}^b = x_{j,u}^{b_1} = x_{j,u}$. Let $x_{j,u}^B = x_{j,u} x_{j,uB}$.

These maps defined on the generators can be extended to endomorphisms of the group T. They all have inverse maps and hence are automorphisms. We define , for example a^{-1} and B^{-1} by

$$x_{j,u}^{a^{-1}} = \begin{cases} x_{j+2,u} & \text{if } j = 1,4,7 \\ x_{j,u}x_{j+1,u} & \text{if } j = 2,5,8 \\ x_{j,u}^{-2}x_{j-1,u}^{-1}x_{j-2,u}x_{j,ua}^{-1} & \text{if } j = 3,6,9 \end{cases}$$

and $B^{-1} = B$. Let us denote by R the subgroup of the group $AutT$ generated by a, a_1, b, b_1, B. Let H be a semidirect product of T and R.

Lemma 4 *The subgroup of H generated by the elements $x_{1,q}, a, b, a_1, b_1, B$ is a homomorphic image of G.*

Proof. The image of x under the homomorphism $G \to H$ is $x_{1,q}$. The images of elements a, b, a_1, b_1, B are denoted by the same letters. Let us verify that relations 1-7 of G are satisfied for the images. Let us check some of the relations 1, for example $[a, b] = 1$. Let us denote

$$\bar{x}_{1,u} = x_{4,u}, \bar{x}_{2,u} = x_{5,u}, \bar{x}_{3,u} = x_{6,u},$$

$$\tilde{x}_{1,u} = x_{7,u}, \tilde{x}_{2,u} = x_{8,u}, \tilde{x}_{3,u} = x_{9,u}.$$

Then $\overline{x_{j,u}^a} = \bar{x}_{j,u}^a, \widetilde{x_{j,u}^a} = \tilde{x}_{j,u}^a$. Suppose u does not contain B. Then

$$x_{1,u}^{ab} = (x_{1,u}x_{2,u}x_{3,u}x_{1,ua})^b =$$

$$(x_{1,u}\bar{x}_{1,u}\tilde{x}_{1,u}x_{1,ub})(x_{2,u}\bar{x}_{2,u}\tilde{x}_{2,u}x_{2,ub})(x_{3,u}\bar{x}_{3,u}\tilde{x}_{3,u}x_{3,ub})(x_{1,ua}\bar{x}_{1,ua}\tilde{x}_{1,ua}x_{1,uab}) =$$

$$(x_{1,u}\bar{x}_{1,u}\tilde{x}_{1,u}x_{1,ub})^a = x_{1,u}^{ba};$$

$$x_{2,u}^{ab} = (x_{1,u}^{-1}x_{2,u})^b =$$

$$(x_{1,u}^{-1}\bar{x}_{1,u}^{-1}\tilde{x}_{1,u}^{-1}x_{1,ub}^{-1})(x_{2,u}\bar{x}_{2,u}\tilde{x}_{2,u}x_{2,ub}) =$$

$$(x_{2,u}\bar{x}_{2,u}\tilde{x}_{2,u}x_{2,ub})^a = x_{2,u}^{ba};$$

$$x_{3,u}^{ab} = (x_{1,u})^b = x_{1,u}\bar{x}_{1,u}\tilde{x}_{1,u}x_{1,ub} = (x_{3,u}\bar{x}_{3,u}\tilde{x}_{3,u}x_{3,ub})^a = x_{3,u}^{ba};$$

$$\bar{x}_{j,u}^{ab} = \overline{(x_{j,u}^a)}^b = \overline{x_{j,u}^a}x_{j,u}^{-a} = (\bar{x}_{j,u}x_{j,u}^{-1})^a = \bar{x}_{j,u}^{ba};$$

$$\tilde{x}_{j,u}^{ab} = (\widetilde{x_{j,u}^a})^b = x_{j,u}^a = \tilde{x}_{j,u}^{ba}.$$

If u contains B we have $x_{j,u}^{ab} = x_{j,u}^a = x_{j,u}^{ba}$.

The other relations (1) can be verified similarly. Now we show that for $z \in \{a, b, B\}$ there is an equality $x_{1,u} * z = x_{1,uz}$ in H. Really, if $z = B$

$$x_{1,u} * B = x_{1,u}^{-1}x_{1,u}^B = x_{1,uB}.$$

If, for example, $z = a$ then

$$x_{1,u} * a = x_{1,u}^{-1}x_{1,u}^a x_{1,u}^{-a^{-1}} x_{1,u}^{a_1^{-1}} = x_{1,u}^{-1}(x_{1,u}x_{2,u}x_{3,u}x_{1,ua})x_{3,u}^{-1}x_{2,u}^{-1} = x_{1,ua}.$$

To complete the proof of Lemma 4 we have to show that H belongs to the variety $\mathcal{N}_2\mathcal{A} \cap \mathcal{B}_p\mathcal{A}$. The derived subgroup H' is generated as a normal subgroup by the elements $[x_{j,u}, a]$, $[x_{j,u}, b]$, $[x_{j,u}, B]$, $[B, b]$ and $[B, b_1]$. We have $[H', H', H'] = 1$ since every commutator of the form $[z_1, \ldots, z_k]$ such that two z-s are B-s or two z-s are x-s is the identity in H.

Now we show that H' has exponent p. Any element in H' has the form xr, where $x \in T$ and r belongs to the normal subgroup generated by $[B, b]$ and $[B, b_1]$. Then $(xr)^p = xx^r x^{r^2} \ldots x^{r^{p-1}} r^p$. But $r^p = 1$, since $B^p = 1$. Really $x_{j,u}^{B^p} = x_{j,u} x_{j,uB}^p = x_{j,u}$ and $x^r = xY$, where Y is a product of elements of the form $x_B^{b^n}$. Hence $x^{r^t} = xY^t$ and $(xr)^p = x^p Y^{p(p-1)/2} = 1$. This proves that H' has exponent p. Lemma 4 and the Proposition have been proved.

Lemma 5 *If N is a normal subgroup of a finite index in G, then $x * B \in N$.*

Proof. Since N has finite index there are m and n such that $m < n$ and $x * a^{(n)} = x * a^{(m)}$ mod N. Then $x * a^{(n)} * b^{(m)} * B = x * a^{(m)} * b^{(m)} * B$ mod N. Lemma 3 and relations (7) imply that $x * a^{(m)} * b^{(m)} * B = x * B$. The same lemma and relations (2) and (7) imply that $x * a^{(n)} * b^{(m)} * B = x * a * B * a^{(n-m)} = 1$. Hence $x * B \in N$. At the same time $x * B \neq 1$, because the image of this element in H is $x_{1,qB}$, and it is nontrivial, since $qB \neq 0$ in the semigroup S.

Lemma 5 and Proposition 1 imply that G is not residually finite. Theorem 2 has been proved.

Problem 1 *Find all minimal varieties of solvable groups which possess relatively finitely presented non-residually finite groups. It is quite possible that these varieties are only $\mathcal{A}_p\mathcal{A}_q\mathcal{A}$ and $\mathcal{N}_2\mathcal{A} \cap \mathcal{B}_p\mathcal{A}$ (p an odd prime) and some subvariety of $\mathcal{N}_2\mathcal{A} \cap \mathcal{B}_4\mathcal{A}$.*

It is interesting that the word problem is decidable for relatively finitely presented groups in the variety $\mathcal{N}_2\mathcal{A}$ and all subvarieties of this variety [3].

The analog of Theorem 2 is also true for Lie algebras.

Theorem 3 *Over a field of characteristic $\neq 2$ there exists a Lie algebra relatively finitely presented in the variety $\mathcal{N}_2\mathcal{A}$ which is not residually finite.*

References

[1] A.Yu. Ol'shanskii, Varieties of finitely approximable groups, *Izv. AN SSSR. Ser. Matem.* 33 (1969), 915-927. MR 41 #3572.

[2] A.V. Jategaonkar, Integral group rings of polycyclic-by-finite groups, *J. Pure and Appl. Algebra* 47 (1974) 337-343. MR 49#9084.

[3] O.G. Kharlampovich, The word problem for subvarieties of the variety $\mathcal{N}_2\mathcal{A}$, *Alg. i Logika* 26, 481-501 (1987); translation in *Algebra and Logic* 26 (1987). MR 89m:20036.

[4] O.G. Kharlampovich and D. Gildenhuys, Locally residually finite metanilpotent varieties of groups, preprint.

[5] R. Bieri and R. Strebel, Valuations and finitely presented metabelian groups, *Proc. London Math. Soc. (3)* (1980), 439–464.

[6] G. Baumslag, Subgroups of finitely presented metabelian groups, *J. Austral. Math. Soc.* 16 (1973), 98–110. MR 48 #11324.

[7] V. N. Remeslennikov, Studies on infinite solvable and finitely approximable groups, Abstracts of the Doctor's Thesis, *Mat. Zametki* 17 (1975) 819–824. MR 52 #14051.

[8] J.R.J. Groves, On some finiteness conditions for varieties of metanilpotent groups, *Arch. Math.* 24 (1973) 252– 268. MR 48 #6253.

[9] P. Hall, On the finiteness of certain soluble groups, *Proc. London Math. Soc.* 3 (1959), 595–622. MR 22 #1618.

O.G. Kharlampovich
Dept. of Mathematics and Statistics
McGill University
Montreal
Quebec
H3A 2K6
Canada

M.V. Sapir
Dept. of Mathematics
University of Nebraska
Lincoln
Nebraska
68588
U.S.A.

Hierarchical decompositions, generalized Tate cohomology, and groups of type $(FP)_\infty$

PETER H. KROPHOLLER [1]

Abstract

We outline a proof that if G is a soluble or linear group of type $(FP)_\infty$ then G has finite virtual cohomological dimension. The proof depends on hierarchical decompositions of soluble and linear groups and also makes use of a recently discovered generalized Tate cohomology theory. A survey of this *complete cohomology* is included. The paper concludes with a review of some open problems.

1 Preface

The first part of this article is based on a lecture delivered at the conference. It concerns the proof that soluble and linear groups of type $(FP)_\infty$ have finite vcd. More general results have been published in [21], but in order to make the key new arguments widely accessible I thought it worthwhile going through the special cases again. Several technical problems can be avoided this way, and I hope that this will make for clarity.

At the conference, a number of people asked about the generalized Tate cohomology theory which plays such a crucial and somewhat miraculous role. For this reason I have included a detailed account in §4.

In the last section, some problems and questions are discussed which I did not have time to cover in the lecture. Some of the results in this section have not been published elsewhere.

---

[1] This research was partially supported by SERC grant GR/F80616

2 Introduction

Let G be a group. This paper studies projective resolutions $P_* \twoheadrightarrow \mathbb{Z}$ of the trivial module $\mathbb{Z}$ over the group ring $\mathbb{Z}G$. Before starting I want to explain the interest in such resolutions. Firstly they are used to define and compute cohomology groups $H^n(G, M)$ of a group G with coefficients in a G-module M. Cohomology groups are important algebraic invariants, and if we agree that these are useful then we must also agree that projective resolutions are necessary. Perhaps I should say *necessary evil*, for although it is trivial that projective resolutions exist and elementary that cohomology does not depend on a particular choice of resolution, it is an order of magnitude harder to lay hands on explicit projective resolutions suitable for calculations.

So where do good projective resolutions come from? By far the best sources are free actions on contractible cell complexes. In this article I shall always consider cellular (or *admissible*) actions, which means to say that the group, in addition to acting as self-homeomorphisms, permutes the cells and each cell stabilizer fixes that cell pointwise. The cell complexes may be simplicial but often it is convenient to allow CW-complexes too. These cell complexes arise in various ways, one of the most important being as Bruhat-Tits buildings. Every cell complex has an augmented cellular chain complex $C_*(X) \twoheadrightarrow \mathbb{Z}$ in which C_j is essentially the free abelian group on the set of j-dimensional cells of X. If X is contractible then this chain complex is an exact sequence. If G admits an action on X then G permutes the j-dimensional cells and each C_j inherits the structure of a permutation module: in this situation one obtains an augmented chain complex of G-modules. If G acts freely on X then each C_j is a free G-module. Thus if G acts freely and X is contractible then the augmented chain complex is a free resolution of $\mathbb{Z}$ over $\mathbb{Z}G$.

Informally then, G-actions on cell complexes are fundamental tools for building good projective resolutions. In turn, projective resolutions are tools for computing cohomology. The schematic diagram (figure 1) illustrates this.

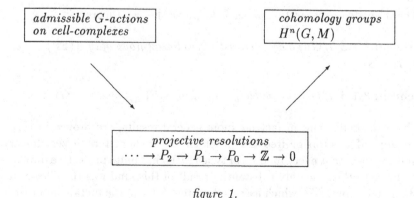

figure 1.

In view of this one should expect that any study of projective resolutions will involve an interplay between group actions on spaces and cohomology.

3 Soluble and linear groups of type (FP)$_\infty$

3.1 Two theorems

There are two key properties of groups which can be defined in terms of projective resolutions.

Definition 3.1.1 *A group G is said to be of type* $(FP)_\infty$ *if and only if there is a projective resolution* $P_* \twoheadrightarrow \mathbb{Z}$ *of finite type: that is, in which every P_j is finitely generated.*

Definition 3.1.2 *A group G has finite cohomological dimension if and only if there is a projective resolution* $P_* \twoheadrightarrow \mathbb{Z}$ *of finite length: that is, in which the P_j are zero from some point on.*

Finite groups are the most obvious examples of groups of type $(FP)_\infty$. Indeed, if G is a finite group one would have to work to avoid projective resolutions of finite type: even the *bar resolution* has all projectives finitely generated.

For further examples of groups of type $(FP)_\infty$ one must look to actions on contractible spaces. Natural examples arise as groups G which admit cocompact and properly discontinuous actions on contractible spaces. When the action is properly discontinuous, all the isotropy groups (or cell stabilizers) are finite, so examples of this kind are generalizations of finite groups. In this way one knows that polycyclic-by-finite groups, arithmetic groups and Coxeter groups are all of type $(FP)_\infty$. Now, these examples have the additional property that there is a subgroup of finite index which has finite cohomological dimension: that is to say, the virtual cohomological dimension vcd(G) is finite.

The immediate goal of this paper is to prove the two theorems:

Theorem 3.1.3 *If G is a characteristic zero linear group of type* $(FP)_\infty$ *then* vcd$(G) < \infty$.

Theorem 3.1.4 *If G is a soluble group of type* $(FP)_\infty$ *then* vcd$(G) < \infty$.

These theorems, and in fact far more general results, are proved in [21]: for example if G is torsion-free and either soluble or characteristic zero linear then it is shown that every module of type $(FP)_\infty$ has finite projective dimension. Prior to this, the only substantial result of this kind was the Theorem of Bieri and Groves, [7], which asserts Theorem 3.1.4 in the metabelian case.

They used sophisticated methods from commutative algebra, and their arguments, which are derived from the valuation-sphere methods invented by Bieri and Strebel in their characterization of finitely presented metabelian groups [8], remain of some interest because they give more insight in the metabelian case than our arguments could possibly give in the more general cases we consider.

The proofs depend on the fact that both soluble and linear groups admit *hierarchical decompositions*. We discuss these decompositions in detail in §3.2. In addition, cohomological methods are needed. It turns out that the right theory is a generalized Tate cohomology. Such a theory has been discovered by Vogel and a detailed account by Goichot can be found in [17]. Independently, and using a different appraoch, a generalized Tate cohomology theory has recently been axiomatized by Mislin [24]. Generalized Tate cohomology groups were also discovered by Benson and Carlson, [5]. In §3.3 we state the properties of this complete cohomology which we need. For the interested reader, a more detailed account is included in §4.

In spirit, the arguments follow the philosophy outlined in §2. Hierarchical decompositions of groups arise when one has useful actions of a group and its subgroups on finite dimensional contractible spaces. For soluble groups one needs little more than Bieberbach's theorem that finitely generated abelian-by-finite groups admit cocompact and properly discontinuous actions on Euclidean spaces. For linear groups one needs more sophisticated technology. Alperin and Shalen were the first to show that finitely generated linear groups always admit hierarchical decompositions, see [1]. They use several methods for constructing actions on contractible spaces. Firstly, if A is a finitely generated commutative ring there are Bruhat-Tits buildings for $SL_n(A)$ associated to discrete valuations on A, and secondly, for special kinds of linear group there are discrete actions on symmetric spaces associated with certain Lie groups. The actions we use are a long way from being free, but nevertheless they provide a good handle to grasp problems about projective resolutions.

Given the philosophy, it is not surprising that cohomology is involved as well. Perhaps it is surprising that we use complete cohomology rather than ordinary cohomology. All the steps of the proof could be done with ordinary cohomology because it really only depends on a few minor axioms like the existence of long exact sequences — it's just that you do not get such a strong conclusion this way

3.2 Hierarchical decompositions of groups

The basic method works for the following class of groups.

Definition 3.2.1 *We write* H𝔉 *for the smallest class of groups which contains all finite groups and which contains a group G whenever there is an*

admissible action of G on a finite dimensional contractible cell complex for which all isotropy groups already belong to $\mathbf{H}\mathfrak{F}$.

One should think of $\mathbf{H}\mathfrak{F}$ as built up in the following way.

- Let $\mathbf{H}_0\mathfrak{F}$ be the class $\mathfrak{F}$ of all finite groups. Now define classes $\mathbf{H}_\alpha\mathfrak{F}$ inductively:

- if α is a successor ordinal then $\mathbf{H}_\alpha\mathfrak{F}$ is the class of all groups G having an admissible action on a finite dimensional contractible cell complex for which all isotropy groups already belong to $\mathbf{H}_{\alpha-1}\mathfrak{F}$:

- if α is a limit ordinal then $\mathbf{H}_\alpha\mathfrak{F} = \bigcup_{\beta<\alpha} \mathbf{H}_\beta\mathfrak{F}$.

Now a group G belongs to $\mathbf{H}\mathfrak{F}$ if and only if it belongs to $\mathbf{H}_\alpha\mathfrak{F}$ for some α. Many groups belong to $\mathbf{H}\mathfrak{F}$:

1. $\mathbf{H}_1\mathfrak{F}$ contains all groups of finite virtual cohomological dimension, because these groups admit actions on finite dimensional contractible spaces with finite isotropy groups.

2. $\mathbf{H}_1\mathfrak{F}$ also contains some groups of infinite virtual cohomological dimension. A striking example is the Burnside group $B(d,e)$ of odd exponent $e \geq 665$ on d generators. It is known that this group admits an action on a contractible 2-dimensional space with all non-trivial isotropy groups cyclic of order e.

3. $\mathbf{H}_5\mathfrak{F}$ contains all finitely generated linear groups. This can be proved by using techniques in [1], (cf. the discussion at the end of §3.1 above and also some further remarks below.)

4. $\mathbf{H}_{2d-1}\mathfrak{F}$ contains all finitely generated soluble groups of derived length d.

All these facts are proved in [21], although we only proved that countable linear groups are in $\mathbf{H}\mathfrak{F}$ rather than the stronger statement above: for a hint on how to get the improved result see Brown's discussion in §3 of Chapter VII of [11] . I will not repeat the arguments here. For linear groups the argument derives from the methods of Alperin and Shalen [1]. In that paper, Alperin and Shalen prove that if G is a finitely generated torsion-free linear group then G has finite cohomological dimension if and only if there is a bound on the ranks of the abelian subgroups. So you might think this would be a good approach to proving Theorem 3.1.3: all one needs to do is show that there is a bound on the ranks of abelian subgroups of linear groups of type $(FP)_\infty$. But no, I do not know a way of carrying this approach through.

As a matter of fact, a form of Theorems 3.1.3 and 3.1.4 holds for any group in $\mathbf{H}\mathfrak{F}$, (see Theorems 3.4.1 and 3.4.2 below). By contrast, note the following example:

Lemma 3.2.2 *There exists a finitely generated torsion-free $\mathbf{H}\mathfrak{F}$-group of infinite cohomological dimension in which every abelian subgroup has rank at most 2.*

Proof. First, for each $n \geq 2$ let H_n be a torsion-free discrete cocompact subgroup of $SO(n,1)$. Thus H_n is the fundamental group of a closed hyperbolic n-manifold, and in particular it has cohomological dimension n and all its abelian subgroups are cyclic. Now let H be the free product of all the H_n. Then H belongs to $\mathbf{H}\mathfrak{F}$, as $\mathbf{H}\mathfrak{F}$ is closed under free products, and H is torsion-free of infinite cohomological dimension. We need to embed H into a finitely generated group. For this, the original approach for embedding countable groups into 2-generator groups using HNN-extensions is best because $\mathbf{H}\mathfrak{F}$ is closed under forming fundamental groups of graphs of groups. Following the proof of ([28], Theorem 11.37) one first forms the free product $H * F$ of H with a free group on two generators and then one forms an HNN-extension $G = (H * F)_{E,t}$ in which two carefully chosen copies of a free group E of countably infinite rank are made conjugate. With the right construction, H is a subgroup of a two generator subgroup of G. Now G is certainly a torsion-free $\mathbf{H}\mathfrak{F}$-group, and since the abelian subgroups of $H * F$ are cyclic, it is guaranteed that the abelian subgroups of G have cohomological dimension ≤ 2 and hence rank ≤ 2. Since subgroups of $\mathbf{H}\mathfrak{F}$-groups also belong to $\mathbf{H}\mathfrak{F}$, the 2-generator subgroup of G which contains H is an example of the kind required.

Regarding the class $\mathbf{H}\mathfrak{F}$ there is one useful elementary trick which is well known.

Lemma 3.2.3 *Let G be a group and let $G_0 < G_1 < G_2 < \cdots$ be a chain of subgroups indexed by the natural numbers such that $G = \bigcup_{n \geq 0} G_n$. Then there is an action of G on a tree such that every vertex and edge stabilizer is conjugate to one of the G_n.*

Proof. Let both the vertex set V and the edge set E be equal to the set of those subsets of G which are cosets $G_n g$ of some G_n. The initial and terminal vertices of a typical edge are defined by $\iota(G_n g) = G_n g$ and $\tau(G_n g) = G_{n+1} g$ respectively. G acts by right multiplication on the resulting graph, with stabilizers as required. The condition $G = \bigcup_{n \geq 0} G_n$ ensures the graph is connected and it is easy to see it has no loops, so it is a tree.

Corollary 3.2.4 *If G is a countable group whose finitely generated subgroups belong to $\mathbf{H}\mathfrak{F}$ then G is in $\mathbf{H}\mathfrak{F}$.*

Proof. The above Lemma can be applied by expressing G as the union of an ascending chain of finitely generated subgroups.

The argument breaks down if G is uncountable and this raises some curious questions about **H𝔉**. Certainly **H𝔉** contains groups of every cardinality because it contains all free groups, but I do not know which abelian groups belong to it. My first guess is that an abelian group belongs to **H𝔉** if and only if it has cardinality $\leq \aleph_\omega$: it would be intriguing to know the truth.

One last remark about **H𝔉**:

What really matters about groups G in **H𝔉** is that there are exact sequences

$$0 \to C_r \to \cdots \to C_1 \to C_0 \to \mathbb{Z} \to 0$$

of $\mathbb{Z}G$-modules, which are finite in length and in which each C_i is a direct sum of modules which are induced from subgroups of G which are simpler than G. Consequently, it is not necessary to have G acting on a contractible space: an acyclic space would be perfectly adequate. One might wonder whether this means we can use a larger class of groups. The following Proposition, pointed out to me by Mladen Bestvina, clarifies the situation..

Proposition 3.2.5 *The class* **H𝔉** *is not changed if one replaces 'contractible' in Definition 3.2.1 by 'acyclic'.*

Proof. Let G be a group with an admissible action on a finite dimensional acyclic space X. View G as a discrete (0-dimensional) space with G acting freely by right multiplication. Then G acts on the join $Y := X \# G$. Moreover, Y is still finite-dimensional, having dimension one more than X, every cell stabilizer for the action of G on Y is either equal to a cell stabilizer of the original action on X or is trivial, and since X is acyclic it follows that Y is contractible. The Proposition clearly follows from this remark.

3.3 Generalized Tate cohomology in brief

Tate cohomology was invented by Tate. He devised a theory for finite groups which subsumed the ordinary cohomology and the ordinary homology into a single cohomological functor and he was originally interested in applications to number theory. See Chapter XII of [13], or [2] for a good survey. From our point of view, Tate cohomology does not seem relevant because most certainly we are concerned only about infinite groups.

Subsequently, Farrell generalized Tate's theory so that it could be applied to any group G with $\mathrm{vcd}(G) < \infty$. As well as Farrell's paper [15], there is also a useful account in [10]. Farrell cohomology looks more relevant to our situation here, but since our theorems *conclude* with finite vcd it seems implausible that Farrell cohomology could be useful *en route*.

Vogel and Mislin have independently discovered a generalized Tate cohomology theory, [17, 24]. Mislin was strongly influenced by the paper [16] of Gedrich and Gruenberg in which the authors develop a theory of *terminal completions* of cohomological functors for certain classes of groups. A different approach to generalized Tate cohomology was discovered by Vogel [17] and by Benson and Carlson, [5]. The paper by Benson and Carlson is ostensibly about Tate cohomology (of finite groups), but they work with definitions which make sense for arbitrary groups and which turn out to yield a theory isomorphic to that of Vogel and Mislin. This theory, which we shall call the *complete cohomology*, turns out to be precisely the tool we need. We shall denote the cohomology groups by $\widehat{H}^j(G, M)$. Like ordinary cohomology, one can have coefficients in any G-module M, and each cohomology group is functorial in M. The theory shares basic properties with ordinary cohomology, but it also enjoys some distinctive features. One point to emphasize at once is that $\widehat{H}^j(G, M)$ is defined and can be non-zero for all integers j, positive and negative.

There are four crucial properties of complete cohomology relevant to this paper. First we state a property which one would expect of any cohomology theory.

Lemma 3.3.1 *There are natural long exact sequences of complete cohomology associated to short exact sequences of coefficient modules.*

Thus, for any short exact sequence $A \rightarrowtail B \twoheadrightarrow C$ of G-modules, there are natural connecting homomorphisms $\delta \colon \widehat{H}^j(G, C) \to \widehat{H}^{j+1}(G, A)$ which, together with functorially induced maps, give rise to a long exact sequence:

$$\cdots \to \widehat{H}^j(G, A) \to \widehat{H}^j(G, B) \to \widehat{H}^j(G, C) \to \widehat{H}^{j+1}(G, A) \to \cdots$$

The next property we need concerns the $(\mathrm{FP})_\infty$ property.

Lemma 3.3.2 *If G is of type $(\mathrm{FP})_\infty$ then the functors $\widehat{H}^j(G, \ )$ commute with direct sums of coefficient modules.*

More precisely, given a group G of type $(\mathrm{FP})_\infty$ and a family $(M_\lambda \mid \lambda \in \Lambda)$ of G-modules, the natural map

$$\bigoplus_{\lambda \in \Lambda} \widehat{H}^j(G, M_\lambda) \to \widehat{H}^j(G, \bigoplus_{\lambda \in \Lambda} M_\lambda)$$

is an isomorphism. As a matter of fact, this holds equally for ordinary cohomology of $(\mathrm{FP})_\infty$-groups.

The remaining two properties distinguish complete cohomology from ordinary cohomology. The first of these is very striking, and of great importance to us.

Lemma 3.3.3 $\widehat{H}^0(G, \mathbb{Z}) = 0$ *if and only if* $\mathrm{cd}(G) < \infty$

This property stands out because it ensures that complete cohomology need not be identically zero. It is wonderfully convenient that this zeroth cohomology group with trivial coefficients carries so much information. In general, not surprisingly, $\widehat{H}^0(G,\mathbb{Z})$ is very hard to compute for groups of infinite cohomological dimension. If G is finite then

$$\widehat{H}^0(G,\mathbb{Z}) = \mathbb{Z}/|G|\mathbb{Z}$$

but even for polycyclic-by-finite groups, no general formula is known.

The last property we need, like the first, is really axiomatic. This is built in to complete cohomology when it is defined, and the theory satisfies a universal property in relation to ordinary cohomology subject to this condition:

Lemma 3.3.4 *For all integers j, and all projective modules P, the complete cohomology groups $\widehat{H}^j(G,P)$ are zero.*

This property does not usually hold for ordinary cohomology. There are exceptions: for example if G is a free abelian group of infinite rank then $H^j(G,P)$ is zero for all j (including $j = 0$) and all projective modules P. For such groups, the complete cohomology and the ordinary cohomology coincide. More generally, if there is an integer j such that the ordinary cohomology vanishes on projectives from dimension j onwards, then the complete cohomology and the ordinary cohomology coincide from that point on.

3.4 Proofs of the theorems

We begin by proving the following result:

Theorem 3.4.1 *Every torsion-free* **HF**-*group of type* $(\mathrm{FP})_\infty$ *has finite cohomological dimension.*

Proof. Let G be a torsion-free **HF**-group of type $(\mathrm{FP})_\infty$. For a contradiction, suppose that G has infinite cohomological dimension. Then $\widehat{H}^0(G,\mathbb{Z})$ is non-zero by Lemma 3.3.3. Let α be least such that G belongs to $\mathbf{H}_\alpha \mathfrak{F}$. Consider the collection $\mathcal{O}$ of ordinals β such that there is an integer j and an $\mathbf{H}_\beta \mathfrak{F}$-subgroup H of G such that $\widehat{H}^j(G,\mathbb{Z}{\uparrow}_H^G)$ is non-zero. We shall reach a contradiction by proving that $\mathcal{O}$ contains zero, for this implies that there is an integer j such that $\widehat{H}^j(G,\mathbb{Z}G)$ is non-zero, contrary to Lemma 3.3.4. There are two steps.

Step 1. $\mathcal{O}$ is non-empty. In fact α belongs to $\mathcal{O}$, as can be seen by taking $H := G$ and $j := 0$.

Step 2. If $\beta > 0$ and $\beta \in \mathcal{O}$ then there exists $\gamma < \beta$ with $\gamma \in \mathcal{O}$. Here, we may suppose given an integer j and an $\mathbf{H}_\beta \mathfrak{F}$-subgroup H of G such that $\widehat{H}^j(G,\mathbb{Z}{\uparrow}_H^G)$ is non-zero. We may assume that β is a successor ordinal, and hence there is an admissible action of H on a finite dimensional contractible cell complex X such that for each cell σ of X, the isotropy group H_σ belongs to $\mathbf{H}_{\beta-1}\mathfrak{F}$. The augmented cellular chain complex of X is an exact sequence of $\mathbb{Z}H$-modules:

$$0 \to C_r \to \cdots \to C_1 \to C_0 \to \mathbb{Z} \to 0$$

Here, r is the dimension of the space X, and C_j is the free abelian group on the set of j-dimensional cells of X. The chain complex is exact because X is contractible. Given that H acts on X we thus have an exact sequence of $\mathbb{Z}H$-modules ending at the right with the trivial module $\mathbb{Z}$. By applying induction from H to G we obtain an exact sequence of G-modules:

$$0 \to C_r{\uparrow}_H^G \to \cdots \to C_1{\uparrow}_H^G \to C_0{\uparrow}_H^G \to \mathbb{Z}{\uparrow}_H^G \to 0$$

This new exact sequence ends at the right with the module $\mathbb{Z}{\uparrow}_H^G$. Since the jth complete cohomology of G is non-zero on this module, it follows by a very easy dimension shifting argument that there is an integer i such that $\widehat{H}^{j+i}(G, C_i{\uparrow}_H^G)$ is non-zero, (cf. Lemma 1 of [20] and (3.1) of [21]). Recall what C_i is: as an H-module, it is the permutation module coming from the action of H on the i-dimensional cells of X. Since the setwise stabilizer of each cell equals its pointwise stabilizer, C_i can be expressed as a direct sum of modules induced from the trivial module:

$$C_i = \bigoplus_\sigma \mathbb{Z}{\uparrow}_{H_\sigma}^H.$$

Here, σ runs through a set of H-orbit representatives of i-dimensional cells. Inducing to G, we have

$$C_i{\uparrow}_H^G = \bigoplus_\sigma \mathbb{Z}{\uparrow}_{H_\sigma}^G.$$

At this point we can use the fact that G is of type $(\mathrm{FP})_\infty$, so that by Lemma 3.3.2 its complete cohomology commutes with direct sums. In particular, since $\widehat{H}^{j+i}(G, C_i{\uparrow}_H^G)$ is non-zero, there must be a cell σ such that $\widehat{H}^{j+i}(G, \mathbb{Z}{\uparrow}_{H_\sigma}^G)$ is non-zero. But this completes the argument for *Step 2* because H_σ belongs to $\mathbf{H}_{\beta-1}\mathfrak{F}$ and hence $\beta - 1$ belongs to $\mathcal{O}$.

These steps complete the transfinite induction showing that $0 \in \mathcal{O}$, and the result is proved.

Now suppose that G is a linear group of type $(\mathrm{FP})_\infty$. By *linear*, we mean that G has a faithful finite dimensional representation over a field of characteristic zero. Since $(\mathrm{FP})_\infty$-groups are finitely generated, G is a finitely generated linear group and it follows from Selberg's Lemma (which is really a result of Minkowski) that G has a torsion-free subgroup H of finite index. Being of finite index, H is still of type $(\mathrm{FP})_\infty$, and it belongs to $\mathbf{H}\mathfrak{F}$, so by Theorem 3.4.1 it has finite cohomological dimension. Hence $\mathrm{vcd}(G) < \infty$. This proves Theorem 3.1.3.

To prove Theorem 3.1.4 requires just a little more work, for the simple reason that finitely generated soluble groups do not necessarily have torsion-free
subgroups of finite index. Because of this, Theorem 3.4.1 is not a good way
to get started, and we need a variation on it which will give some information
about non-torsion-free groups. In fact, by very similar arguments one can
prove

Theorem 3.4.2 *Every* **H𝔉**-*group of type* (FP)$_\infty$ *has finite cohomological dimension over* $\mathbb{Q}$.

In proving this one needs the complete cohomology criterion that a group
G has finite cohomological dimension over $\mathbb{Q}$ if and only if $\widehat{H}^0(G, \mathbb{Q}) = 0$, to
be used in place of Lemma 3.3.3. Otherwise the proof is essentially the same,
except that one works with coefficient ring $\mathbb{Q}$ in place of $\mathbb{Z}$ throughout.

One further result is needed:

Proposition 3.4.3 *Let* G *be a group of type* (FP)$_\infty$ *which has finite cohomological dimension over* $\mathbb{Q}$. *Then there is a bound on the orders of the finite
subgroups of* G.

Proof. Since G has finite cohomological dimension over $\mathbb{Q}$, we have
$\widehat{H}^0(G, \mathbb{Q}) = 0$. Using the (FP)$_\infty$ property, one can show that $\widehat{H}^0(G, \mathbb{Q}) =
\widehat{H}^0(G, \mathbb{Z}) \otimes \mathbb{Q}$. Combining these two, it follows that $\widehat{H}^0(G, \mathbb{Z})$ is torsion.
Now this zeroth complete cohomology group has a natural ring structure,
and being torsion, there is a positive integer n such that the ring homomor-
phism $\mathbb{Z} \to \widehat{H}^0(G, \mathbb{Z})$ has kernel $n\mathbb{Z}$. If H is any subgroup there is a ring
homomorphism $\widehat{H}^0(G, \mathbb{Z}) \to \widehat{H}^0(H, \mathbb{Z})$, (called the restriction map). So if H
is finite we obtain a ring homomorphism

$$\mathbb{Z}/n\mathbb{Z} \to \widehat{H}^0(G, \mathbb{Z}) \to \widehat{H}^0(H, \mathbb{Z}) = \mathbb{Z}/|H|\mathbb{Z}$$

and hence $|H|$ divides n.

Now suppose that G is a soluble group of type (FP)$_\infty$. By Theorem 3.4.2,
G has finite cohomological dimension over $\mathbb{Q}$. The structure of such groups
is very well understood. There is a chain of subgroups

$$N = G_0 \le G_1 \le \cdots \le G_n = H$$

where each G_i is normal in G, N is locally finite, H has finite index in G, and
G_{i+1}/G_i is torsion-free abelian of finite rank for $0 \le i < n$. This structure
for soluble groups of finite torsion-free rank was originally established by
Mal'cev [23]. It follows from Stammbach's work [29] that soluble groups of
finite cohomological dimension over $\mathbb{Q}$ also have this structure. Details can
also be found in [6] and [26]. Clearly we have that vcd(G/N) is finite. The
Proposition above shows that N is finite, and now it follows from abstract
group theoretic arguments that G is virtually torsion-free because it is not
difficult in this situation to find a subgroup of finite index in G which has
trivial intersection with N. Thus vcd$(G) < \infty$ as required.

4   Complete cohomology in detail

Throughout this section, let $\mathcal{C}$ and $\mathcal{D}$ be abelian categories. For convenience we always assume that $\mathcal{C}$ has both *enough projectives* and *enough injectives*. This means to say that for any object M of $\mathcal{C}$, there is an epimorphism $P \twoheadrightarrow M$ with P projective and a monomorphism $M \rightarrowtail J$ with J injective. We shall be concerned with cohomological functors from $\mathcal{C}$ to $\mathcal{D}$. The principal example to keep in mind is the case where G is a group, $\mathcal{C}$ is the category of $\mathbb{Z}G$-modules and $\mathcal{D}$ is the category of abelian groups. Then one has the familiar cohomological functors $H^n(G, \quad)$. More generally, if R is any ring and M is a fixed R-module, then we can take $\mathcal{C}$ to be the category of R-modules, and we can consider the cohomological functors $\mathrm{Ext}_R^n(M, \quad)$.

The term *cohomological functor* has a technical meaning which we recall in §4.1. Our goal is to describe Mislin's recent work [24] which associates a *completion* $\widehat{U}^*$ to any cohomological functor U^* from $\mathcal{C}$ to $\mathcal{D}$. Here, we call this the Mislin completion. We outline its construction, and indicate how to establish its key properties. We also include an outline of the different approach to the same theory discovered by Benson and Carlson [5].

4.1   Axioms

A *cohomological functor* from $\mathcal{C}$ to $\mathcal{D}$ consists of a family $(U^n \mid n \in \mathbb{Z})$ of additive functors satisfying the following two axioms:

Axiom 4.1.1 *For each $n \in \mathbb{Z}$ and each short exact sequence $A \rightarrowtail B \twoheadrightarrow C$ in $\mathcal{C}$, there is a natural connecting homomorphism $\delta \colon U^n(C) \to U^{n+1}(A)$.*

Naturality here means that if you have a commutative diagram

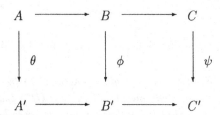

in which the rows are short exact sequences, then the resulting squares

$$
\begin{array}{ccc}
U^n(C) & \xrightarrow{\ \ \delta\ \ } & U^{n+1}(A) \\
\downarrow{\scriptstyle \psi_*} & & \downarrow{\scriptstyle \theta_*} \\
U^n(C') & \xrightarrow[\ \ \delta\ \ ]{} & U^{n+1}(A')
\end{array}
$$

commute. The second axiom is

Axiom 4.1.2 *For each short exact sequence $A \overset{\iota}{\rightarrowtail} B \overset{\pi}{\twoheadrightarrow} C$, the resulting long sequence*

$$\cdots \overset{\delta}{\to} U^n(A) \overset{\iota_*}{\to} U^n(B) \overset{\pi_*}{\to} U^n(C) \overset{\delta}{\to} U^{n+1}(A) \overset{\iota_*}{\to} \cdots$$

is exact.

Given cohomological functors (U^*) and (V^*) there is a notion of morphism $\nu: (U^*) \to (V^*)$. This consists of a family $(\nu^n \mid n \in \mathbb{Z})$ of natural transformations, $\nu^n: U^n \to V^n$ satisfying the following:

Axiom 4.1.3 *For each integer n and each short exact sequence $A \rightarrowtail B \twoheadrightarrow C$, the square*

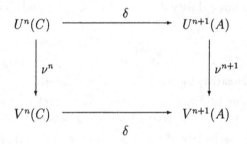

commutes.

The following general definition can be made for the Mislin completion.

Definition 4.1.4 *Given a cohomological functor (U^*) from C to D, its Mislin completion consists of a cohomological functor $(\widehat{U}^*)$ together with a morphism $(U^*) \to (\widehat{U}^*)$ such that the following conditions hold:*

1. *for all integers n and all projective modules P, $\widehat{U}^n(P) = 0$:*

2. *if (V^*) is any other cohomological functor which vanishes on projective modules then any morphism $(U^*) \to (V^*)$ factors uniquely through the given morphism $(U^*) \to (\widehat{U}^*)$.*

It is clear from general nonsense that complete cohomology is uniquely determined up to isomorphism by this definition. However, it is not yet clear that it exists.

4.2   A naive definition of generalized Tate cohomology groups

Here, we describe a simple method of defining generalized Tate cohomology groups which has the advantage that many properties are transparent. For group cohomology this definition was first mentioned by Benson and Carlson in [5]. In the same paper they also mention a more sophisticated formulation using hypercohomology, which we discuss in §4.4 below. At this stage it will not be clear that the cohomology groups fit together to form a cohomological functor.

For simplicity, we take $\mathcal{C}$ to be the category of modules over a ring R. Let M and N be right R-modules. The subset

$$\mathcal{P}_R(M, N)$$

of $\text{Hom}_R(M, N)$ consisting of homomorphisms which factor through a projective module is an additive subgroup and we shall denote the quotient by $[M, N]_R$ or simply $[M, N]$.

$$[M, N] := \text{Hom}_R(M, N)/\mathcal{P}_R(M, N).$$

It is easy to check that $[\ ,\ ]$ inherits a bifunctor structure from $\text{Hom}_R(\ ,\ )$.

For any R-module M, let FM denote the free module on the underlying set of M. Then F is a functor from R-modules to R-module, and moreover there is an obvious natural map $FM \to M$ determined by sending each free generator of FM to the corresponding element of M. Given an R-module map $\phi: M \to N$ we obtain a commutative diagram

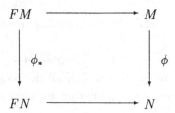

We write ΩM for the kernel of the map $FM \to M$. Then Ω is also a functor taking R-modules to R-modules, and the above square extends to the following commutative diagram:

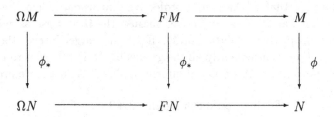

Although Ω itself is not an additive functor, it is still true and easy to check that the induced map

$$\mathrm{Hom}_R(M,N) \to \mathrm{Hom}_R(\Omega M, \Omega N),$$

is an additive homomorphism. (In the special case when $M = N$, this map is actually a ring homomorphism of the endomorphism rings.) Since it is clear that ΩP is projective whenever P is projective, the homomorphism carries $\mathcal{P}_R(M,N)$ into $\mathcal{P}_R(\Omega M, \Omega N)$ and hence it induces a map

$$[M,N] \to [\Omega M, \Omega N].$$

This process can be iterated so that one obtains a whole direct limit system

$$[M,N] \to [\Omega M, \Omega N] \to [\Omega^2 M, \Omega^2 N] \to [\Omega^3 M, \Omega^3 N] \to \cdots.$$

The zeroth Mislin Ext-group can now be defined.

Definition 4.2.1 $\widehat{\mathrm{Ext}}_R^0(M,N) := \varinjlim_i [\Omega^i M, \Omega^i N].$

In fact, we can write down a formula for $\widehat{\mathrm{Ext}}_R^n(M,N)$ where n is any integer:

Definition 4.2.2 $\widehat{\mathrm{Ext}}_R^n(M,N) := \widehat{\mathrm{Ext}}_R^0(M, \Omega^n N) = \varinjlim_i [\Omega^i M, \Omega^{i+n} N].$

Notice that although $\Omega^n N$ has no meaning for $n < 0$, the above definition makes perfect sense because in the direct limit, only a finite number of initial terms are undefined. Notice that for any integer i,j,k we have

$$\widehat{\mathrm{Ext}}_R^i(\Omega^j M, \Omega^k N) = \widehat{\mathrm{Ext}}_R^{i-j+k}(M,N) = \widehat{\mathrm{Ext}}_R^0(\Omega^{-i+j-k} M, N)$$
$$= \widehat{\mathrm{Ext}}_R^0(M, \Omega^{i-j+k} N).$$

For a group G and $\mathbb{Z}G$-module M, the complete cohomology can be defined in the obvious way:

$$\widehat{H}^n(G,M) := \widehat{\mathrm{Ext}}_{\mathbb{Z}G}^n(\mathbb{Z},M).$$

At this stage we cannot easily establish all the properties of the new cohomology theory. However, some properties are transparent. The first is the general form of Lemma 3.3.4.

Lemma 4.2.3 *If either one of M and N has finite projective dimension over R then $\widehat{\mathrm{Ext}}_R^*(M,N) = 0$. In particular, complete cohomology vanishes on projective modules.*

Proof. Suppose that M has finite projective dimension. Then there is an integer i such that $\Omega^i M$ is projective, because the $\Omega^i M$ arise as successive kernels in a projective resolution of M. If j is any larger integer then $\Omega^j M$ is still projective. Consequently all the groups $[\Omega^j M, \Omega^j N]$ are zero for $j \geq i$ and the result follows. A similar argument applies if N has finite projective dimension.

It is also not difficult to establish Lemma 3.3.3.

Lemma 4.2.4 *The following are equivalent.*

1. $\widehat{\mathrm{Ext}}_R^0(M, M) = 0$:

2. M has finite projective dimension.

Proof. That $(2) \Rightarrow (1)$ follows from the previous Lemma. Now suppose that (1) holds. Then there must be an integer i such that the identity map on M becomes zero in $[\Omega^i M, \Omega^i M]$ otherwise it would survive to be a non-zero element in the direct limit. But this simply means that the identity map on $\Omega^i M$ factors through a projective and so $\Omega^i M$ itself is projective. Hence M has projective dimension at most i.

Lemma 3.3.3 now follows directly because a group G has finite cohomological dimension if and only if the trivial module $\mathbb{Z}$ has finite projective dimension over $\mathbb{Z}G$, and $\widehat{H}^0(G, \mathbb{Z}) = \widehat{\mathrm{Ext}}_{\mathbb{Z}G}^0(\mathbb{Z}, \mathbb{Z})$.

4.3 Mislin's approach via satellites

There are two natural approaches to complete cohomology. We first consider that taken by Mislin [24] and this section is closely based Mislin's paper. We refer the reader to [24] for further details, and in particular for the proof that the complete cohomology groups agree with the naive definition discussed above. The theory depends on the manufacture of a cohomological functor from a single half-exact functor. A functor U from $\mathcal{C}$ to $\mathcal{D}$ is called *half-exact* if it is additive and if, for any short exact sequence $A \rightarrowtail B \twoheadrightarrow C$ in $\mathcal{C}$, the induced sequence

$$U(A) \to U(B) \to U(C)$$

is exact at $U(B)$. Given a half-exact functor U, one can define a cohomological functor $(S^n U \mid n \in \mathbb{Z})$ by means of *satellites*. For $n < 0$, the functors $S^n U$ are called left satellites and for $n > 0$ they are called right satellites. The case $n = 0$ is covered by defining $S^0 U = U$. For a good exposition, see Chapter III of [13]. It is important not to confuse the concept of left and right satellites with that of left and right derived functors — the latter can be quite different. It is perhaps worth noting that left and right satellites are defined for any additive functor U, even if it is not half-exact.

To set up complete cohomology, it turns out that one only needs left satellites, but for clarity, I will recall the definition of the right satellites as well. To construct the left satellites one needs that $\mathcal{C}$ has enough projectives. To construct the right satellites one needs that $\mathcal{C}$ has enough injectives. First the left satellites: for any R-module M, choose a short exact sequence $K \rightarrowtail P \twoheadrightarrow M$ in which P is projective. Then define

$$S^{-1}U(M) = \ker(U(K) \to U(P)).$$

One has to show that this definition does not depend on the choice of short exact sequence, and that $S^{-1}U$ can be made into an additive functor in a natural way. This is straightforward: see Chapter III, §1 of [13]. Now the sequence of left satellites $S^{-2}U, S^{-3}U, \ldots$ can be defined inductively, each being the left satellite, as defined above, of the preceding one.

The zeroth satellite is simply defined by

$$S^0 U := U.$$

For the right satellites, choose a short exact sequence $M \rightarrowtail J \twoheadrightarrow L$ in which J is injective and set

$$S^1 U(M) := \operatorname{coker}(U(J) \to U(L)).$$

Iterating this process one obtains the sequence of right satellites. Taken as a whole, it is possible to define connecting homomorphisms so that Axiom 4.1.1 and half of Axiom 4.1.2 hold. We refer the reader to [13] for details, but to give a hint, one can define the map $\delta \colon S^{-1}U(C) \to U(A)$ associated to a short exact sequence $A \rightarrowtail B \twoheadrightarrow C$ by first choosing a short exact sequence $K \rightarrowtail P \twoheadrightarrow C$ with P projective, then defining maps $P \to B$ and $K \to A$ to make a commutative diagram

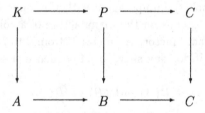

and then defining δ to be the composite map $S^{-1}U(C) \to U(K) \to U(A)$. It takes some diagram chasing to check that this is well-defined. Similar definitions can be made for the connecting homomorphisms in other dimensions. The satellites together with the connecting homomorphisms not only satisfy Axiom 4.1.1 but also satisfy half of Axiom 4.1.2: namely, in the long sequence

$$\cdots \to S^n U(A) \to S^n U(B) \to S^n U(C) \to S^{n+1} U(A) \to \cdots,$$

the composite of any two consecutive maps is zero. A system of functors satisfying one and a half of the axioms in this way is called a *connected sequence of functors*, (cf. Chapter III, §4 of [13]). We recall one of the fundamental results in this theory, ([13], Theorem 3.1 of Chapter III).

Theorem 4.3.1 *If U is a half-exact functor then the connected sequence of functors $(S^n U \mid n \in \mathbb{Z})$ described above is a cohomological functor.*

If all this seems very abstract, keep in mind that the familiar cohomology functors $H^n(G, \quad)$ of a group G are all half-exact. Ultimately we shall be interested in what happens when Theorem 4.3.1 is applied to each of these.

In general, suppose we are given a cohomological functor (U^*) from $\mathcal{C}$ to $\mathcal{D}$. Then there is automatically a supply of half-exact functors, namely the U^n. Let k be a fixed integer and consider the cohomological functor obtained by forming the satellites $S^n U^k$ of U^k. Theorem 4.3.1 shows that together these form a cohomological functor $(S^* U^k)$ from $\mathcal{C}$ to $\mathcal{D}$. It is convenient to re-index these, so we consider the zeroth functor to be $S^{-k} U^k$. With this indexing, notice that the kth functor $S^0 U^k$ is equal to the kth functor we had to begin with, namely U^k. Because of this we can define a perfectly good cohomological functor which coincides with (U^*) in dimensions $\geq k$ and where we make use of the satellites only in dimensions $\leq k$. Thus we define

$$
\begin{aligned}
U_{(k)}^n &= S^{-k+n} U^k, \ n \leq k \\
&= U^n, \ n \geq k.
\end{aligned}
$$

Now given a cohomological functor (U^*) it is easy to define a natural transformation $U^0 \to S^{-1} U^1$. Briefly, for any module M, choose a short exact sequence $K \rightarrowtail P \twoheadrightarrow M$ with P projective. The connecting homomorphism $U^0(M) \to U^1(K)$ carries $U^0(M)$ onto the kernel of the map $U^1(K) \to U^1(P)$, and since this kernel is, by definition, $S^{-1} U^1(M)$ we get a map $U^0(M) \to S^{-1} U^1(M)$. It is easy to check that this defines a natural transformation as required.

One can iterate this process to obtain a whole sequence of natural transformations

$$U^0 \to S^{-1} U^1 \to S^{-2} U^2 \to S^{-3} U^3 \to \cdots.$$

The direct limit gives a definition of $\hat{U}^0$, the zeroth functor for the Mislin completion of (U^*). But this on its own is not satisfactory. One wants to get all the functors together with the Axioms 4.1.1, 4.1.2, and the requirements of Definition 4.1.4. To do this we need to appeal to the following crucial Theorem, which is a special case of Proposition 5.2, Chapter III of [13]. Perhaps it is worth emphasizing now that it is precisely because there are theorems of this kind that there is a real advantage in setting up the theory of cohomological functors with the technical Axioms and Definitions in §4.1.

Theorem 4.3.2 *Let (W^*) and (V^*) be cohomological functors from $\mathcal{C}$ to $\mathcal{D}$. Let k be an integer. Suppose that V^n vanishes on projective modules for all $n < k$. Let $\nu^j : W^j \to V^j$ be natural transformations defined for $j \geq k$ which commute with connecting homomorphisms as in Axiom 4.1.3 where this makes sense. Then the ν^j extend uniquely to a morphism $\nu : (W^*) \to (V^*)$.*

For each integer k there is a natural map $U^k \to S^{-1}U^{k+1}$. Using the Theorem above we can extend this to a morphism $(U_{(k)}^*) \to (U_{(k+1)}^*)$, first by using the identity natural transformations $U_{(k)}^n \to U_{(k+1)}^n$ when $n > k$ and then appealing to the Theorem for $n < k$. Thus we now have a direct limit system of cohomological functors:

$$(U_{(0)}^*) \to (U_{(1)}^*) \to (U_{(2)}^*) \to (U_{(3)}^*) \to \cdots,$$

and the direct limit inherits the structure of a cohomological functor. (Note: it certainly inherits the structure of a connected sequence of functors, but because direct limits preserve exactness, one sees that the limit satisfies Axiom 4.1.2 in full.) The Mislin completion is defined by

$$(\hat{U}^*) := \varinjlim_{k} (U_{(k)}^*).$$

This can now be shown to satisfy the conditions of Definition 4.1.4. One can appeal to Theorem 4.3.2 in order to check that the universal property (2) of 4.1.4 holds. Actually it is easy to see that any morphism $(U^*) \to (V^*)$ of cohomological functors induces a morphism $(\hat{U}^*) \to (\hat{V}^*)$ of their Mislin completions so that one obtains a commutative square

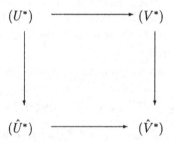

giving a nice generalized version of the universal property.

It is not difficult to show that if R is a ring and M is a fixed R-module then the functors $\widehat{\mathrm{Ext}}_R^*(M, \ )$ obtained by applying this satellite approach to $(\mathrm{Ext}_R^*(M, \ ))$ coincide with the functors described in §4.2. I leave this as an exercise.

The satellite functor approach provides a good way of seeing that the complete cohomology of a group of type $(FP)_\infty$ commutes with direct sums, so proving Lemma 3.3.2. The point is that if an additive functor commutes with direct sums then so do its left satellites. Since the Mislin completion is built as a direct limit of left satellites it inherits the property. Thus Lemma 3.3.2 follows directly from the fact that the corresponding result holds for ordinary cohomology.

4.4 Hypercohomology and the Benson-Carlson-Vogel approach

Here, we sketch a different view of complete cohomology based on hypercohomology of chain complexes. This approach was first described by Goichot in [17] and was also discovered by Benson and Carlson in [5]. It has the advantage that some properties are more transparent. On the other hand it is less obvious that it satisfies the universal property of Definition 4.1.4.

Further information about hypercohomology can be found at the end of [13]. It is also mentioned in [3] and [4], and a nice account can be found in the section on bicomplexes in Chapter 11 of [27]. The approach is related to the more sophisticated theory of triangulated and derived categories which we do not touch on: for an excellent treatment of this topic, see Hartshorne's book [18]. Here, we can proceed without all the general machinery of derived categories essentially because we are working with abelian categories which have enough projectives.

Throughout this section we work with the category of modules over a ring R, and consider chain complexes of R-modules which are bounded below.

Given two chain complexes $\mathbf{C}$ and $\mathbf{D}$, we define three chain complexes called the *total complex*, the *hypercohomology complex* and the *Vogel complex*.

First we define the hypercohomology complex. This is written $\mathrm{Hom}_{\mathbf{R}}(\mathbf{C}, \mathbf{D})$, the group of n-chains is defined by

$$\mathrm{Hom}_{\mathbf{R}}(\mathbf{C}, \mathbf{D})_{\mathbf{n}} := \prod_{\mathbf{p+q=n}} \mathrm{Hom}_{\mathbf{R}}(\mathbf{C}_{-\mathbf{p}}, \mathbf{D}_{\mathbf{q}}),$$

and the differential $\mathbf{d} : \mathrm{Hom}_{\mathbf{R}}(\mathbf{C}, \mathbf{D})_{\mathbf{n}} \to \mathrm{Hom}_{\mathbf{R}}(\mathbf{C}, \mathbf{D})_{\mathbf{n-1}}$ is defined by

$$\mathbf{d}(\phi) := d' \circ \phi - (-1)^n \phi \circ d,$$

where d and d' denote the differentials in $\mathbf{C}$ and $\mathbf{D}$ respectively.

The total complex $\mathrm{Tot}(\mathbf{C}, \mathbf{D})$ is the subcomplex of the hypercohomology complex in which the nth chain group is given by

$$\mathrm{Tot}(\mathbf{C}, \mathbf{D})_n := \bigoplus_{p+q=n} \mathrm{Hom}_R(C_{-p}, D_q).$$

The Vogel complex is defined to be the quotient of the hypercohomology complex by the total complex. Thus one has a short exact sequence of chain complexes:

$$\mathrm{Tot}(\mathbf{C}, \mathbf{D}) \rightarrowtail \mathrm{Hom}_{\mathbf{R}}(\mathbf{C}, \mathbf{D}) \twoheadrightarrow \widehat{\mathrm{Hom}}_{\mathbf{R}}(\mathbf{C}, \mathbf{D}),$$

with the Vogel complex at the right. The key result about this is as follows:

Theorem 4.4.1 *Let* $\mathbf{P}$ *and* $\mathbf{Q}$ *be non-negative chain complexes of projective R-modules, and suppose that both have homology concentrated in degree zero, with* $H_0(\mathbf{P}) = M$ *and* $H_0(\mathbf{Q}) = N$. *(In effect,* $\mathbf{P}$ *and* $\mathbf{Q}$ *are projective resolutions of M and N respectively.) Then*

1. the nth homology group of $\mathbf{Hom_R(P,Q)}$ *is equal to* $\text{Ext}_R^{-n}(M,N)$*; and*

2. the nth homology group of $\widehat{\mathbf{Hom}}_R(\mathbf{P},\mathbf{Q})$ *is equal to* $\widehat{\text{Ext}}_R^{-n}(M,N)$*.*

One advantage of this approach is that it enables one to establish properties of the Mislin Ext groups as functors of the left hand variable. For example it becomes clear that there are long exact sequences just as for the ordinary Ext. Thus if $A \rightarrowtail B \twoheadrightarrow C$ is a short exact sequence of R-modules, there is a long exact sequence

$$\cdots \to \widehat{\text{Ext}}_R^n(C,N) \to \widehat{\text{Ext}}_R^n(B,N) \to \widehat{\text{Ext}}_R^n(A,N) \to \widehat{\text{Ext}}_R^{n+1}(C,N) \to \cdots$$

which can be established by resolving the short exact sequence $A \rightarrowtail B \twoheadrightarrow C$ with a short exact sequence of projective resolutions.

The short exact sequence of chain complexes mentioned above sheds interesting light on complete cohomology because, as observed by Goichot [17], on passage to homology it yields a long exact sequence:

$$\cdots \to X_{-n} \to \text{Ext}_R^n(M,N) \to \widehat{\text{Ext}}_R^n(M,N) \to X_{-n+1} \to \cdots$$

where the X_* are the homology groups of the total complex. The linking maps from ordinary cohomology to complete cohomology coincide with the natural map described in Definition 4.1.4, and the long exact sequence shows that the homology of the total complex is critical in understanding the difference between complete cohomology and ordinary cohomology.

5   Conjectures, problems and recent developments

5.1   H$\mathfrak{F}$-groups of type (FP)$_\infty$

Although it is tempting to ask the question, (FP)$_\infty$-groups do not always have finite vcd. Important examples in [9], [12] of Brown and Geoghegan demonstrate this forcefully: these examples are torsion-free and of infinite cohomological dimension over $\mathbb{Z}$. It makes more sense to ask the question if one restricts to the class of H$\mathfrak{F}$-groups. However, even here there are counterexamples. A key example [25] of Raghunathan is of type (FP)$_\infty$, since it has a finite central subgroup with arithmetic quotient, but it is not virtually torsion-free and hence cannot have finite vcd. This example suggests that in general one should not try to demand finite vcd. Instead, let $\mathbf{K_1\mathfrak{F}}$ denote the class of groups which admit properly discontinuous cocompact actions on finite dimensional contractible cell-complexes. Then the strongest conjecture one could make is as follows:

Conjecture 5.1.1 *Every* H$\mathfrak{F}$*-group of type* (FP)$_\infty$ *belongs to* $\mathbf{K_1\mathfrak{F}}$*.*

If proved, this would complete the story, for the converse certainly holds: every $\mathbf{K_1\mathfrak{F}}$ both belongs to $\mathbf{H\mathfrak{F}}$ and is of type $(\mathrm{FP})_\infty$. However, with present methods it is far too strong for there to be anything useful one can say. Notice that it implies, for example, that every group of type (FP) is finitely presented and of type (FL), and this in turn is completely unresolved. There are reasonable prospects for proving Conjecture 5.1.1 for soluble-by-finite groups.

Here is more tractable looking conjecture.

Conjecture 5.1.2 *Every* $\mathbf{H\mathfrak{F}}$*-group of type* $(\mathrm{FP})_\infty$ *belongs to* $\mathbf{H_1\mathfrak{F}}$.

To me, this seems a very natural generalization of Theorems 3.1.3 and 3.1.4 because $\mathbf{H_1\mathfrak{F}}$ is a natural generalization of the class of groups of finite vcd. Indeed there is a perfectly sensible *algebraic dimension* for groups in this class, defined to be the greatest integer n for which there is a projective $\mathbb{Z}G$-module P such that $H^n(G, P)$ is non-zero. The following Lemma shows that such an n always exists, and that it coincides with the vcd when the latter is finite.

Lemma 5.1.3 *If G has an admissible properly discontinuous action on a contractible space X of dimension d then*

1. *there is a projective $\mathbb{Z}G$-module P and a non-negative integer $n \leq d$ such that $H^n(G, P)$ is non-zero:*

2. *for all integers $m > d$ and all projective modules P, $H^m(G, P) = 0$:*

3. *if G has finite vcd then the greatest integer satisfying* (1) *is equal to* $\mathrm{vcd}(G)$.

Proof.

1. By Theorem 2 of [22] there is a $\mathbb{Z}G$-module I of projective dimension at most d such that $H^0(G, I)$ is non-zero. Let

$$0 \to P_d \to \cdots \to P_1 \to P_0 \to I \to 0$$

be a projective resolution of I. Since $H^0(G, I)$ is non-zero, Lemma 1 of [20] shows that for some n with $0 \leq n \leq d$, $H^n(G, P_n)$ is non-zero.

2. Let P be a projective module and let

$$0 \to C_d \to \cdots \to C_1 \to C_0 \to \mathbb{Z} \to 0$$

be the augmented cellular chain complex of X. Then there is a first quadrant spectral sequence

$$E_1^{p,q} = \mathrm{Ext}_{\mathbb{Z}G}^q(C_p, P) \Rightarrow H^{p+q}(G, P).$$

(This is the cohomology version of the spectral sequence which Brown mentions as (7.10), Chapter VII of [10].) In our situation, it follows from the Eckmann-Shapiro Lemma that $E_1^{p,q} = 0$ whenever $q > 0$ because all isotropy groups are finite and P is projective. Hence the spectral sequence collapses and the result follows.

3. Let H be a subgroup of finite index in G with finite cohomological dimension equal to n. Note that we now have $\text{vcd}(G) = n$. For any free $\mathbb{Z}H$-module F, $H^j(G, F \otimes_{\mathbb{Z}H} \mathbb{Z}G)$ is isomorphic to $H^j(H, F)$. From this it follows that $H^j(G, \ )$ vanishes on free modules (and hence also on projective modules) for $j > n$ and also that there exists a free G-module E with $H^n(G, E)$ non-zero. Thus the integer determined by (1) is the same as n.

In view of this Lemma, it is clear that Conjecture 5.1.2 implies the following:

Conjecture 5.1.4 *If G is an $\mathbf{H}\mathfrak{F}$-group of type* $(\text{FP})_\infty$ *then there is an integer m such that for all $n \geq m$, $H^n(G, \mathbb{Z}G) = 0$.*

My efforts so far to prove Conjecture 5.1.4 have been unsuccessful. If true one would be able to address a problem of Bieri for $\mathbf{H}\mathfrak{F}$-groups of type $(\text{FP})_\infty$. Of the four conjectures mentioned in this section, the following is the weakest:

Conjecture 5.1.5 *If G is an $\mathbf{H}\mathfrak{F}$-group of type* $(\text{FP})_\infty$ *then every torsion-free subgroup of G has finite cohomological dimension.*

Here is an outline of the proof that Conjecture 5.1.4 implies Conjecture 5.1.5. Let G and m be chosen according to Conjecture 5.1.4. Let H be a torsion-free subgroup of G. Let n be any integer greater than m and suppose that M is an H-module such that $H^n(H, M)$ is non-zero. By the Eckmann-Shapiro Lemma, $H^n(H, M) \cong H^n(G, \text{Coind}_H^G(M))$. A variation on the proof of Theorem 3.4.1 can be used to deduce from the non-vanishing of $H^n(G, \text{Coind}_H^G(M))$ that there is an integer $j \geq n$ and a finite subgroup F of G with $H^j(G, (\text{Coind}_H^G(M)) \otimes_{\mathbb{Z}F} \mathbb{Z}G)$. Since H is torsion-free, Mackey decomposition shows that $\text{Coind}_H^G(M) \otimes_{\mathbb{Z}F} \mathbb{Z}G$ is of the form $V \otimes \mathbb{Z}G$ for some torsion-free abelian group V. Using the fact that G is an $(\text{FP})_\infty$-group it follows that $H^j(G, \mathbb{Z}G)$ is non-zero, contrary to the choice of m. Thus $H^n(H, M)$ vanishes for all modules M and H has cohomological dimension less than n.

Thus the four conjectures are related:

$$(5.1.1) \Rightarrow (5.1.2) \Rightarrow (5.1.4) \Rightarrow (5.1.5)$$

but there is no proof of even the weakest of the four.

5.2 Centres and central factor groups

For various reasons it is interesting to ask whether the quotient of an **H𝔉**-group by a central subgroup belongs to **H𝔉**. Here, we consider just one of the reasons: namely a long-standing problem of Robert Bieri. For simplicity I will state this as a conjecture, though it has to be said that the evidence is rather flimsy.

Conjecture 5.2.1 *If G is an* $(FP)_\infty$*-group then the centre of G is finitely generated.*

This is known to be true if G is soluble or linear. We show here that it is true for any poly-linear group: that is any group G with a series of finite length in which the factors are linear. The key to this, and the most substantial evidence for the conjecture is as follows:

Proposition 5.2.2 *Let G be a group of type* $(FP)_\infty$*. If Z is a central subgroup such that G/Z belongs to* **H𝔉** *then Z is finitely generated.*

Proof. The class **H𝔉** contains all countable abelian groups and in particular it contains Z. Since **H𝔉** is extension closed and G/Z belongs to **H𝔉**, it follows that G is in **H𝔉**. By Theorem 3.4.2, G has finite cohomological dimension over $\mathbb{Q}$. In particular, Z has finite cohomological dimension over $\mathbb{Q}$ and hence there is a finitely generated subgroup Z_0 of Z such that Z/Z_0 is torsion. It can be shown that the quotient of an $(FP)_\infty$-group by a finitely generated central subgroup inherits the $(FP)_\infty$ property. Hence G/Z_0 is of type $(FP)_\infty$. Moreover, being an extension of Z/Z_0 by G/Z, the group G/Z_0 belongs to **H𝔉**. Theorem 3.4.2 shows that G/Z_0 has finite cohomological dimension over $\mathbb{Q}$ and it follows from Proposition 3.4.3 that there is a bound on the orders of the finite subgroups of G/Z_0. Hence Z/Z_0 is finite and Z is finitely generated.

Given this, the following suffices to prove 5.2.1 for poly-linear groups.

Proposition 5.2.3 *If G is a countable poly-linear group and Z is a central subgroup then G/Z belongs to* **H𝔉**.

Proof. Suppose first that G is linear. Then the centre $\zeta(G)$ is closed in the Zariski topology so $G/\zeta(G)$ is also linear. Hence $G/\zeta(G)$ belongs to **H𝔉** and it follows that G/Z belongs to **H𝔉**, because it is an extension of $\zeta(G)/Z$ by $G/\zeta(G)$.

In general, let $\mathfrak{X}$ denote the class of groups G with the property that whenever Z is a central subgroup then G/Z is in **H𝔉**. Since **H𝔉** is extension closed, so is $\mathfrak{X}$. We have proved that linear groups belong to $\mathfrak{X}$, and hence so do poly-linear groups. This proves the Proposition.

5.3 Note added in March 1994

Recent joint work [14] of the author and Jonathan Cornick has led to improvements of several results discussed here. In particular, Conjecture 5.1.5 is proved. The situation regarding linear groups is clarified: results of [14] confirm that countable linear groups belong to **H$\mathfrak{F}$** regardless of characteristic. This is convenient because both in this article and in [21] the term *linear* is used to mean linear *in characteristic zero*. In fact it now turns out that the main results discussed here hold for all linear groups. One of the key results of [14] is the following improvement of Theorem 3.1.3:

Theorem 5.3.1 *If G is a residually finite* **H$\mathfrak{F}$**-*group of type* (FP)$_\infty$ *then* vcd(G) $< \infty$.

Since finitely generated linear groups are residually finite we can state the following Corollary in which *linear* can be interpreted in the conventional way, allowing for any characteristic.

Corollary 5.3.2 *If G is a linear group of type* (FP)$_\infty$ *then* vcd(G) $< \infty$.

6 Acknowledgements

I am indebted to Karl Gruenberg for drawing my attention to Mislin's work [24]. I thank Karl, and also Jon Carlson, Jeremy Rickard and Guido Mislin for useful discussions about complete cohomology.

References

[1] R. C. Alperin and P. B. Shalen, 'Linear Groups of Finite Cohomological Dimension', *Invent. Math.*, **66** (1982), 89–98.

[2] M. F. Atiyah and C. T. C. Wall, 'Cohomology of groups', **in:** (eds. J. W. S. Cassels and A Frölich) *Algebraic Number Theory*, (Academic Press, London, 1967).

[3] D. J. Benson, *Representations and Cohomology I: Basic representation theory of finite groups and associative algebras*, Cambridge Studies in Advanced Mathematics 30, (Cambridge U.P. 1991).

[4] D. J. Benson, *Representations and Cohomology II: Cohomology of groups and modules*, Cambridge Studies in Advanced Mathematics 31, (Cambridge U.P. 1991).

[5] D. J. Benson and J. Carlson, 'Products in negative cohomology', *J. Pure Appl. Algebra* **82** (1992), 107–130.

[6] R. Bieri, *Homological Dimension of Discrete Groups*, Mathematics Notes (Queen Mary College, London, 2nd ed. 1981).

[7] R. Bieri and J. R. J. Groves, 'Metabelian groups of type $(FP)_\infty$ are virtually of type (FP)', *Proc. London Math. Soc.* (3) **45** (1982), 365–384.

[8] R. Bieri and R. Strebel, 'Valuations and finitely presented metabelian groups', *Proc. London Math. Soc.* (3) **41** (1980), 439–464.

[9] K. S. Brown, 'Finiteness properties of groups', *J. Pure Appl. Algebra* **44** (1987), 45–75.

[10] K. S. Brown, *Cohomology of Groups,* Graduate Texts in Math. **87** (Springer, Berlin, 1982).

[11] K. S. Brown, *Buildings,* (Springer, Berlin, 1989).

[12] K. S. Brown and R. Geoghegan, 'An infinite dimensional $(FP)_\infty$-group', *Invent. Math.* **77** (1984), 367–381.

[13] H. Cartan and S. Eilenberg, *Homological Algebra,* (Oxford U.P. 1956).

[14] J. Cornick and P. H. Kropholler, *Some cohomological properties of a new class of groups,* in preparation.

[15] F. T. Farrell, 'An extension of Tate cohomology to a class of infinite groups', *J. Pure Appl. Algebra* **10** (1977), 153–161.

[16] T. V. Gedrich and K. W. Gruenberg, 'Complete cohomological functors on groups', *Topology and its Applications* **25** (1987), 203–223.

[17] F. Goichot, *Homologie de Tate-Vogel équivariante*, J. Pure Appl. Algebra **82** (1992), 39–64.

[18] R. Hartshorne, *Residues and Duality*, Lecture Notes in Mathematics 20 (Springer, Berlin, 1966).

[19] P. J. Hilton and U. Stammbach, *A Course in Homological Algebra*, Graduate Texts in Mathematics 4, (Springer, Berlin, 1970).

[20] P. H. Kropholler, 'Soluble groups of type $(FP)_\infty$ have finite torsion-free rank', *Bull. London Math. Soc.* to appear.

[21] P. H. Kropholler, 'On groups of type $(FP)_\infty$', *J. Pure Appl. Algebra* to appear.

[22] P. H. Kropholler and O. Talelli, 'On a property of fundamental groups of graphs of finite groups', *J. Pure Appl. Algebra* **74** (1991), 57-59.

[23] A. I. Mal'cev, 'On certain classes of infinite soluble groups', *Math. Sb.* **28** (1951), 567–588 = *Amer. Math. Soc. Translations* (2) **2** (1956), 1–21.

[24] G. Mislin, 'Tate cohomology for arbitrary groups via satellites', to appear in *Topology and its Applications*.

[25] M. S. Raghunathan, 'Torsion in cocompact lattices in coverings of $Spin(2, n)$', *Math. Ann.* **266** (1984), 403–419.

[26] D. J. S. Robinson, *A Course in the Theory of Groups*, Graduate Texts in Mathematics **80** (Springer, Berlin, 1982).

[27] J. J. Rotman, *An introduction to homological algebra,* Pure and Applied Mathematics **85**, (Academic Press, London, 1970).

[28] J. J. Rotman, *The Theory of Groups*, (Allyn and Bacon, Boston 1973).

[29] U. Stammbach, 'On the weak (homological) dimension of the group algebra of solvable groups', *J. London Math. Soc.* (2) **2** (1970), 567–570.

School of Mathematical Sciences
Queen Mary and Westfield College
Mile End Road
London E1 4NS
U.K.
E-mail: P.H.Kropholler@qmw.ac.uk

Tree-lattices and lattices
in Lie groups

ALEXANDER LUBOTZKY [1]

Let G be a locally compact group. A discrete subgroup Γ of G is called a *lattice* of G if G/Γ carries a G-invariant finite measure. Γ is a *uniform lattice* if G/Γ is compact.

The theory of lattices in semi-simple Lie groups has been a central focus of interest in the last three decades accumulating into a remarkable theory. In recent years the first steps toward a theory of lattices in the automorphism group $A = Aut(X)$ of a locally finite tree X have been carried out.

An intermediate case between G and A is played by a simple algebraic K-group H of K-rank one over a local non-archimedean field K. Examples of such H are $SL_2(\mathbb{Q}_p)$ or $SL_2(\mathbb{F}_p((t)))$. The group H acts on its associated Bruhat-Tits tree T. H is not the full group of automorphisms of T but it is cocompact in $Aut(T)$. This implies that every lattice of H is also a lattice of $A = Aut(T)$, but not vice versa. An interesting feature of the theory of lattices of A is that lattices coming from H behave quite differently from the general lattices. See for example section 2.

In this paper we survey and compare properties of lattices in G, H and $A = Aut(X)$. We consider the aspects of existence (and arithmeticity), finite generation, covolumes, the geometry and combinatorics of the fundamental domains, residual finiteness, the commensurability group, super-rigidity and the congruence subgroup problem.

There are additional topics for which it is interesting to compare A to G and H and some work has been done in those directions; e.g., classifying the irreducible representations (cf. [FTN]), the Howe-Moore property ([LM]) and mixing of all orders ([Mz1], [LM]). While these topics have consequences for lattices, we limit ourselves in this survey to those aspects which directly deal with discrete subgroups of A, H and G.

This survey is based on a lecture given at the conference on combinatorial group theory which took place in Edinburgh in March 1993. We are grateful to the organizers for the excellent conference and the help in preparing the paper.

We introduce the following conventions:

- G is a non-compact simple real Lie group (e.g., $G = PSL_n(\mathbb{R})$)

- H is the K-points of a simple K-rank one algebraic group $\mathbf{H}$ defined over a local non-archimedean field K (e.g., $H = PSL_2(\mathbb{F}_p((t)))$).

[1] Partially sponsored by the Edmund Landau Center for Research in Mathematical Analysis.

- *A is the full group of automorphisms of a locally finite tree X. We will assume that $A\backslash X$ is finite. This implies that a lattice in A is also an X-lattice (in the sense of [BK] and [BL]) and vice versa. An important example is $A = Aut(X_k)$ where X_k is the k-regular tree.*

By our assumptions G and H are simple groups (as abstract groups). A is not necessarily a simple group but by a result of Tits [Ti] it has a large simple normal subgroup A^+. For example for a k-regular tree $(A : A^+) = 2$ so A is almost simple.

When G is a simple Lie group and K a maximal compact subgroup then G/K is a symmetric space on which Γ acts properly discontinuously. The quotient orbifold $\Gamma\backslash G/K$ is very useful for the study of a lattice Γ of G. Such Γ always has a finite index torsion free subgroup Δ and so $\Delta\backslash G/K$ is a manifold which covers $\Gamma\backslash G/K$ with finite fibers.

The analogous object for H is the Bruhat-Tits building T associated with H. Since H is of rank one, T is a tree and H acts on it with two orbits of vertices. In particular it is a biregular tree of two possible degrees $q_1 + 1$ and $q_2 + 1$.

By our assumption on A it acts on the tree X with finitely many orbits of vertices. In either case, a lattice of H or A is conveniently described using the quotient graph of groups $\Gamma\backslash\backslash T$ or $\Gamma\backslash\backslash X$ (see [Se2] and [BK]). As Γ is discrete, the stabilizers in Γ of vertices and edges are finite. Γ is a uniform lattice of H (resp. A) if and only if $\Gamma\backslash\backslash T$ (resp. $\Gamma\backslash\backslash X$) is a finite graph. A discrete subgroup of A is a lattice if and only if $\sum_x (1/|\Gamma_x|)$ converges, where Γ_x denotes the stabilizer in Γ of the vertex x and the sum runs over a fundamental domain for the action of Γ on X, or equivalently, Γ_x is the group attached to a vertex of the quotient graph of groups $\Gamma\backslash\backslash X$ and the sum is over all the vertices of $\Gamma\backslash\backslash X$.

Note that the tree T is a special case of X and H is therefore a (closed) subgroup of $A = Aut(T)$. Moreover, since $H\backslash T$ has only two vertices, H is a cocompact subgroup of A and hence, as mentioned before, every lattice of H is also a lattice in $A = Aut(T)$.

Recall that the graph of groups $\Gamma\backslash\backslash X$ gives complete information on Γ and its action on X (up to conjugation). Constructing lattices in A is therefore translated into a problem of constructing graphs of groups with some desirable properties. This is usually a problem which mixes combinatorics and finite group theory. To get a lattice out of a graph of groups one uses the "inverse Bass-Serre theory" as described in [Se2] I5.3. This chapter of the theory of groups acting on trees is less well known than the other direction, but it is the crucial tool for the construction of lattices in A. A more delicate theory was developed by Bass [Ba] which, among other things, enables one to construct lattices of A within H.

Finally, I would like to express my gratitude to Hyman Bass whose work plays a central role in the theory of trees and their automorphism groups, and from whom I have learnt most of what I know about the subject.

Thanks are due also to Shahar Mozes for help at several points of the paper.

1. Existence of lattices

The first question is, of course, whether G, H and A have lattices.

(1G) Theorem

(a) Borel [B1]. *G has uniform and non-uniform arithmetic lattices.*

(b) Margulis [Ma] Gromov-Schoen [GS]. *Unless G is locally isomorphic to $SO(n, 1)$ or $SU(n, 1)$, every lattice in G is arithmetic.*

(c) Gromov-Piatetski-Shapiro [GPS]. *For every $n \geq 2$, $SO(n, 1)$ has non-arithmetic uniform and non-uniform lattices.*

(d) Mostow [Mo1], Deligne-Mostow [DM]. *$SU(2, 1)$ and $SU(3, 1)$ have uniform and non-uniform non-arithmetic lattices.*

The problem of existence of non-arithmetic lattices in $SU(n, 1)$ $(n \geq 4)$ is still open.

(1H) Theorem

(a) *If $char(K) = 0$ then:*

(i) Tamagawa [Ta]. *Every lattice of H is uniform.*

(ii) Borel-Harder [BH]. *H has a uniform arithmetic lattice.*

(iii) Lubotzky [L2]. *H has non-arithmetic uniform lattices.*

(b) *If $char(K) > 0$ then:*

(i) Lubotzky [L2]. *H has uniform and non-uniform non-arithmetic lattices.*

(ii) Borel-Harder [BH]. *H has an arithmetic non-uniform lattice. It has an arithmetic uniform lattice if and only if G is of type A_n (i.e., over the algebraic closure of K, the algebraic group $\mathbf{H}$ is isomorphic to SL_{n+1} for some n).*

(1A) Theorem (Bass-Kulkarni [BK]).
A has a uniform lattice if and only if A is unimodular (i.e., the left Haar measure is equal to the right Haar measure).

The unimodularity of A can be reformulated via a combinatorial condition on the quotient $A \backslash X$ (see [BK] §1). It is a necessary, but not sufficient, condition for the existence of non-uniform lattices. For example, [BK] Example

(4.12)2, furnishes an example when A itself is discrete, hence a uniform lattice, and so non-uniform lattices do *not* exist. The question of existence of a non-uniform lattice in A is open.

The notion of arithmetic lattice does not exist in A, but recall Margulis' Theorem (cf. [Ma] IX, 1.12., and see (6G)): A lattice Γ in G is arithmetic if and only if the commensurability group $C(\Gamma)$ of Γ is not discrete. In light of this theorem one is tempted to *define* a lattice of A to be arithmetic if $C(\Gamma)$ is not discrete. It turns out (Liu [Li]) that every cocompact lattice in A is "arithmetic", in fact $C(\Gamma)$ is dense in A (see (6A) below). For non-uniform lattices the situation is not yet clear (see [BL] for some partial results).

We should also mention the following important theorem:

(1A') Theorem (Bass-Kulkarni [BK]).   *Any two uniform lattices Γ_1 and Γ_2 of A are commensurable up to conjugation (i.e., there exists $g \in A$ such that $\Gamma_2 \cap g^{-1}\Gamma_1 g$ is of finite index in Γ_2 and in $g^{-1}\Gamma_1 g$).*

Theorem (1A') does not hold for non-uniform lattices of A, nor does it hold for lattices in G or in H. In a different direction we have:

(1G") Theorem (Weil [We], Mostow [Mo1] Margulis [Ma]).   *Unless $G = PSL_2(\mathbb{R})$, G has only countably many conjugacy classes of lattices. If $G = PSL_2(\mathbb{R})$ it has uncountably many conjugacy classes of uniform lattices and uncountably many of non-uniform ones.*

The first statement is a corollary of the local rigidity of lattices in G ($G \neq PSL_2(\mathbb{R})$) proved by Weil [We] for uniform lattices and by Mostow, Prasad and Margulis for non-uniform ones. (For $G = PSL_2(\mathbb{C})$ and Γ non-uniform, local rigidity does not hold in the strict sense – i.e., there might be small deformators of Γ – but they are not lattices: one can therefore still deduce that there are only countably many conjugacy classes of lattices).

The second statement of the Theorem is classical; it follows from the positive dimensionality of the Teichmüller space classifying the conformal structures of a given surface.

H behaves like $PSL_2(\mathbb{R})$:

(1H") Theorem (Lubotzky [L2]).   *H has uncountably many conjugacy classes of uniform lattices, and, if $\operatorname{char}(K) > 0$, uncountably many non-uniform ones.*

In this aspect, A behaves neither like G nor like H:

(1A") Theorem

(a) Bass-Kulkarni [BK]. *A has only countably many conjugacy classes of uniform lattices.*

(b) Bass-Lubotzky [BL]. *If $A = Aut(X_k)$ (where X_k is the k-regular tree, $k \geq 3$), then A has uncountably many conjugacy classes of non-uniform lattices. But for some trees X, $A = Aut(X)$ has no non-uniform lattices.*

Proof of (1A"(a)). Fix a uniform lattice Δ in A. The finitely generated group Δ has only countably many finite index subgroups, say, $H_1, H_2, \ldots$ etc. If Γ is an arbitrary uniform lattice, then a conjugate of it Γ' contains one of the H_i's as a subgroup of index say m. According to [BK] Theorem 6.5, a given uniform lattice Λ in A (e.g., $\Lambda = H_i$) is contained in only finitely many lattices as a subgroup of a given index m. This implies that A has only countably many conjugacy classes of uniform lattices.

A stronger version of (1A"(b)) is given below in (4A).

2. The geometry and combinatorics of lattices

Let G be as before, a non-compact simple real Lie group and K a maximal compact subgroup of it. The symmetric space G/K is contractible. If Γ is a cocompact lattice it contains a finite index torsion-free group Δ and $\Delta\backslash G/K$ is a compact manifold. If Γ is a uniform lattice of H or A, then $\Gamma\backslash X$ is a finite graph and $\Gamma\backslash X$ is a finite graph of groups. Thus Γ is virtually free.

The geometry and combinatorics of non-uniform lattices are far more complicated:

(2G(i)) Theorem (Margulis [Ma], Borel [B2]). *Assume* rank $(G) \geq 2$. *Every (non-uniform) lattice of G is arithmetic. As such, reduction theory implies that a fundamental domain for Γ in G is covered by a union of finitely many Siegel sets.*

Remark. Without going into details, we mention that a Siegel set is a product of a compact set and a sector of a single torus. Moreover, every point in it is at a bounded distance from a point on that sector. So going to infinity in a Siegel set is essentially equivalent to going to infinity within that sector. Thus in $\Gamma\backslash G$ there are finitely many such sectors along which one can go to infinity. These sectors, however, might have unbounded intersections. On the other hand when rank $(G) = 1$, Garland and Raghunathan [GR] established a similar result for arbitrary lattices (not necessarily arithmetic). Here, different sectors have only bounded intersections. Moreover, every sector is asymptotic to a unique point at infinity. Thus, G/Γ has only finitely many ends. This is usually expressed as follows:

(2Gii) Theorem (Garland-Raghunathan [GR]). *If* rank $(G) = 1$ *and Γ is a lattice in G, then $\Gamma\backslash G/K$ is a union of a compact set plus finitely many "cusps".*

For H, which is a simple rank one group over a non-archimedean field, we have a similar result. The geometry however is replaced by combinatorics:

(2H) Theorem (Raghunathan [R2], Lubotzky [L2]).   *Let Γ be a (non-uniform) lattice in H. Then the quotient $\Gamma\backslash X$ is a union of a finite graph and finitely many infinite rays.*

The situation for A is however different:

(2A) Theorem (Bass-Lubotzky [BL]).   *Let $X = X_k$ and $A = Aut(X)$. Then A has a non-uniform lattice Γ such that $\Gamma\backslash X$ is of infinite homology and with infinitely many "cusps" (i.e., infinitely many different directions to go to infinity on $\Gamma\backslash X$).*

3. Finite generation

Let Γ be a uniform lattice of G (resp. H or A): then Γ has a finite index torsion-free subgroup Δ. Such a subgroup Δ is thus a fundamental group of a compact Riemannian manifold (resp. finite graph) obtained by dividing the symmetric space associated with G (resp. the tree associated with H or A) by Δ. Thus Δ and Γ are finitely generated and finitely presented.

We will therefore restrict our discussion in this section to non-uniform lattices.

(3G) Theorem (Garland-Raghunathan [GR], Kazhdan [Ka])   *Every (non-uniform) lattice of G is finitely presented.*

Line of proof.   If $\operatorname{rank}(G) = 1$, Garland and Raghunathan ([GR]) show how to construct for Γ a suitable fundamental domain from which one can write a finite presentation for Γ. If $\operatorname{rank}(G) \geq 2$, then by Kazhdan ([Ka], [Ma]) G and Γ have property (T). Discrete groups with property (T) and in particular Γ, are finitely generated. Moreover, by the Margulis Arithmeticity Theorem, Γ is arithmetic. Reduction theory for arithmetic groups (cf. [B2]) implies that Γ is finitely presented. (The fact that Γ is arithmetic implies it is also finitely generated - but the proof of the arithmeticity require the apriori knowledge that Γ is finitely generated).

(3H) Theorem (Behr [Be], Lubotzky [L2]).   *A non-uniform lattice of H is **not** finitely generated.*

(3A) Theorem ([BL]).   *A non-uniform lattice Γ of A is **not** finitely generated.*

Proof.   As described in the introduction, Γ is the fundamental group of an infinite graph of finite groups of unbounded order. If Γ is finitely generated it must be isomorphic to the fundamental group of a finite subgraph of groups.

This implies that Γ is virtually free, hence virtually torsion-free and the size of its finite subgroups is bounded. As Γ has unbounded finite subgroups we get a contradiction.

Note that (3A) implies (3H).

4. Covolumes

Quite a lot is known about the possible covolumes of lattices in G, H and A.

(4G) Theorem

(a) Kazhdan-Margulis ([KM], [R1]). *There exists an $\varepsilon = \varepsilon(G) > 0$ such that $\mu(\Gamma\backslash G) \geq \varepsilon$ for every lattice Γ in G (here μ is a fixed Haar measure of G).*

(b) Borel [B3], Wang [W], Thurston [TH]. *If $G \neq PSL_2(\mathbb{R})$, $PSL_2(\mathbb{C})$ then the set $V(G)$ consisting of the real numbers $\mu(\Gamma\backslash G)$, as Γ runs over all lattices of G, is discrete. On the other hand $V(PSL_2(\mathbb{R}))$ and $V(PSL_2(\mathbb{C}))$ are countable non-discrete subsets of $\mathbb{R}$.*

(c) Siegel [Si]. *When $G = PSL_2(\mathbb{R})$ and μ is the standard Haar measure of G (i.e., the one induced by the measure of $\mathbb{H}^2 = G/SO(2)$ corresponding to the curvature (-1)-Riemannian metric of $\mathbb{H}^2$) then $\varepsilon(PSL_2(\mathbb{R})) = \frac{\pi}{21}$ and this minimum is obtained for the uniform $(2,3,7)$-triangle group.*

(d) Meyerhoff [Me]. *For $G = PSL_2(\mathbb{C})$, the lattice $\Gamma_0 = PSL_2(\mathcal{O}_3)$ obtains the minimal covolume* **among the non-uniform** *lattices of G. Here $\mathcal{O}_3$ is the ring of integers of the quadratic extension $\mathbb{Q}(\sqrt{-3})$ and $\mu(\Gamma_0\backslash G) \cong 0.0863$ (when μ is the Haar measure induced by the measure of $\mathbb{H}^3 = G\backslash PSU(2)$ corresponding to the curvature (-1)-Riemannian metric of $\mathbb{H}^3$). There are, however, uniform lattices in $G = PSL_2(\mathbb{C})$ of smaller covolume, but the absolute minimum is not known.*

(4H) Theorem

(a) (Raghunathan [R2]). *There exists $\varepsilon = \varepsilon(H) > 0$ such that $\mu(\Gamma\backslash H) \geq \varepsilon$ for every lattice Γ in H.*

(b) (Lubotzky [L2]). *$V(H)$, the set of covolumes of lattices in H, is a discrete subset of $\mathbb{R}$.*

(c) (Lubotzky [L3]). *For $H = SL_2(\mathbb{F}_q((t)))$, we have*

$$\mu(\Gamma \backslash H) \geq 2(q-1)^{-2}(q+1)^{-1}, \ (\mu(\Gamma \backslash H) \geq (q-1)^{-2}(q+1)^{-1} \text{ if } q \text{ is even }),$$

when μ is normalized so that all maximal compact subgroups of H have measure 1. This minimum is obtained for the non-uniform (!) lattice $\Gamma_0 = PSL_2(\mathbb{F}_q[\frac{1}{t}])$. (If $q = 2$, then the same minimum is also obtained by some uniform lattices).

(d) (Lubotzky-Weigel [LW]). *Let $H = SL_2(K)$ where K is of characteristic zero, so K is a finite extension of $\mathbb{Q}_p$. In all cases where $p \neq 2$, the minimum possible covolume was determined and a lattice with that covolume was presented. (The reader is referred to [LW] for details).*

Remark. Part (b) is not explicitly proved in [L2] but it can be deduced from the discussion there in sections 5 and 6. See also [L3, §2] for a more detailed discussion for the case $H = SL_2$.

The situation for (regular) trees is dramatically different:

(4A) Theorem (Bass (see [BL])). *Let $X = X_k$ be the k-regular tree (k $\geq$ 3), $A = Aut(X_k)$ and μ a Haar measure of A. Then for every $0 < r \in \mathbb{R}$ there exists a lattice Γ of A with $\mu(\Gamma \backslash A) = r$. In particular there is no minimum.*

The result does not hold for general X: for some X the full automorphism group A is a discrete finitely generated group. In this case A has only countably many finite index subgroups. Hence the set of covolumes of lattices in such A is a countable discrete subset of $\mathbb{R}$.

Another aspect of covolumes for which G and A are different is the following theorem which was proved by Borel [B3] for SL_2 and in general by Margulis and Rohlfs [MR]:

(4G') Theorem *Let Γ be a lattice in G. Then there exists a constant $0 < c \in \mathbb{R}$ such that $c\mu(\Delta \backslash G)$ is an integer for any lattice Δ commensurable with Γ, where μ is the Haar measure of G.*

The same result probably holds also for H, but this has not been established yet. In A, however, the result is not true. This follows from:

(4A') Theorem (Bass-Kulkarni [BK]) *For $X = X_k$ (k $\geq$ 3), there exists in A an infinite ascending chain of uniform lattices $\Gamma_1 < \Gamma_2 < \Gamma_3 < \ldots$.*

As every two uniform lattices of A are commensurable (after conjugation) (see 1A'), Theorem 4A' implies that, for $X = X_k$ (k $\geq$ 3), no uniform lattice

Γ in A satisfies the assertion of Theorem 4G'. It is less clear what is the situation for non-uniform lattices of A.

5. Residual finiteness

A group Γ is said to be residually finite if the intersection of its finite index subgroups is the identity. Finitely generated linear groups are residually finite. As G and H are linear, this remark applies to their finitely generated subgroups. Hence:

(5G) Proposition *Every lattice of G is residually finite.*

Proof. By (3G) every lattice is finitely generated.

(5H) Proposition *Every lattice Γ of H is residually finite.*

Proof. If Γ is uniform in H then it is finitely generated and the above remark applies. If Γ is non-uniform, then it is not finitely generated (3H), but as is shown in [L2], it can be expressed as a fundamental group of a finite graph of groups whose vertex groups are residually finite and whose edge groups are finite. Such a group is residually finite by [Se] Prop. 12, p. 122.

(5A) Proposition (Bass-Lubotzky [BL]).

(a) *Every uniform lattice of $A = Aut(X)$ is virtually free and in particular residually finite.*

(b) *When $X = X_k$ (the k-regular tree, $k \geq 3$), $A = Aut(X_k)$ has a non-uniform lattice with no finite index subgroups, while it also has non-uniform residually finite lattices.*

To understand (b) look at the following two graphs of groups:

$$Y = \overset{\Gamma_0}{\bullet}\underline{\qquad}\overset{\Lambda_1}{\bullet}\underline{\qquad}\overset{\Lambda_2}{\bullet}\underline{\qquad}\bullet\cdots\cdots \qquad (i)$$

where $\Gamma_0 = SL_2(\mathbb{F}_p)$, and for $n \geq 0$,

$$\Lambda_n = \begin{pmatrix} a & d^0 \leq n \\ 0 & a^{-1} \end{pmatrix} = \left\{ \begin{pmatrix} a & b \\ 0 & a^{-1} \end{pmatrix} \mid a \in \mathbb{F}_p^*,\ b \in \mathbb{F}_p[t], deg(b) \leq n \right\}.$$

So $\Lambda_n \simeq \mathbb{F}_p^{n+1} \rtimes \mathbb{F}_p^*$, and $\Lambda_0 = \Gamma_0 \cap \Lambda_1$.

If follows from [Se 2], p. 87, that the fundamental group of Y is $SL_2(\mathbb{F}_p[t])$ and $PSL_2(\mathbb{F}_p[t])$ is a lattice in $Aut(X_{p+1})$, which is clearly residually finite.

$$Y' = \overset{\Gamma_0'}{\bullet}\underline{\qquad}\overset{\Lambda_1'}{\bullet}\underline{\qquad}\overset{\Lambda_2'}{\bullet}\underline{\qquad}\bullet\cdots\cdots \qquad (ii)$$

where $\Gamma_0' = SL_2(\mathbb{F}_p)$ and $\Lambda_n' = V_n \rtimes \mathbb{F}_p^*$, where

$$V_n = p^{-(n+2)}\mathbb{Z}_p/\mathbb{Z}_p \subset \mathbb{Q}_p/\mathbb{Z}_p,$$

and $t \in \mathbb{F}_p^*$ acts on V by lifting to a $(p-1)$–th root of unity $w(t) \in \mathbb{Z}_p^\times$, and multiplying by $w(t)^2$. Then Λ_0' ($\subset \Lambda_1'$) can be identified with $\Lambda_0 \subset \Gamma_0 = \Gamma_0'$, thus defining Y'.

It is not difficult to check that Γ' (the fundamental group of Y') acts on X_{p+1} as a non-uniform lattice. The group Γ' is not residually finite. Indeed, Γ' contains $\mathbb{Q}_p/\mathbb{Z}_p = \bigcup_n V_n$, which is a p-divisible p-group and hence has no non-trivial finite quotients. Since the triangular group $\Lambda_0' \subset \Gamma_0'$ generates Γ_0' as a normal subgroup, it follows that Γ' has no non-trivial finite quotients.

6. The commensurability group; super rigidity

Let Γ be a lattice in G (resp. H, A). The *commensurability group of Γ* (also called the commensurator of Γ), denoted $C(\Gamma)$, is equal to the set of all $g \in G$ (resp. H, A) such that $g^{-1}\Gamma g \cap \Gamma$ is of finite index in Γ and in $g^{-1}\Gamma g$. This is a subgroup of G (resp. H, A).

When $G = \underset{\sim}{G}(\mathbb{R})$, where $\underset{\sim}{G}$ is an algebraic group defined over $\mathbb{Q}$ and $\Gamma = \underset{\sim}{G}(\mathbb{Z})$, one has $C(\Gamma) = \underset{\sim}{G}(\mathbb{Q})$. So we think of $C(\Gamma)$, also in the context of trees, as the group of "rational points" of A. In particular a question of interest is whether $C(\Gamma)$ is dense in A.

(6G) Theorem (Margulis [Ma] IX 1.12.)   *Let Γ be a lattice in G. Then the following are equivalent:*

(i) $C(\Gamma)$ *is dense in G.*

(ii) $C(\Gamma)$ *is not discrete.*

(iii) Γ *is an arithmetic lattice.*

(6H) Theorem (Margulis cf. [Ma], IX 1.12, 1.13).   *If Γ is a uniform lattice of H then (i), (ii), and (iii) of (6G) are equivalent.*

It does not seem to be known whether (6H) also holds for non-uniform lattices of H.

The equivalence of (i) and (ii) in (6G) and (6H) is easy. But in A, these two condition are probably not equivalent. One might be tempted to take one of them as a definition of an arithmetic lattice in A. In this sense all uniform lattices are arithmetic:

(6A) Theorem (Liu [Li]).   *Let Γ be a uniform lattice in $A = Aut(X)$. Then $C(\Gamma)$ is dense in A.*

The fact that all uniform lattices of A behave the same way is not surprising in the light of Theorem (1A') which asserts that they are all commensurable up to conjugation. The situation for non-uniform lattices is less clear. It seems that non-uniform lattices can behave quite differently. But very little has been proved so far in this direction.

Margulis' celebrated super rigidity theorem asserts that when rank $(G) \geq 2$, then essentially every linear representation of Γ can be extended, on a finite index subgroup, to a representation of G. This does not hold for rank one groups. As A is similar to rank one groups we can not expect it to hold for A, either. On the other hand the following super-rigidity result of Margulis is "rank-free":

(6G'H') Theorem (Margulis [Ma] VII 5.4.).   *Let Γ be a lattice in G (resp. H). Let k be a local field, F a connected adjoint k-simple algebraic group defined over k, and $\varphi : \Lambda \to F(k)$ a homomorphism, where Λ is a subgroup of $C(\Gamma)$ containing Γ and dense in G (resp. H). Assume $\varphi(\Gamma)$ is Zariski dense in F, but not relatively compact in $F(k)$. Then φ extends, uniquely, to a continuous homomorphism from G (resp. H) to $F(k)$.*

The following analogue has been established for regular trees by Lubotzky-Mozes-Zimmer [LMZ] and in general by Burger and Mozes [BM]:

(6A') Theorem   *Let Γ be a lattice in A, Λ a subgroup of $C(\Gamma)$ containing Γ and D the closure of Λ in A. Let $\varphi : \Lambda \to Aut(X')$ be a homomorphism, where X' is some locally finite tree. Assume $\varphi(\Gamma)$ is not relatively compact in $Aut(X')$ and $\varphi(\lambda)$ acts minimally on X'. Then φ can be extended, uniquely, to a homomorphism from D to $Aut(X')$.*

7. The congruence subgroup problem

One of the central problems about arithmetic groups is the congruence subgroup problem. Namely, let $\mathbf{G}$ be an algebraic group defined over a global field F with ring of integers $\mathcal{O}$. Let $\Gamma = \mathbf{G}(\mathcal{O})$ and I a non-zero ideal of $\mathcal{O}$. We denote by $\Gamma(I)$ the kernel of the canonical map from $\mathbf{G}(\mathcal{O})$ to $\mathbf{G}(\mathcal{O}/I)$. A subgroup Λ of Γ is called a *congruence subgroup* if Λ contains $\Gamma(I)$ for some I. Two topologies are naturally defined on Γ: the profinite topology (resp. the congruence topology) - i.e., the one for which the family of all finite index subgroups (resp. the congruence subgroups) serves as a basis of neighborhoods of the identity. Let $\hat{\Gamma}$ (resp. $\bar{\Gamma}$) be the completion of Γ with respect to the profinite (resp. congruence) topology. As the first is finer than the later, the identity map of Γ can be extended to a surjective homomorphism π from $\hat{\Gamma}$ to $\bar{\Gamma}$. The arithmetic group $\Gamma = \mathbf{G}(\mathcal{O})$ is said to have the *congruence subgroup property (CSP)* if $\ker(\pi)$ is finite.

(7G)   The congruence subgroup problem has been intensively studied for arithmetic groups which appear as lattices in G. It was conjectured by Serre

[Se1] that if Γ is an arithmetic lattice of G, then Γ has CSP if and only if $\mathrm{rank}_{\mathbb{R}}(G) \geq 2$. The positive part of this conjecture is by now almost completely proved - with some exceptions of uniform lattices. We refer the reader to Rapinchuk [Ra] for the detailed history. For rank one groups the story is as follows. For lattices in $SO(n,1)$ or $SU(n,1)$ it was proved in almost all cases that CSP fails (see [RV], [V] and the references therein). But for the other rank one real Lie groups $Sp(n,1)$ and F_4, nothing is known. In recent years it was shown that these rank one groups are in many ways similar to higher rank (i.e., rank ≥ 2) groups: they have Kazhdan's property (T) and all lattices there are arithmetic ([GS]). One may thus suggest a modification of Serre's conjecture from [Se1] to include $Sp(n,1)$ and F_4 with the groups whose lattices have CSP. Note, however, that lattices in these groups, like all lattices in rank one groups and unlike lattices in higher rank groups, do have infinite normal subgroups of infinite index. Also recall that while it is strongly believed that the CSP for Γ depends only on G and not on Γ (this is the main point of Serre's conjecture), this has never been proved, so in principal it is possible that some lattices in $Sp(n,1)$ or F_4 satisfy CSP and some do not.

For H however, Serre's conjecture is fully confirmed. This was proved by Serre ([Se1] and [Se2]) for SL_2 and by Lubotzky [L2] in general:

(7H) Theorem   *Let Γ be an arithmetic lattice in H. Then Γ does not have the congruence subgroup property.*

In the context of A, there is no classical meaning for a lattice to be arithmetic. It was suggested in chapter 6 to define a lattice to be arithmetic if its commensurability group is not discrete. In this sense, however, every uniform lattice is arithmetic if the tree X is regular (see chapters 1 and 6). Furthermore, we can use the commensurability group of a lattice Γ of A to define the congruence topology of Γ. Namely, we say that a subgroup Λ of Γ is a congruence subgroup if there exist $g_1,\ldots,g_n \in C(\Gamma)$ such that Λ contains $\bigcap_{i=1}^{n} g_i^{-1}\Gamma g_i$. Clearly a congruence subgroup of Γ is of finite index and this defines a family of neighborhoods of the identity of Γ. Let us call the topology determined by this family the congruence topology and $\bar{\Gamma}$ will be the completion with respect to that topology. We say that Γ has the congruence subgroup property if $\ker(\pi) : \hat{\Gamma} \to \bar{\Gamma}$ is finite. It is possible to show that applying this for a lattice in G or H recovers the congruence subgroup problem there.

(7A) Theorem (Mozes [Mz2]).   *Let X be a k-regular tree and Γ a uniform lattice in $A = Aut(X)$. Then every finite index subgroup of Γ is a congruence subgroup.*

This is a somewhat surprising result as we are trained to think that lattices in A are similar to (or sometimes worse than) lattices in H and we would expect them to behave as lattices in rank one groups. Somehow, the richness of the commensurability group of Γ defines a very strong congruence topology on Γ - as strong as the profinite topology. This is especially impressive if we keep in mind that Γ is virtually free so has "a lot" of finite index subgroups. (Just for comparison, let us mention that for $\Gamma = PSL_2(\mathbb{Z})$ which is a virtually free lattice in $G = PSL_2(\mathbb{R})$, the number of subgroups of index at most n in Γ grows super exponentially as a function of n. On the other hand the number of congruence subgroups of index at most n, grows as $n^{\log n / \log \log n}$, i.e., just slightly faster than polynomial. See [L4] for more.)

Anyway, this last surprise illustrates the charm of the study of lattices in A: Not always do we know what to expect!

References

[Ba] H. Bass, Covering theory for graphs of groups, *J. Pure Appl. Alg.* 89 (1993), 3-47.

[BK] H. Bass and R. Kulkarni, Uniform tree lattices, *J. Amer. Math. Soc.* 3 (1990), 843–902.

[BL] H. Bass and A. Lubotzky, Non-uniform tree lattices, in preparation.

[Be] H. Behr, Finite presentability of arithmetic groups over global function fields , *Proc. Edinburgh Math. Soc.* 30 (1987), 23–39.

[B1] A. Borel, Compact Clifford-Klein forms of symmetric spaces, *Topology* 2 (1963), 111–122.

[B2] A. Borel, *Introduction aux Groupes Arithmetiques*, Herman, Paris (1969).

[B3] A. Borel, Commensurability classes and volumes of hyperbolic 3 - manifolds, *Ann. Scuola Norm. Sup. Pisa Cl. Sci.* 8 (1981), 1–23.

[BH] A. Borel and G. Harder, Existence of discrete cocompact subgroups of reductive groups over local field, *J. Reine Angew. Math.* 298, (1978) 53–64.

[BM] M. Burger and S. Mozes, CAT(-1) spaces, divergence groups and their commensurators, preprint.

[DM] P. Deligne and G.D. Mostow, Monodromy of hypergeometric functions and non-lattice integral monodromy, *Publ. Math. IHES* 63 (1986), 5–89.

[FTN] A. Figa-Talamanca and C. Nebbia, *Harmonic Analysis and Representation Theory for Groups Acting on Trees*, Lond. Math. Soc. Lecture Notes 162, Cambridge University Press (1991).

[GR] H. Garland and M.S. Raghunathan, Fundamental domains for lattices in ℝ-rank one semisimple Lie groups, *Ann. of Math.* 92 (1970), 279–326.

[GPS] M. Gromov, I. Piatetski-Shapiro, Non-arithmetic groups in Lobochevsky spaces, *Publ. Math. IHES* 66 (1988), 93–103.

[GS] M. Gromov and R. Schoen, Harmonic maps into singular spaces and *p*-adic superrigidity for lattices in groups of rank one, *Publ. Math. IHES* 76 (1992), 165–246.

[Ka] D.A. Kazhdan, On the connection between the dual space of a group and the structure of its closed subgroups, *Funct. Anal. Appl.* 1 (1967), 63–65.

[KM] D. Kazhdan and G. Margulis, A proof of Selberg's hypothesis, *Mat. Sbornik* 75 (1968), 162–168.

[Li] Y.S. Liu, Density of the commensurability groups of uniform tree lattices, *J. Algebra* 165 (1994), 346-359.

[L1] A. Lubotzky, Trees and discrete subgroups of Lie groups over local fields, *Bull. Amer. Math. Soc.* 20 (1989), 27–31.

[L2] A. Lubotzky, Lattices in rank one Lie groups over local fields, *Geometric and Functional Analysis* 1 (1991), 405–431.

[L3] A. Lubotzky, Lattices of minimal covolume in SL_2: A nonarchimedean analogue of Siegel's Theorem $\mu \geq \pi/21$, *J. Amer. Math. Soc.* 3 (1990), 961–975.

[L4] A. Lubotzky, Subgroup growth and congruence subgroups, Invent. Math., to appear.

[LM] A. Lubotzky and S. Mozes, Asymptotic properties of unitary representations of tree automorphism, in: *Harmonic Analysis and Discrete Potential Theory*, M.A. Picardello (ed.) , Plenum Press, N.Y. (1992).

[LMZ] A. Lubotzky, S. Mozes and R.J. Zimmer, Superrigidity for the commensurability group of tree lattices, to appear.

[LW] A. Lubotzky and T. Weigel, Lattices of minimal covolume in SL_2 over nonarchimedean fields, in preparation.

[Ma] G.A. Margulis, *Discrete Subgroups of Semi-Simple Lie Groups,* Springer-Verlag, New York (1991).

[MR] G.A. Margulis and J. Rohlfs, On the proportionality of covolumes of discrete subgroups, *Math. Ann.* 275 (1986), 197–205.

[Me] R. Meyerhoff, The cusped hyperbolic 3-orbifold of minimum volume, *Bull. Amer. Math. Soc.* 13 (1985), 154–156.

[Mo1] G.D. Mostow, *Strong Rigidity of Locally Symmetric Spaces*, Ann. of Math. Studies 78, Princeton Univ. Press (1973).

[Mo2] G.D. Mostow, On a remarkable class of polyhedra in complex hyperbolic spaces, *Pacific J. Math.* 86 (1980), 171–276.

[Mz1] S. Mozes, Mixing of all orders of Lie groups actions, Invent. Math. 107 (1992), 235–241.

[Mz2] S. Mozes, in preparation.

[R1] M.S. Raghunathan, *Discrete Subgroups of Lie Groups*, Springer Verlag, New York (1972).

[R2] M.S. Raghunathan, Discrete subgroups of algebraic groups over local fields of positive characteristic, *Proc. of the Indian Academy of Sciences (Math. Sci.)* 99 (1989), 127–146.

[RV] M.S. Raghunathan and T.N. Venkataramana, The first Betti number of arithmetic groups and the congruence subgroup problem, in: *Proc. Conf. held in honour of R. Steinberg*, UCLA (1992), Contempoarary Math. 153 (1993), 95-107.

[Ra] A.S. Rapinchuk, Congruence subgroup problem for algebraic groups: old and new, *Astérisque*, 209 (1992), 73–84.

[Se1] J.P. Serre, Le problème des groupes de congruences pour SL_2, *Ann. of Math.* 92 (1970), 489–527.

[Se2] J.P. Serre, *Trees*, Springer Verlag, New York (1980).

[Si] C.L. Siegel, Some remarks on discontinuous groups, *Ann. of Math.* 46 (1945), 708–718.

[Ta] T. Tamagawa, On discrete subgroups of p-adic algebraic groups, in: *Arithmetical Algebraic Geometry*, O.F.G. Schilling (ed.), Harper and Row, New York (1965).

[Ti] J. Tits, Sur le groupe des automorphisms d' un arbre, in: *Essays on topology and related topics: Memoirs dedié à George de Rham*, A. Haefliger and R. Narasimhan, (eds.) Springer Verlag (1970).

[Th] W. Thurston, *The Geometry and Topology of 3-manifolds*, Mimeographed notes, Princeton University.

[V] T.N. Venkataramana, On the first cohomology of arithmetic groups, preprint.

[W] H.C. Wang, *Topics on totally discontinuous groups in symmetric spaces*, W. Boothey (ed.) M. Dekker (1972), 460–487.

[We] A. Weil, Discrete subgroups of Lie groups II, *Ann. of Math.* 75 (1962), 578–602.

[Zi] R.J. Zimmer, *Ergodic Theory and Semisimple Groups*, Birkhäuser, Boston (1984).

Institute of Mathematics
Hebrew University
Jerusalem 91904
Israel.

Generalisations of Fibonacci numbers, groups and manifolds

C. MACLACHLAN

1 Introduction

The Fibonacci groups are defined by

$$F_n = < a_1, a_2, \ldots, a_n \mid a_i a_{i+1} = a_{i+2} \; i = 1, 2, \ldots, n > \qquad (1)$$

where the suffixes are taken *mod n*. The nomenclature follows from the inductive definition of the Fibonacci numbers

$$f_i + f_{i+1} = f_{i+2}. \qquad (2)$$

Much of the interest in these groups, and their generalisations, has focussed on whether or not these groups are finite. They are known to be finite precisely when $n = 3, 4, 5, 7$ (see [12] for a survey).

For $n = 2m$, $m \geq 3$, there is a compact 3-manifold, obtained by face-pairing on a polyhedral 3-cell, whose fundamental group is isomorphic to F_n. For $m \geq 4$, the polyhedron can be realised hyperbolically to give a tesselation of hyperbolic 3-space and hence a complete compact hyperbolic manifold of constant negative curvature (see [5]). This shows, independently, that for these values of n, the groups F_n are infinite . Furthermore, these manifolds are m-fold cyclic covers of S^3, branched over the figure 8 knot (e.g. [6]).

For a finite number of values of m only, these manifolds are arithmetic in the sense that their covering groups are discrete arithmetic subgroups of $PSL(2, \mathbb{C})$ ([5],[6]).

In this paper, these hyperbolic manifolds are considered from a slightly different viewpoint (cf. [6]) to show that there is a direct link between Fibonacci numbers and hyperbolic Fibonacci manifolds, which is not merely a happy accident of nomenclature.

Furthermore, this relationship is emphasised by considering certain generalisations of the Fibonacci numbers, correponding to which we construct groups and manifolds, showing that, for all but a finite number, the manifolds corresponding to even values are hyperbolic. The corresponding groups are thus infinite. It will also be shown that only finitely many of these hyperbolic manifolds for each family are arithmetic.

The author would like to thank Alan Reid for helpful conversations on the subject matter of this paper.

2   Numbers

It is well-known that the Fibonacci numbers as defined at (2), with $f_1 = f_2 = 1$, can be generated by the matrix

$$A_1 = \begin{pmatrix} 1 & 1 \\ 1 & 0 \end{pmatrix} \tag{3}$$

in the sense that

$$A_1^n = \begin{pmatrix} f_{n+1} & f_n \\ f_n & f_{n-1} \end{pmatrix}. \tag{4}$$

If λ_1, λ_2 are the eigenvalues of A_1, the slopes of the corresponding eigenvectors are $\lambda_1^{-1}, \lambda_2^{-1}$ so that diagonalising A_1, the alternative description of the Fibonacci numbers is obtained using (4) as

$$f_n = \frac{\lambda_1^n - \lambda_2^n}{\lambda_1 - \lambda_2} \tag{5}$$

where $\lambda_1, \lambda_2 = (1 \pm \sqrt{5})/2$.

The generalisations of the Fibonacci numbers considered here are defined, for each $k \geq 1$, by

$$f_i^{(k)} + k f_{i+1}^{(k)} = f_{i+2}^{(k)} \tag{6}$$

with $f_1^{(k)} = 1, f_2^{(k)} = k$.

These numbers can be generated in the sense given at (4) above by the matrix

$$A_k = \begin{pmatrix} k & 1 \\ 1 & 0 \end{pmatrix}. \tag{7}$$

The alternative description of these numbers is exactly as at (5) with in this case the eigenvalues being given by $(k \pm \sqrt{(k^2 + 4)})/2$.

3   Groups

The Fibonacci groups are defined at (1). Corresponding to the generalised numbers as defined at (6), we define the following generalisations of the Fibonacci groups

$$F_n^{(k)} = < a_1, a_2, \ldots, a_n \mid a_i a_{i+1}^k = a_{i+2} \quad i = 1, 2, \ldots, n > \tag{8}$$

where the suffixes are taken *mod n*.

Remark. There are various generalisations of the Fibonacci groups, many of them more popular than those just defined (e.g.[12], [8]).

If $X_n^{(k)} = \dfrac{F_n^{(k)}}{[F_n^{(k)}, F_n^{(k)}]}$, then $X_n^{(k)}$ is defined by the relation matrix $A_k^n - I$, and so its order is given by

$$o(X_n^{(k)}) = (\lambda_1^n + \lambda_2^n) - 1 - (-1)^n = f_{n+1}^{(k)} + f_{n-1}^{(k)} - 1 - (-1)^n$$

which is exactly analogous to the Fibonacci case where $k = 1$ (see [8]).

We will eventually only be concerned with the cases where n is even, as the fundamental groups of the manifolds which are constructed give rise to the generalised Fibonacci groups $F_{2m}^{(k)}$. In the case $k = 1$ also, only the fundamental groups F_{2m} arise. The following result deals in part with the case where n is odd.

Theorem 3.1 *If n is odd, the groups F_n cannot be the fundamental groups of hyperbolic 3-orbifolds of finite volume.*

Proof. Suppose that $\rho : F_n \to Isom\,\mathbb{H}^3$ is a faithful representation such that $\rho(F_n)$ is discrete of finite covolume. Now F_n admits an automorphism α of order n permuting the generators cyclically. Thus, by Mostow rigidity, $\exists\, t \in Isom\,\mathbb{H}^3$ such that $t\rho(\gamma)t^{-1} = \rho(\alpha(\gamma))$ for every $\gamma \in F_n$. Let $\Gamma =< t, \rho(F_n) >$. Now $t^n \in Z(\rho(F_n))$ so that $t^n = 1$ since the limit set of t^n would be $\partial\mathbb{H}^3$. The group Γ then has a presentation

$$< a_1, t \mid t^n = 1, \ (a_1 t)^2 = t^2 a_1 > .$$

Since n is odd, $\Gamma^{(2)} = \Gamma$ where $\Gamma^{(2)} =< \gamma^2 \mid \gamma \in \Gamma >$. This implies that $\Gamma \subset PSL(2,\mathbb{C})$, the orientation-preserving subgroup of $Isom\,\mathbb{H}^3$. After conjugation we can assume that $t = P \begin{pmatrix} \xi & 0 \\ 0 & \xi^{-1} \end{pmatrix}$ where ξ is a $2n$'th root of unity. Let $a_1 = P \begin{pmatrix} x & y \\ z & w \end{pmatrix}$ with $xw - yz = 1$ so that $yz \neq 0$, since $\rho(F_n)$ has finite covolume. Substituting in the second defining relation for Γ forces $\xi^6 = 1$. Thus the result follows for $n \neq 3$ and for $n = 3$ since F_3 is finite.

4 Manifolds

The construction of the manifolds is by Dehn filling on once-punctured torus bundles and includes the hyperbolic Fibonacci manifolds as the case $k = 1$ (cf. [6]).

Let ϕ be an element of the mapping class group M of the once-punctured torus S. The once-punctured torus bundle MS_ϕ is obtained by taking the mapping torus of a homeomorphism of S representing ϕ. Now MS_ϕ depends on the conjugacy class of $\phi^{\pm 1}$ in $M \cong SL(2,\mathbb{Z})$ (e.g. [4]). Furthermore by results of Thurston and Jorgensen the manifold has a complete hyperbolic structure of finite volume if and only if ϕ determines a hyperbolic element of $PSL(2,\mathbb{Z})$ [11]. A representative ϕ can be chosen of the form $(-1)^\epsilon L^{n_1} R^{n_2} \ldots L^{n_{2r-1}} R^{n_{2r}}$ where $\epsilon = 0,1$ and $L = \begin{pmatrix} 1 & 1 \\ 0 & 1 \end{pmatrix}, R = \begin{pmatrix} 1 & 0 \\ 1 & 1 \end{pmatrix}$ (e.g. [9]).

Let θ be an automorphism of $\Pi_1(S) = F_2(a,b)$, the free group on two generators, inducing $\phi \in SL(2,\mathbb{Z})$. Then the fundamental group of $\Pi_1(MS_\phi)$ is given by

$$< a, b, t \mid tat^{-1} = \theta(a), \ tbt^{-1} = \theta(b) > . \tag{9}$$

Now the squares of the matrices defined at (7) determine hyperbolic elements of the group $PSL(2, \mathbb{Z})$. Indeed $A_k^2 = L^k R^k$. Thus the corresponding once-punctured torus bundles, denoted $M^{(k)}$, are complete hyperbolic manifolds of finite volume. The automorphism θ_k where

$$\theta_k(a) = (a^k b)^k a \quad \theta_k(b) = a^k b \tag{10}$$

induces A_k^2 so that

$$\Pi_1(M^{(k)}) = < a, b, t \mid tat^{-1} = (a^k b)^k a, \ tbt^{-1} = a^k b > . \tag{11}$$

Now on the toroidal boundary $\partial M^{(k)}$ we can choose the meridian and longitude to be represented by t and $[a, b]$ (cf. [1]). Then if we carry out (p, q) Dehn filling, by the hyperbolic Dehn Surgery theorem [10], for all but a finite number of (p, q) we obtain an orbifold which has a complete hyperbolic structure of finite covolume. For large enough n, let $H_n^{(k)}$ denote the fundamental group of the orbifold obtained by $(n, 0)$ filling on $M^{(k)}$. Thus

$$H_n^{(k)} = < a, b, t \mid t^n = 1, \ tat^{-1} = (a^k b)^k a, \ tbt^{-1} = a^k b > . \tag{12}$$

Now $H_n^{(k)}$ contains a torsion-free normal subgroup with cyclic quotient of order n. In the case $k = 1$, this normal subgroup is the fundamental group of the n-fold cyclic cover of S^3 branched over the figure 8 knot complement i.e. the Fibonacci groups F_{2n}. More generally for any $k \geq 1$, a presentation for this normal subgroup is obtained by Reidemeister-Schreier rewriting. By setting

$$a_{2i-1} = t^{n-i-1} b^{-1} t^{-(n-i-1)} \quad a_{2i} = t^{n-i-1} a^{-1} t^{-(n-i-1)}$$

for $i = 1, 2, \ldots, n$, it is easily seen that this group is isomorphic to the group $F_{2n}^{(k)}$ defined at (8).

Theorem 4.1 *For each $k \geq 1$, there exists a positive integer N_k such that for all $n \geq N_k$, there is a hyperbolic manifold $M_n^{(k)}$ of finite volume such that $\Pi_1(M_n^{(k)}) \cong F_{2n}^{(k)}$.*

Corollary 4.2 *For those values of k, n described in the Theorem, $F_{2n}^{(k)}$ is infinite.*

5 Arithmeticity

From [5],[6], it is known that the hyperbolic manifolds $M_n^{(1)}$ are arithmetic for only finitely many values of n; precisely for the values $n = 4, 5, 6, 8, 12, \infty$ where $M_\infty^{(1)} = M^{(1)}$. (For similar results using $(n, 0)$ surgery on hyperbolic 2-bridge knots, see [7]). Here a similar result is proved for the manifolds $M_n^{(k)}$. Note however, that only for the values $k = 1, 2$ are the manifolds $M_\infty^{(k)}$ arithmetic (see [3]). A finite covolume hyperbolic 3-orbifold is referred to as being arithmetic if its covering group is an arithmetic Kleinian group. For information on arithmetic groups see [2], [13].

Theorem 5.1 *For each $k \geq 1$, there are only finitely many values of n such that the hyperbolic manifolds $M_n^{(k)}$ are arithmetic.*

Proof. If $M_n^{(k)}$ is arithmetic, then the groups $H_n^{(k)}$ at (12) will also be arithmetic. But this is the fundamental group obtained by Dehn filling on the manifold $M^{(k)}$. Thus the covolume of $H_n^{(k)}$ is bounded above by the volume of $M^{(k)}$ [10]. But the number of conjugacy classes of arithmetic Kleinian groups of bounded covolume is finite [2]. The result now follows.

Remarks. The general methods used in the proofs of the above results leave some intriguing questions unanswered: The precise values of N_k for which, for all $n \geq N_k$, the manifolds $M_n^{(k)}$ are hyperbolic, are not determined, except as we know, in the case $k = 1$, where $N_1 = 4$. For small values of n, it is of interest to determine the manifolds $M_n^{(k)}$. Of course, $M_1^{(1)} = S^3$. More generally, $\Pi(M_1^{(k)})$ is the free product of two cyclic groups of order k so that $M_1^{(k)}$ is a connected sum. Finally, the groups $H_n^{(1)}$ appear as subgroups of index 2 in generalised triangle groups (see [12]) and a matrix representation of the generators in these extended groups is easily determined. From this it is straightforward to deduce (cf. [3] Section 4) for precisely which values of n these groups are arithmetic. The methods used above do not expose an obvious general approach to this problem for $H_n^{(k)}$ and this problem, together with those just outlined, will be explored elsewhere.

References

[1] S.Betley, J.Przytycki and T.Zukowski. *Hyperbolic structures on Dehn filling of some punctured torus bundles over S^1*. Kobe J. Math. 3 (1987) 117-147.

[2] A.Borel. *Commensurability classes and volumes of hyperbolic 3-manifolds*. Ann. Scuola Norm. Pisa 8 (1981) 1 - 33.

[3] B.Bowditch, A.W.Reid and C.Maclachlan. *Arithmetic hyperbolic surface bundles*. To apppear.

[4] M.Culler,W.Jaco and H.Rubinstein. *Incompressible surfaces in once-punctured torus bundles*. Proc. London Math. Soc 45 (1982) 385 - 419.

[5] H.Helling, A.C.Kim and J.L.Mennicke. *On Fibonacci groups*. To appear.

[6] H.M.Hilden, M.T.Lozano and J.M.Montesinos-Amilibia. *The arithmeticity of the figure eight knot orbifolds*. Topology '90. Proceedings of Research Semester at Ohio State University. De Gruyter 1992 (169 - 183).

[7] H.M.Hilden, M.T.Lozano and J.M.Montesinos-Amilibia. *On the arithmetic 2-bridge knot and link orbifolds* To appear.

238 Fibonacci numbers, groups and manifolds

[8] D.L.Johnson, J.W.Wamsley and D.Wright. *The Fibonacci groups.* Proc. London Math. Soc. 29 (1974) 577 - 592.

[9] C.Series. *The geometry of Markoff numbers.* Math. Intelligencer 7 (1985) 20 - 29.

[10] W.P.Thurston. *The geometry and Topology of 3-manifolds.* Mimeographed notes, Princeton University 1977.

[11] W.P.Thurston. *Hyperbolic structures on 3-manifolds II: surface groups and 3-manifolds that fiber over the circle* To appear in Annals of Maths.

[12] R.M.Thomas. *The Fibonacci groups revisited.* Groups St Andrews 1989 L.M.S.Lecture Notes Series 160 (445 - 454).

[13] M-F.Vigneras. *Arithmetique des algebres de quaternions.* Lecture Notes in Maths 800, Springer-Verlag 1980.

Department of Mathematical Sciences,
University of Aberdeen,
Dunbar Street,
Aberdeen AB9 2TY,
Scotland.

Knotted surfaces in the 4-sphere with no minimal Seifert manifolds

TORU MAEDA

Let M^n be an n-dimensional smooth manifold in the $(n+2)$-sphere S^{n+2}, and let $S^{n+2} - N(M^n)$ be the complement of an open regular neighbourhood of M^n in S^{n+2}. A *Seifert manifold* H of M^n is a compact orientable $(n+1)$-dimensional submanifold of S^{n+2} such that the boundary of H is M^n. Moreover, if the induced homomorphism from $\pi_1(H)$ into $\pi_1(S^{n+2} - N(M^n))$ is one-to-one, then the Seifert manifold H is called *minimal*.

One important case is when M^n is a sphere, i.e. a knot. The existence of Seifert manifolds of knots of any dimension is known. It is well known that the Seifert surfaces of 1-knots with minimal genus are minimal. Gutierrez [1] asserted the existence of minimal Seifert manifolds for knots of any dimension. However, his proof has a gap. For $n \geq 3$, Silver [6] has given examples of n-dimensional knots with no minimal Seifert manifolds. Necessary and sufficient conditions for an n-knot ($n \geq 3$) to have a minimal Seifert manifold have been given in [7].

In this paper we prove the following result:

Theorem. *For any integer $g \geq 1$, there exist closed orientable surfaces of genus g in S^4 which have no minimal Seifert manifolds and no trivial 1-handles.*

This result was presented to KOOK topology seminar in Osaka in September 1989, and at KNOT 90 in Osaka in 1990. The existence of 2-knots with no minimal Seifert manifolds is still an open question.

1 Preliminaries

Let H be a Seifert manifold of a knot K in S^{n+2}. Then the knot group $G(K)$ of K has an HNN-decomposition with base group the image of $\pi_1(Cl(S^{n+2} - N(K)) - H)$ in $G(K)$ under the natural homomorphism. The existence of a minimal Seifert manifold of K implies that both $\pi_1(Cl(S^{n+2} - N(K)) - H)$ and $\pi_1(H)$ are finitely presented. Therefore $G(K)$ has an HNN-decomposition with stable letter representing a meridian of $N(K)$, a finitely presented base group $\pi_1(Cl(S^{n+2} - N(K)) - H)$, and finitely presented associated subgroups $\pi_1(H)$.

Here we will give an example of a group G which is not a higher-dimensional knot group as characterized by Kervaire [4], but close to it.

239

Proposition. *The group G given by the presentation*

$$\langle x, a, b : b^2 = 1, x^{-1}ax = a^2, x^{-1}bx = [b, a] \rangle, \qquad (1.1)$$

where $[b, a] = b^{-1}a^{-1}ba$, has no HNN-decomposition with finitely presented base groups.

Proof of Proposition. The presentation (1.1) shows that G has an HNN-decomposition with base group $B = gp\{a, b\}$, the group generated by a and b, and stable letter x. Associated to this HNN-decomposition, the commutator subgroup $[G, G]$ of G has the structure of a limit group given by the following tree product:

$$[G, G] = \cdots \underset{B_{i-1}}{*} B_{i-1} \underset{B_i}{*} B_i \underset{B_{i+1}}{*} \cdots,$$

where $B_i = x^{-i}Bx^i$, $i = 0, \pm 1, \ldots$. Since $x^{-1}Bx \le B$,

$$[G, G] = \cdots \underset{B_{i-1}}{*} B_{i-1} \underset{B_i}{*} B_i.$$

Lemma 1. *The group B has a presentation*

$$\langle a, b : W_i(a, b)^2, \ i = 0, 1, \ldots \rangle \qquad (1.2)$$

where $W_0(a, b) = b$ and (for $i > 0$),

$$W_i(a, b) = [b, a, a^2, a^{2^2}, \ldots, a^{2^{i-1}}]$$
$$(= [[b, a, a^2, a^{2^2}, \ldots, a^{2^{i-2}}], a^{2^{i-1}}] = x^{-i}bx^i)$$

(we remark that the relators $x^{-i}b^2x^i$ for $i < 0$ do not induce any nontrivial relators in $B = B_0 \in [G, G]$).

Proof of Lemma 1. By the Embedding Theorem for HNN-extensions, it is enough to show that the group B^* presented by (1.2) is isomorphic to its subgroup $B_1^* = gp\{a^2, [b, a]\}$ under the map that sends a and b to a^2 and $[b, a]$ respectively. Let $L^* = \ll B_1^* \gg^{B^*}$, the normal closure of B_1^* in B^*. The factor group B^*/L^* is the abelian group $\langle a : a^2 \rangle \oplus \langle b : b^2 \rangle$, so to get a presentation of L^* we apply the Reidemeister-Schreier method to (1.2) with the minimal Schreier system $S = \{1, a, b, ab\}$ (see [5], Theorem 2.9, p. 94). Then we have the following presentation of L^*:

$$L^* \cong \langle S_{a,a}, S_{b,a}, S_{b,b}, S_{ab,a}, S_{ab,b} : \ \begin{aligned} &S_{b,b}, S_{ab,b} \\ &W_i(S_{a,a}, S_{ab,a}^{-1}S_{a,a})^2, \\ &W_i(S_{a,a}, S_{b,a}^{-1})^2, \\ &W_i(S_{b,a}S_{ab,a}, S_{a,a}^{-1}S_{ab,a})^2, \\ &W_i(S_{ab,a}S_{b,a}, S_{b,a})^2 \ (i = 0, 1, \ldots) \rangle \end{aligned}$$

where $S_{X,Y} = XY(\overline{XY})^{-1}$, and $\overline{XY}$ is the right coset representative of XY in S.

Now we take the factor group L of L^* by the normal closure $\ll S_{b,a} \gg^{L^*}$ of $S_{b,a}$ in L^*.

$$L \cong \langle S_{a,a}, S_{ab,a} \; : \; W_i(S_{a,a}, S_{ab,a}^{-1}S_{a,a})^2, W_i(S_{ab,a}, S_{a,a}^{-1}S_{ab,a})^2 \; (i = 0, 1, \ldots) \rangle.$$

Since $S_{a,a}$ and $S_{ab,a}$ represent a^2 and $a^2[b,a]^{-1}$ respectively, L is a factor group of B_1^*. For convenience, we replace the generating system $\{S_{a,a}, S_{ab,a}\}$ of L by the set consisting of $\alpha = S_{a,a}$ and $\beta = S_{ab,a}^{-1}S_{a,a}$. Then

$$L \cong \langle \alpha, \beta \; : \; W_i, (\alpha, \beta)^2, W_i(\alpha\beta^{-1}, \beta^{-1})^2 \; (i = 0, 1, \ldots) \rangle.$$

Here, we show that

$$(\alpha\beta^{-1})^{2^i} = \alpha^{2^i} W_i(\alpha, \beta) \quad \text{modulo } \{W_k(\alpha, \beta)^2 \;, \; k = 0, 1, \ldots, i-1\}$$

by induction on i.

$$
\begin{aligned}
(\alpha\beta^{-1})^2 \; &= \alpha\beta^{-1}\alpha\beta^{-1} \\
&= \alpha^2 \cdot \alpha^{-1}\beta^{-1}\alpha\beta \\
&\qquad \text{modulo } \{W_0(\alpha, \beta)^2 \; (= \beta^2), \; W_1(\alpha, \beta)^2 \; (= [\beta, \alpha]^2)\} \\
&= \alpha^2[\alpha, \beta] \\
&= \alpha^2 W_1(\alpha, \beta)^{-1} \\
&= \alpha^2 W_1(\alpha, \beta) \quad \text{modulo } W_1(\alpha, \beta)^2.
\end{aligned}
$$

$$
\begin{aligned}
(\alpha\beta^{-1})^{2^i} \; &= ((\alpha\beta^{-1})^{2^{i-1}})^2 \\
&= (\alpha^{2^{i-1}} W_{i-1}(\alpha, \beta))^2 \quad \text{by inductive assumption} \\
&= \alpha^{2^i} \cdot \alpha^{-2^{i-1}} W_{i-1}(\alpha, \beta)\alpha^{2^{i-1}} W_{i-1}(\alpha, \beta) \\
&= \alpha^{2^i} \cdot \alpha^{-2^{i-1}} W_{i-1}(\alpha, \beta)^{-1}\alpha^{2^{i-1}} W_{i-1}(\alpha, \beta) \\
&\qquad \text{modulo } W_{i-1}(\alpha, \beta)^2 \\
&= \alpha^{2^i}[\alpha^{2^{i-1}}, W_{i-1}(\alpha, \beta)] \\
&= \alpha^{2^i} W_i(\alpha, \beta)
\end{aligned}
$$

Then, we can show that $W_i(\alpha\beta^{-1}, \beta^{-1}) = W_i(\alpha, \beta)$ modulo $\{W_k(\alpha, \beta), \; k = 0, 1, \ldots, i\}$ by induction on i.

$$
\begin{aligned}
W_0(\alpha\beta^{-1}, \beta^{-1}) \; &= \beta^{-1} \\
&= \beta \quad \text{modulo } W_0(\alpha, \beta)^2 \\
&= W_0(\alpha, \beta)
\end{aligned}
$$

$$
\begin{aligned}
W_i(\alpha\beta^{-1},\beta^{-1}) &= [W_{i-1}(\alpha\beta^{-1},\beta^{-1}),(\alpha\beta^{-1})^{2^{i-1}}]\\
&= [W_{i-1}(\alpha\beta^{-1},\beta^{-1}),\alpha^{2^{i-1}}\cdot W_{i-1}(\alpha,\beta)]\\
&= [W_{i-1}(\alpha,\beta),\alpha^{2^{i-1}}\cdot W_{i-1}(\alpha,\beta)] \quad \text{by induction}\\
&= [W_{i-1}(\alpha,\beta),W_{i-1}(\alpha,\beta)]\cdot[W_{i-1}(\alpha,\beta),\alpha^{2^{i-1}}]\\
&\qquad\qquad\qquad\qquad \cdot[W_{i-1}(\alpha,\beta),\alpha^{2^{i-1}},W_{i-1}(\alpha,\beta)]\\
&= [W_{i-1}(\alpha,\beta),\alpha^{2^{i-1}}]\cdot[W_{i-1}(\alpha,\beta),\alpha^{2^{i-1}},W_{i-1}(\alpha,\beta)]\\
&= W_{i-1}(\alpha,\beta)^{-1}\cdot[W_{i-1}(\alpha,\beta),\alpha^{2^{i-1}}]\cdot W_{i-1}(\alpha,\beta)\\
&= [\alpha^{2^{i-1}},W_{i-1}(\alpha,\beta)] \quad \text{modulo } W_{i-1}(\alpha,\beta)^2\\
&= W_i(\alpha,\beta)^{-1}\\
&= W_i(\alpha,\beta) \quad \text{modulo } W_i(\alpha,\beta)^2
\end{aligned}
$$

Therefore $W_i(\alpha\beta^{-1},\beta^{-1})^2 = 1$ modulo $\{W_k(\alpha,\beta)^2,\ k=0,1,\ldots,i\}$ for any $i = 0,1,\ldots$. Hence

$$
L \cong \langle \alpha,\beta\ :\ W_i(\alpha,\beta)^2,\ i=0,1,2,\ldots\rangle.
$$

Since α and β are the images of $S_{a,a}$ and $S_{a,a}^{-1}S_{ab,a}$, that is, the generating system $\{a^2,[a,b]\}$ of $B_1^\star$, under the epimorphism from $L^\star$ onto L, and since $W_i(S_{a,a},S_{a,a}^{-1}S_{ab,a})^2 = 1$ in $B_1^\star$, we have $B_1^\star \cong L \cong B^\star$. This completes the proof of Lemma 1.

Lemma 2. *The group B has no finite presentation.*

Proof of Lemma 2. Suppose that B is finitely presented. Since $H_1(B) \cong \langle a\ :\ \rangle \oplus \langle b\ :\ b^2\rangle$, it follows that B has an HNN-decomposition with finitely generated base group, X say, finitely generated associated subgroups C and D, and stable letter a. The kernel E of the natural epimorphism from B onto $\langle a\ :\ \rangle$ is $gp\{b_p; p \in \mathbb{Z}\}$, where $b_p = a^{-p}ba^p$. A presentation of E is

$$
\langle b_p\ (p \in \mathbb{Z})\ :\ (b_p b_{p+1}\cdots b_{p+2^k-1})^2 = 1\ (k \in \mathbb{Z}, k \geq 0)\rangle.
$$

The first homology group $H_1(E)$ of E is

$$
\bigoplus_{-\infty}^{\infty} \langle b_b\ :\ b_p^2 = 1\rangle \tag{1.3}
$$

Besides this, E is a limit group

$$
\cdots \quad \underset{C_{-2}=D_{-1}}{*} \quad \overset{X_{-1}}{} \quad \underset{C_{-1}=D_0}{*} \quad \overset{X_0}{} \quad \underset{C_0=D_1}{*} \quad \overset{X_1}{} \quad \underset{C_1=D_2}{*} \quad \cdots \tag{1.4}
$$

where $X_p = a^{-p}Xa^p$, $C_p = a^{-p}Ca^p$ and $D_p = a^{-p}Da^p$.

We consider E as a group acting on the tree associated to (1.4). Let us denote the element $b_p\cdots b_{p+2^k-1}$ by $B_p(k)$. Since $B_p(k)$ is of finite order ($=2$) for any k, $B_0(k)$ is in a conjugate of X_{u_k} for some integer u_k. Then

$B_{-u_k}(k) = a^{u_k}B_0(k)a^{-u_k}$ is in a conjugate of $X_0 (= X)$. Therefore the image of X in $H_1(E)$ contains all of $\{B_{-u_k}(k) \; ; \; k \in \mathbb{Z}, \; k \geq 0\}$. Since the word length of $B-u_k(k)$ in (1.3) is 2^k, the image of X is not finitely generated. This is a contradiction. This completes the proof of Lemma 2.

The statement and proof of the following Lemma 3, implicit in an earlier version of this paper, were supplied by the referee. Assume that $\phi \; : \; B \hookrightarrow B$ is a self-embedding of a finitely generated group B. Let G be the group presented as an HNN-extension $\langle x, B \; : \; x^{-1}Bx = \phi(B) \rangle$. Let $\psi \; : \; G \to \mathbb{Z}$ be the natural epimorphism such that $\psi(x) = 1$ and $\psi(B) = 0$.

Lemma 3. *Assume that some (and hence any) presentation of the subgroup $\phi(B)$ extends to a presentation of B by adjoining only finitely many generators and relators. The group G has an HNN-decomposition with finitely presented base contained in the kernel of ϕ if and only if B itself is finitely presented.*

Proof of Lemma 3. Suppose that G has an HNN-decomposition with stable letter $\overline{x}$ and finitely presented base $\overline{B}$ contained in the kernel of ϕ. After replacing B with the base $gp(\bigcup_{|i| \leq p} x^{-i}\overline{B}x^i)$ for sufficiently large p (see for example the proof of Lemma 2.6 of [2]), we can assume without any loss of generality that $B \subset \overline{B}$. Then $B \subset \overline{B} \subset x^qBx^{-q}$ for sufficiently large q. By the hypothesis and induction, there exists a presentation $\langle X, y_1, \ldots, y_k \; : \; R, s_1, \ldots, s_\ell \rangle$ for x^qBx^{-q} such that $\langle X \; : \; R \rangle$ is a presentation of B. Suppose that $\langle Z \; : \; T \rangle$ is some finite presentation for $\overline{B}$. Then x^qBx^{-q} has a presentation of the form $\langle X, y_1, \ldots, y_k, Z \; : \; R, s_1, \ldots, s_\ell, T, x_i = W_i(X, y_j) \rangle$. Since $B \subset \overline{B}$, the relators R are a consequence of T and so can be eliminated from the presentation. Similarly, each of the generators X can be expressed as a word in Z, and so can be removed. What remains is a finite presentation for x^qBx^{-q}. Since x^qBx^{-q} is isomorphic to B, the base B is finitely presented.

This completes the proof of Lemma 3. By the Corollary to Lemma 2, The Proposition holds.

2 Proof of the Theorem

Let F_g be a compact orientable surface in S^4. Then the fundamental group $G(F_g) = \pi_1(S^4 - F_g)$ has a finite Wirtinger presentation with first homology group $H_1(G) \cong \mathbb{Z} = \langle t \; : \; \rangle$, and the generators represent meridians of the boundary of $N(F_g)$. Conversely, if a group W has a Wirtinger presentation with p generators and q relators, where p and q are finite, and $H_1(W) \cong \mathbb{Z}$, then there exists a compact orientable surface of genus $q - p + 1$ in S^4 such that $G(F_g) \cong W$ and its generators represent meridians of the boundary of $N(F_g)$. Because, using only a suitable set of $p - 1$ of the q defining relators, we obtain a Wirtinger presentation of deficiency 1. This corresponds to a ribbon 2-knot in symmetric position in S^4. Adding $q - p + 1$ fusion-bands to this ribbon knot according to the remaining $q - p + 1$ defining relators, we have a surface of genus $q - p + 1$ in S^4. (See for example [3].)

Firstly, we will consider the case of a torus in S^4. The existence of a group G_1 with the following properties implies the existence of a surface F of genus at most 1 in S^4 with no minimal Seifert manifold:

(1) G_1 has a Wirtinger presentation of deficiency zero;

(2) $H_1(G_1) \cong \mathbb{Z}$;

(3) G_1 has no HNN-decomposition with finitely presented base group, in which the stable letter consists of one of the generators in the Wirtinger presentation in (1). (G_1 always has some HNN-decomposition by (2).)

Adding one more condition:

(4) the second homology group $H_2(G_1) \neq 0$,

we have that G_1 cannot be a knot group. The F in S^4 is a torus which is not given from any 2-knot by attaching a trivial 1-handle. (Note that, even if $H_2(G_1)$ vanished, we would not be able to decide whether or not G_1 was a 2-knot group.) We define G_1 as follows:

$$G_1 = \langle x, a, b \ : \ x^{-1}ax = a^2, x^{-1}bx = [b, a], x^{-1}b^2x = b^2 \rangle.$$

We now verify conditions (1) to (4) for G_1.

(1) G_1 has a Wirtinger presentation of deficiency zero.

Since

$$\begin{aligned} a &= x^{-1} \cdot axa^{-1} \text{ and} \\ b &= x[b, a]x^{-1} = xb^{-1}a^{-1}bx^{-1} \cdot xax^{-1} \\ &= xb^{-1}ax^{-1}a^{-1}bx^{-1} \cdot xb^{-1}xbx^{-1} \cdot xax^{-1} \end{aligned}$$

we have

$$\begin{aligned} G_1 \cong \langle x, a, b \ : \ &a = x^{-1} \cdot axa^{-1}, \\ &b = xb^{-1}ax^{-1}a^{-1}bx^{-1} \cdot xb^{-1}xbx^{-1} \cdot xax^{-1}, \\ &x = b^{-2}xb^2 \rangle. \end{aligned}$$

Add three new generators y, z, w by $y = axa^{-1}$, $z = xb^{-1}xbx^{-1}$ and $w = xb^{-1}axa^{-1}bx^{-1}$, then eliminate a and b by $a = x^{-1}y$ and $b = w^{-1}zyx^{-1}$. Thus

$$\begin{aligned} G_1 \cong \langle x, y, z, w \ : \ &y = x^{-1}yxy^{-1}x, \\ &z = x^2y^{-1}z^{-1}wxw^{-1}zyx^{-2}, \\ &w = x^2y^{-1}z^{-1}wyx^{-1}xxy^{-1}w^{-1}zyx^{-2}, \\ &x = (w^{-1}zyx^{-1})^{-2}x(w^{-1}zyx^{-1})^2 \rangle. \end{aligned}$$

(2) $H_1(G_1) \cong \mathbb{Z}$ is immediate from the presentation of G_1.

(3) G_1 has an HNN-decomposition with base group $B^* = gp\{a, b\}$, associated subgroups B^* and $gp\{a^2, [b, a]\}$, and stable letter x. The group B is the image of B^* under the natural epimorphism from G_1 onto G, i.e. $B \cong B^*/(\ll b^2 \gg^{G_1} \cap B^*)$. Suppose that B^* is finitely presented. Then B^* has a finite presentation with generating system $\{a, b\}$, say $\langle a, b : Y(1), \ldots, Y(m) \rangle$. These defining relators induce the relations

$$W_i(a, b)^2 = W_{i+1}(a, b)^2 \text{ for } i = 0, 1, \ldots$$

Since B has the presentation (1.2), it also has a finite presentation $\langle a, b : Y(1), \ldots, Y(m), b^2 \rangle$. This contradicts Lemma 2, so B^* is not finitely presented. By the Corollary to Lemma 2, G_1 has no finitely presented base group with stable letter x.

(4) Let us use the Hopf formula to investigate $H_2(G_1)$. Since

$$G_1 = \langle x, a, b : x^{-1}ax = a^2, x^{-1}bx = [b, a], x^{-1}b^2x = b^2 \rangle,$$

it is enough to show that $b^{-2}x^{-1}b^2x \neq 1$ in the group $G^\dagger$ presented by

$$\langle x, a, b : x^{-1}ax = a^2, x^{-1}bx = [b, a], [[b^2, x], x] = [[b^2, x], a] = [[b^2, x], b] = 1 \rangle.$$

Consider the factor group $G^\ddagger$ of $G^\dagger$:

$$
\begin{aligned}
G^\ddagger \cong \langle x, a, b : \quad & x^{-1}ax = a^2, x^{-1}bx = [b, a], [b, x^2], \\
& [[b^2, x], x] = [[b^2, x], a] = [[b^2, x], b] = 1, \\
& b^8, b^4 = a^{-1}b^4a = x^{-1}b^4x, \\
& [b^2, a^{-k}ba^k] = [x^{-1}b^2x, a^{-k}ba^k] = 1 \ (k \in \mathbb{Z}) \rangle
\end{aligned}
$$

$$
\begin{aligned}
\cong \langle x, a, b_k \ (k \in \mathbb{Z}) : \quad & x^{-1}ax = a^2, x^{-1}b_0x = b_0^{-1}b_1, \\
& x^{-1}b_0^{-1}b_1x = b_0, a^{-k}b_0a^k = b_k, \\
& (b_0^{-2}(b_0^{-1}b_1)^2)^{-1}((b_0^{-1}b_1)^{-2}b_0^2) = 1, \\
& (b_0^{-2}(b_0^{-1}b_1)^2)^{-1}(b_1^{-2}(b_1^{-1}b_2)^2) = 1, \\
& (b_0^{-2}(b_0^{-1}b_1)^2)^{-1}b_0^{-1}(b_0^{-2}(b_0^{-1}b_1)^2)b_0 = 1, \\
& b_0^8, b_0^4 = b_1^4 = (b_0^{-1}b_1)^4, [b_0^2, b_k] = [(b_0^{-1}b_1)^2, b_k] = 1 \rangle.
\end{aligned}
$$

Since

$$a^{-2}x^{-1}b_0xa^2 = a^{-2}b_0^{-1}b_1a^2 = b_2^{-1}b_3,$$

we have

$$
\begin{aligned}
b_3 &= b_2 a^{-2} x^{-1} b_0 x a^2 \\
&= b_2 x^{-1} a^{-1} b_0 a x \\
&= b_2 x^{-1} b_1 x \\
&= b_2 \cdot x^{-1} b_0 x \cdot x^{-1} b_0^{-1} b_1 x \\
&= b_2 b_0^{-1} b_1 b_0.
\end{aligned}
$$

Hence
$$G^\ddagger \cong \langle x, a, b_0, b_1, b_2 \ : \ x^{-1}ax = a^2, x^{-1}b_0x = b_0^{-1}b_1,$$
$$x^{-1}b_1x = b_1^{-1}b_0^2, x^{-1}b_2x = b_0^{-1}b_1^{-1}b_0b_2^{-1},$$
$$a^{-1}b_0a = b_1, a^{-1}b_1a = b_2, a^{-1}b_2a = b_2b_0^{-1}b_1b_0,$$
$$b_0^{-4}(b_0^{-1}b_1)^{-4} = 1,$$
$$b_0^2(b_0^{-1}b_1)^{-2}b_1^{-2}(b_1^{-1}b_2)^2 = 1,$$
$$b_0^8, b_0^4 = b_1^4 = (b_0^{-1}b_1)^4,$$
$$[b_0^2, b_1] = [b_1^2, b_0] = [(b_0^{-1}b_1)^2, b_0] = 1 \rangle.$$

The epimorphism $G^\dagger \to G^\ddagger$ is given by $x \mapsto x$, $a \mapsto a$ and $b \mapsto b_0$. The subgroup $gp\{b_0, b_1\}$ of $G^\ddagger$ is a finite group

$$\langle b_0, b_1 \ : \ b_0^8 = b_1^8 = 1, b_0^4 = b_1^4 = (b_0^{-1}b_1)^4,$$
$$[b_0^2, b_1] = [b_1^2, b_0] = [(b_0^{-1}b_1)^2, b_0] = [(b_0^{-1}b_1)^2, b_1] = 1 \rangle$$

of order 64. In this group, $b_0^{-2}x^{-1}b_0^2x = b_0^{-2}(b_0^{-1}b_1)^2 \neq 1$.

To complete the proof of the Theorem, one constructs a suitable surface of genus g by taking the connected sum of g copies of F. We omit the details.

References

[1] M. A. Gutierrez, An exact sequence calculation for the second homotopy of a knot, *Proc. Amer. Math. Soc.* **32** (1972) 571-577.

[2] O. Kakimizu, Combinatorial distance between HNN decompositions of a group, *J. Pure Appl. Algebra* **82** (1992) 273-288.

[3] A. Kawauchi, T. Shibuya and S. Suzuki, Descriptions on surfaces in four-space I, *Mathematics Seminar Notes* **10** (1982) 75-125.

[4] M. Kervaire, On higher dimensional knots, in: *Diff. and Comb, Topology*, (S. S. Cairns, ed.), Princeton University Press (1965) 105-119.

[5] W. Magnus, A. Karrass and D. Solitar, *Combinatorial Group Theory*, Interscience Publications, New York-London-Sydney (1966).

[6] D. S. Silver, Examples of 3-knots with no minimal Seifert manifolds, *Math. Proc. Cambridge Phil. Soc.* **110** (1991) 417-420.

[7] D. S. Silver, On the existence of minimal Seifert manifolds, *Math. Proc. Cambridge Phil. Soc.* **114** (1993) 103-109.

Department of Mathematics,
Faculty of Engineering,
Kansai University,
Suita, Osaka 564,
Japan.

The higher geometric invariants of modules over Noetherian group rings

HOLGER MEINERT

1 Introduction

For every finitely generated group G let $V(G)$ denote the finite dimensional $\mathbb{R}$-vector space $\mathrm{Hom}(G; \mathbb{R})$ of all homomorphisms from G into the additive group of the reals. If A is a G-module[1] and $m \in \mathbb{N}_0$ or $m = \infty$ one defines

$$\Sigma^m(G; A) := \{ \chi \in V(G) \mid A \text{ is of type } \mathrm{FP}_m \text{ over } \mathbb{Z}G_\chi \} ,$$

where G_χ is the submonoid $\{g \in G \mid \chi(g) \geq 0\}$ of G. Recall that a module A over a ring R with $1 \neq 0$ is said to be of type FP_m (resp. FP_∞) over R if A admits a free resolution which is finitely generated in all dimensions $\leq m$ (resp. in all dimensions) [3]. A monoid M is of type FP_m if the trivial M-module $\mathbb{Z}$ is of type FP_m over the monoid ring $\mathbb{Z}M$.

The present definition of $\Sigma^m(G; A)$ differs from the original one insofar as we have endowed it with the "singular" point $0 \in \Sigma^m(G; A)$ if and only if A is of type FP_m over $\mathbb{Z}G$. On the other hand it follows from [7] that A is of type FP_m over $\mathbb{Z}G$ if $\Sigma^m(G; A) \neq \emptyset$.

There are two major results on the invariants justifying the effort of investigating them: 1) If N is a normal subgroup of G with Abelian quotient, then by one of the main results in [7] A is of type FP_m over $\mathbb{Z}N$ if and only if $\Sigma^m(G; A)$ contains the set $\{\chi \in V(G) \mid \chi(N) = \{0\}\}$. As an application one can characterize the FP_m-type of all normal subgroups above the derived subgroup of direct products of free groups [14]. 2) In 1980 Bieri and Strebel [8] showed that if G is an extension $A \rightarrowtail G \twoheadrightarrow Q$ of Abelian groups A and Q, then G is finitely presented if and only if G is of type FP_2 if and only if $\Sigma^0(Q; A) \cup -\Sigma^0(Q; A) = V(Q)$. It is conceivable that the invariant $\Sigma^0(Q; A)$ is strong enough to characterize all finiteness properties of G. The precise statement of the so-called FP_m-conjecture can be found in [3] or [4]; for further results in that direction the reader is referred to [1], [5], [15].

The first main result of the present paper shows that for groups G with (left) Noetherian group ring $\mathbb{Z}G$ the invariant $\Sigma^0(G; A)$ determines all the higher invariants in a very strong sense.

---

[1]Here and in the following modules will always be considered as left modules.

Theorem A. *If the group ring $\mathbb{Z}G$ of a group G is Noetherian, then $\Sigma^{\infty}(G; A) = \Sigma^{0}(G; A)$ for all G-modules A.*

Note that every finitely generated G-module is of type FP_{∞} if $\mathbb{Z}G$ is Noetherian, in particular, G itself is of type FP_{∞}. On the other hand, it is not difficult to see that in this situation the monoid ring $\mathbb{Z}G_{\chi}$ is Noetherian if and only if $\chi(G) \subseteq \mathbb{R}$ is cyclic (see Section 3 below). It is not known whether the converse of Theorem A is true. However, using the result of Kropholler [12], that all soluble-by-finite groups of type FP_{∞} are constructible in the sense of Baumslag and Bieri [2], we prove:

Theorem B. *A soluble-by-finite group G with the property that $\Sigma^{\infty}(G; \mathbb{Z}) = V(G)$ is polycyclic-by-finite.*

It is a well-known result of P. Hall [11] that a soluble-by-finite group is polycyclic-by-finite if and only if its group ring is Noetherian. Hence we obtain:

Corollary AB. *Let G be a soluble-by-finite group. Then the following assertions are equivalent:*

1. $\mathbb{Z}G$ is Noetherian.

2. $\Sigma^{\infty}(G; A) = \Sigma^{0}(G; A)$ for all G-modules A.

3. $\Sigma^{\infty}(G; \mathbb{Z}) = V(G)$.

4. G is polycyclic-by-finite.

Finally, we will observe that the main idea in the proof of Theorem B together with Kropholler's theorem and a result of Bieri and Strebel yields the following group theoretical information.

Proposition C. *Suppose that G is a soluble-by-finite group of type FP_{∞}. If G is not polycyclic-by-finite, then the Abelianization G/G' is infinite.*

2   A criterion

The main step in the proof of Theorem A is to construct a monoid satisfying the assumptions of the following criterion. It is a special case of a more general unpublished result of R. Strebel. Because we are in a more special situation we can use a theorem of Bieri and Renz to replace the second, difficult part of Strebel's original proof by an easier one.

Proposition. *Suppose that G is a finitely generated group, that A is a G-module, and that $\chi \in V(G)$. Let M be a monoid such that $G' \subseteq M \subseteq G_{\chi}$ and such that $G = \bigcup_{j \geq 0} M \cdot t^{-j}$ for some $t \in M$ with $\chi(t) > 0$. If A is of type FP_{m} over $\mathbb{Z}M$ then $\chi \in \Sigma^{m}(G; A)$.*

Proof. Note first that $M \cdot t^k = t^k \cdot M$ for all $k \in \mathbb{Z}$ since $G' \subseteq M$. Now, there exists a $\mathbb{Z}M$-free resolution $\mathbf{P} \xrightarrow{\partial_0} A$ which is finitely generated in all dimensions $\leq m$. Let Y_i be a $\mathbb{Z}M$-basis of P_i for all $i \geq 0$. Without loss of generality we may assume that $\mathbf{P} \twoheadrightarrow A$ is *admissible* in the sense that $\partial_i(y) \neq 0$ for all $y \in Y_i$ and all $i \geq 0$.

Claim. There is a $\mathbb{Z}M$-chain endomorphism $\psi : \mathbf{P} \to \mathbf{P}$ lifting the identity of A such that $\psi_i(Y_i)$ is contained in the $\mathbb{Z}M$-submodule $t \cdot P_i$ of P_i for all $i \geq 0$.

Proof of the claim (R. Strebel). We argue by induction on i. If $i = 0$ then we can find an element $a_y \in A$ for every $y \in Y_0$ with $t \cdot a_y = \partial_0(y)$ because t is a unit in $\mathbb{Z}G$. Let $p_y \in P_0$ be a preimage of a_y under ∂_0. Then $y \mapsto t \cdot p_y$ for all $y \in Y_0$ defines a $\mathbb{Z}M$-homomorphism $\psi_0 : P_0 \to P_0$ lifting the identity of A and such that $\psi_0(Y_0) \subseteq t \cdot P_0$.

Suppose that $\psi_0, \psi_1, \ldots, \psi_{i-1}$ are already defined for some $i > 0$ and let $y \in Y_i$. Using the fact that $\mathbb{Z}M \cdot t = t \cdot \mathbb{Z}M$ it is easy to see that $(\psi_{i-1} \circ \partial_i)(y) = t \cdot c_y$ for some $c_y \in P_{i-1}$. Since $\partial_{i-1} \circ \psi_{i-1} = \psi_{i-2} \circ \partial_{i-1}$ it follows that $t \cdot c_y$ is a cycle, and since t is a unit in $\mathbb{Z}G$ this implies that c_y is a cycle. Choosing a preimage $q_y \in P_i$ of c_y under ∂_i for every $y \in Y_i$ and defining $\psi_i : P_i \to P_i$ by $y \mapsto t \cdot q_y$ for all $y \in Y_i$ completes the proof of the claim.

Next, our assumptions imply that $\mathbb{Z}G = \bigcup_{j \geq 0} \mathbb{Z}M \cdot t^{-j} = \bigcup_{j \geq 0} t^{-j} \cdot \mathbb{Z}M$. Consequently, $\mathbb{Z}G$ is flat as right $\mathbb{Z}M$-module and $\mathbb{Z}G \otimes_{\mathbb{Z}M} A \cong A$ as $\mathbb{Z}G$-modules as can be seen by the calculation $t^{-j} \otimes a = t^{-j} \otimes t^j(t^{-j}a) = 1 \otimes t^{-j}a$ if $a \in A$. We put $\mathbf{F} := \mathbb{Z}G \otimes_{\mathbb{Z}M} \mathbf{P}$, $X_i := 1 \otimes Y_i$ for all $i \geq 0$, and $\phi := 1 \otimes \psi$. We can summarize this by saying that $\mathbf{F} \twoheadrightarrow \mathbb{Z}G \otimes_{\mathbb{Z}M} A \cong A$ is an admissible $\mathbb{Z}G$-free resolution of A which is finitely generated in all dimensions $\leq m$ and $\phi : \mathbf{F} \to \mathbf{F}$ is a $\mathbb{Z}G$-chain endomorphism lifting the identity of A such that $\phi_i(X_i)$ is contained in the $\mathbb{Z}M$-submodule $t \cdot (\mathbb{Z}M \cdot X_i)$ of F_i for every $i \geq 0$.

Associated with $\chi : G \to \mathbb{R}$ there is a certain valuation $v_\chi : \mathbf{F} \to \mathbb{R}_\infty$ (see [7], §2) on the resolution $\mathbf{F}$ with the following properties: $v_\chi(f + f') \geq \inf\{v_\chi(f), v_\chi(f')\}$ and $v_\chi(g \cdot f) = \chi(g) + v_\chi(f)$ for all $f, f' \in \mathbf{F}$ and all $g \in G$. Moreover, $v_\chi(f) = \infty$ if and only if $f = 0$. Now, $\chi \in \Sigma^m(G; A)$ if and only if there exists a $\mathbb{Z}G$-chain endomorphism $\varphi : \mathbf{F} \to \mathbf{F}$ lifting the identity of A, such that $v_\chi(\varphi_i(x)) > v_\chi(x)$ for all $x \in X_i$ and all $i \leq m$ ([7], Theorem 4.1)[2].

Next, we choose $k \in \mathbb{N}$ with the property that

$$k \cdot \chi(t) > \max\{v_\chi(x) \mid x \in X_i, i \leq m\} - \min\{v_\chi(x) \mid x \in X_i, i \leq m\}.$$

Then the k-fold iterate $\varphi := \phi^k$ of ϕ is still a $\mathbb{Z}G$-chain endomorphism lifting the identity of A.

Claim. $v_\chi(\varphi_i(x)) > v_\chi(x)$ for all $x \in X_i$ and all $i \leq m$.

---

[2]We can not extend this to dimensions greater than m because we can not be sure that the F_i are still finitely generated if $i > m$.

Proof of the claim. Since $\phi(X_i) \subseteq t \cdot (\mathbb{Z}M \cdot X_i)$ and since $\mathbb{Z}M \cdot t = t \cdot \mathbb{Z}M$ it is not difficult to show $\varphi(X_i) \subseteq t^k \cdot (\mathbb{Z}M \cdot X_i)$. Let $x \in X_i$ for some $i \leq m$. Then $\varphi_i(x) = t^k \cdot \sum_{z \in X_i} \lambda_{xz} z$ with $\lambda_{xz} \in \mathbb{Z}M \subseteq \mathbb{Z}G_\chi$. Using the formulae above we see that $v_\chi(\sum \lambda_{xz} z) \geq \min\{v_\chi(z) \mid z \in X_i, \lambda_{xz} \neq 0\} =: \mu$. Hence $v_\chi(\varphi_i(x)) \geq k \cdot \chi(t) + \mu > v_\chi(x)$ by the choice of k.

This completes the proof of the proposition.

3 Proof of Theorem A

For the convenience of the reader we offer a proof of the following fact before we embark on the proof of Theorem A.

Lemma. *Let G be a group with Noetherian group ring and let $\chi \in V(G)$. Then the monoid ring $\mathbb{Z}G_\chi$ is Noetherian if and only if $\chi(G) \subseteq \mathbb{R}$ is cyclic, or equivalently, discrete.*

Proof. If $\chi(G)$ is not discrete we can find a sequence $(g_n)_{n \in \mathbb{N}}$ of group elements such that $0 < \chi(g_{n+1}) < \chi(g_n)$ for all $n \in \mathbb{N}$. Then $\mathbb{Z}G_\chi \cdot g_1 < \mathbb{Z}G_\chi \cdot g_2 < \cdots$ is a strictly increasing sequence of submodules of $\mathbb{Z}G_\chi$. Consequently $\mathbb{Z}G_\chi$ is not Noetherian.

Suppose that $\chi \neq 0$ and that $\chi(G)$ is (infinite) cyclic. Then $\chi(G)$ is generated by an element $\chi(t) > 0$, $t \in G$, and $\mathbb{Z}G_\chi$ is a twisted polynomial ring $\mathbb{Z}K[t]^\sigma$ over the group ring $\mathbb{Z}K$ of the kernel K of χ. Since $\mathbb{Z}G$ Noetherian implies $\mathbb{Z}K$ Noetherian ([16], Lemma 10.2.2) it follows that $\mathbb{Z}G_\chi$ is Noetherian by a generalised version of Hilbert's basis theorem. More precisely, we find that $t \cdot \mathbb{Z}K = \mathbb{Z}K \cdot t$ since $G' \subseteq K$ and that the subring $S = \langle \mathbb{Z}K, t \rangle$ of $\mathbb{Z}G_\chi$ generated by $\mathbb{Z}K$ and t is Noetherian ([16], Theorem 10.2.6). But $S = \mathbb{Z}G_\chi$.

Proof of Theorem A. Let G be a group with the property that $\mathbb{Z}G$ is Noetherian. If H is a subgroup of G then its group ring $\mathbb{Z}H$ is Noetherian, too ([16], Lemma 10.2.2). In particular, H is of type FP_∞, hence finitely generated.

This shows that the derived subgroup G' is finitely generated and [10], III, Theorem 3.9, applies to yield [3]

$$\Sigma^0(G; A) = \{\, \chi \in V(G) \mid A \text{ is finitely generated over some} \atop \text{finitely generated submonoid of } G_\chi \,\} .$$

Let $\chi \in \Sigma^0(G; A)$. If $\chi = 0$ then $G_\chi = G$ and A is of type FP_∞ over the Noetherian ring $\mathbb{Z}G$. Hence $0 \in \Sigma^\infty(G; A)$ and we may assume $\chi \neq 0$. Using the characterization of $\Sigma^0(G; A)$ above we see that there exists a finite set $\mathcal{X} = \{x_1, \ldots, x_k\} \subseteq G_\chi$ such that A is finitely generated over the monoid

---

[3]See also [6], Proposition 2.4. However, Bieri, Neumann, Strebel always assume that G' acts by inner automorphisms on A, but their arguments go through without this assumption. Bieri and Strebel make this statement precise in [10].

generated by $\mathcal{X}$. Moreover, by enlarging this set and inverting generators if necessary, we can assume that $\mathcal{X}$ maps onto a generating system of G/G'. In other words, $G = \mathrm{gp}(G', x_1, \ldots, x_k)$. Since χ is non-zero there is at least one x_i with $\chi(x_i) > 0$.

The next step is to introduce a sequence $G' =: M_0 \leq M_1 \leq \cdots \leq M_k =: M$ of submonoids of G_χ by putting $M_i := \mathrm{md}(M_{i-1}, x_i)$, the monoid generated by M_{i-1} and x_i, for all $1 \leq i \leq k$.

Claim. $\mathbb{Z}M_i$ is Noetherian for all $0 \leq i \leq k$.

Proof of the claim (by induction on i). Since $M_0 = G'$ is a subgroup of G, the group ring $\mathbb{Z}M_0$ is Noetherian. If $i > 0$ one observes that $x_i \cdot \mathbb{Z}M_{i-1} = \mathbb{Z}M_{i-1} \cdot x_i$ because $G' \subseteq M_{i-1}$. Then [16], Theorem 10.2.6, asserts that the subring $S_i = \langle \mathbb{Z}M_{i-1}, x_i \rangle$ of $\mathbb{Z}M_i$ generated by $\mathbb{Z}M_{i-1}$ and x_i is Noetherian. But it's obvious that $M_i \subseteq S_i$ and hence $\mathbb{Z}M_i = S_i$.

Since A is finitely generated over $\mathbb{Z}M$ we infer that A is of type FP_∞ over $\mathbb{Z}M$. We set $t := x_1 \cdots x_k \in M$. Then $G' \subseteq M \subseteq G_\chi$ and $\chi(t) > 0$. Now, for every element $g \in G$ there exist $j_1, \ldots, j_k \in \mathbb{Z}$ such that $g \in x_1^{j_1} \cdots x_k^{j_k} \cdot G'$. If all j_i are non-negative then $g \in M$. Otherwise we put $j := \max\{-j_1, \ldots, -j_k\} > 0$ and conclude that $t^j \cdot x_1^{j_1} \cdots x_k^{j_k} \cdot G' = x_1^{j_1+j} \cdots x_k^{j_k+j} \cdot G' \subseteq M$. In other words, $G = \bigcup_{j \geq 0} t^{-j} \cdot M = \bigcup_{j \geq 0} M \cdot t^{-j}$. So the assumptions of the criterion of Section 2 are fulfilled and we find $\chi \in \Sigma^\infty(G; A)$.

4 Proofs of Theorem B and Proposition C

The proof of Theorem B is based on a result on the higher geometric invariants of soluble-by-finite groups of type FP_∞ which is part of my thesis. Because its proof is too long to present it here I confine myself to the statement of the result and to some words of explanation and hope to come back to a detailed proof somewhere else.

Theorem D. ([13], Satz F) *Let G be a soluble-by-finite group of type FP_∞. Then the union $\bigcup_{m \geq 0} \Sigma^m(G; \mathbb{Z})^c$ of the complements of the geometric invariants $\Sigma^m(G; \mathbb{Z})$ in $V(G)$ is contained in an open half space of $V(G) \cong \mathbb{R}^n$ and the inclusion*

$$\operatorname*{conv}_{\leq m} \Sigma^1(G; \mathbb{Z})^c \subseteq \Sigma^m(G; \mathbb{Z})^c$$

is valid for all $m \in \mathbb{N}$.

Here $\mathrm{conv}_{\leq m} \Sigma^1(G; \mathbb{Z})^c$ is the union of the convex hulls (in $V(G) \cong \mathbb{R}^n$) of all subsets of $\Sigma^1(G; \mathbb{Z})^c$ with at most m elements. Below we will give some examples of metabelian-by-finite groups of type FP_∞ where the inclusion above fails to be an equality. However, one of the main results of my thesis [13] states that $\mathrm{conv}_{\leq m} \Sigma^1(G; \mathbb{Z})^c = \Sigma^m(G; \mathbb{Z})^c$ if G is metabelian and of type FP_∞ (or, more general, if G is metabelian, of finite Prüfer rank and of type FP_m).

Proof of Theorem B. We have to show that a soluble-by-finite group G is polycyclic-by-finite if $\Sigma^\infty(G;\mathbb{Z}) = V(G)$. As $0 \in \Sigma^\infty(G;\mathbb{Z})$, G is of type FP_∞ hence constructible in the sense of [2] by Kropholler's theorem [12]. Now, constructible soluble-by-finite groups are of finite Prüfer rank [2] and by a result of Mal'cev (see e.g. [17], Proof of Theorem 10.38, Part (a)) they are nilpotent-by-Abelian-by-finite. Let N be a nilpotent-by-Abelian normal subgroup of G with finite quotient $F := G/N$. Then N is of type FP_∞ and the action of F on N induces an action on $V(N)$, written as $\chi \mapsto \chi^f$ for $f \in F$ and $\chi \in V(N)$. Using [7], Corollary 6.3, we see that F acts on $\Sigma^1(N;\mathbb{Z})$ and on its complement $\Sigma^1(N;\mathbb{Z})^c$.

Claim 1. A homomorphism $\chi : N \to \mathbb{R}$ lifts to a homomorphism $\chi^G : G \to \mathbb{R}$ if and only if χ is a fixed point under the action of F.

Proof of Claim 1. Saying that χ is a fixed point under the F-action is equivalent to saying that $\chi([G,N]) = 0$. Applying the exact functor $\mathrm{Hom}_\mathbb{Z}(-;\mathbb{R})$ to the 5-term exact sequence

$$\mathrm{H}_2(G;\mathbb{Z}) \longrightarrow \mathrm{H}_2(F;\mathbb{Z}) \longrightarrow N/[G,N] \longrightarrow G/G' \longrightarrow F/F' \longrightarrow 0$$

(see e.g. [18], 11.4.7) yields an exact sequence

$$0 \longleftarrow \mathrm{Hom}_\mathbb{Z}(N/[G,N];\mathbb{R}) \longleftarrow \mathrm{Hom}_\mathbb{Z}(G/G';\mathbb{R}) \longleftarrow 0$$

because F is finite. Using the fact that $\mathrm{Hom}_\mathbb{Z}(G/G';\mathbb{R}) \cong V(G)$ this proves the claim.

Claim 2. $\Sigma^1(N;\mathbb{Z}) = V(N)$.

Proof of Claim 2. Suppose that $\tilde{\chi} \in \Sigma^1(N;\mathbb{Z})^c$. Then $\chi := \sum_{f \in F} \tilde{\chi}^f$ is a fixed point under the action of F, hence can by lifted to a homomorphism $\chi^G : G \to \mathbb{R}$ by Claim 1. On the other hand, $\Sigma^1(N;\mathbb{Z})^c$ is by definition closed under multiplication by positive real numbers. From this we infer that $\chi = \sum_{f \in F} \frac{1}{m}(m \cdot \tilde{\chi}^f) \in \mathrm{conv}_{\leq m} \Sigma^1(N;\mathbb{Z})^c$ if $m := |F|$. Using Theorem D we find $\chi \in \Sigma^m(N;\mathbb{Z})^c$. Since N has finite index in G and since $\chi^G|_N = \chi$ this implies $\chi^G \in \Sigma^m(G;\mathbb{Z})^c$ ([10], B4.13), a contradiction.

Now, the proof of Theorem B is easily completed. From Claim 2 we infer that N' is finitely generated (use the result of Bieri and Renz quoted in the introduction, [7], Theorem B). Since N' is nilpotent it must be polycyclic. Consequently N is polycyclic and G is polycyclic-by-finite.

The last arguments of the proof show that a *nilpotent-by-Abelian group G is polycyclic if and only if $\Sigma^1(G;\mathbb{Z}) = V(G)$* (cf. [9], Theorem 5.4). Now, one might suspect that for soluble-by-finite groups the latter condition implies that G is polycyclic-by-finite. However, the following counterexamples were shown to me by Peter Kropholler.

Examples. Let the symmetric group S_k $(k \geq 2)$ act on the direct product $H = H_1 \times \cdots \times H_k$ of k copies of the metabelian group $\langle\, a,t \mid tat^{-1} = a^2 \,\rangle$

by permuting the factors. Define $\chi_i : H_i \to \mathbb{Z}$ by putting $a \mapsto 0$, $t \mapsto 1$ and $\chi_H : H \to \mathbb{Z}$ by $\chi_H(h_1, \ldots, h_k) := \chi_1(h_1) + \cdots + \chi_k(h_k)$. Then χ_H lifts to a homomorphism $\chi : G \to \mathbb{Z}$ of the semi-direct product $G := H \rtimes S_k$. Now, G is metabelian-by-finite, of type FP_∞, not polycyclic-by-finite, and using [13], Satz E, one can compute the complements of the higher geometric invariants: $\Sigma^m(G; \mathbb{Z})^c = \emptyset$ for all $m < k$ and $\Sigma^m(G; \mathbb{Z})^c = \{ r \cdot \chi \mid r > 0 \}$ for $m \geq k$.

Finally, we turn to Proposition C. Our original proof of it was based on Theorem B and used the higher geometric invariants. However, the referee has pointed out that Proposition C follows from Kropholler's theorem and a result of Bieri and Strebel ([9], Theorem 5.2) on the invariant $\Sigma^1(G; \mathbb{Z})^c$ of constructible nilpotent-by-Abelian groups G: $\Sigma^1(G; \mathbb{Z})^c$ is contained in a rationally defined open half-space of $V(G)$, i.e. there exists an element $g \in G$ such that $\chi(g) > 0$ for all $\chi \in \Sigma^1(G; \mathbb{Z})^c$ (for these groups the invariant $\sigma(G)$ defined in [9] can be identified with $\Sigma^1(G; \mathbb{Z})^c$, see [10]). The latter result is an important ingredient in the proof of Theorem D (see [13]).

Proof of Proposition C. As in the proof of Theorem B we find a constructible nilpotent-by-Abelian normal subgroup $N \trianglelefteq G$ with finite quotient F. Now, a non-zero homomorphism $\tilde{\chi} \in \Sigma^1(N; \mathbb{Z})^c$ must exist because N is finitely generated and not polycyclic. Since $\Sigma^1(N; \mathbb{Z})^c$ is contained in a rationally defined open half-space of $V(N)$ the sum $\chi := \sum_{f \in F} \tilde{\chi}^f$ cannot be trivial. As χ extends to $\chi^G : G \to \mathbb{R}$, the group G/G' must be infinite.

References

[1] Åberg, H.: *Bieri-Strebel valuations (of finite rank).* Proc. London Math. Soc. (3) **52**, 269–304 (1986)

[2] Baumslag, G. and Bieri, R.: *Constructable solvable groups.* Math. Z. **151**, 249–257 (1976)

[3] Bieri, R.: *Homological dimension of discrete groups.* Queen Mary College Mathematics Notes, London (1976, 2nd ed. 1981)

[4] Bieri, R.: *The geometric invariants of a group – a survey with emphasis on the homotopical approach.* In: Geometric group theory, Vol. 1 (Edited by G.A. Niblo & M.A. Roller). London Math. Soc. Lecture Note Series **181**, Cambridge University Press, Cambridge; 24–36 (1993)

[5] Bieri, R. and Groves, J.R.J.: *Metabelian groups of type (FP)$_\infty$ are virtually of type (FP).* Proc. London Math. Soc. (3) **45**, 365–384 (1982)

[6] Bieri, R., Neumann, W.D. and Strebel, R.: *A geometric invariant for discrete groups.* Invent. Math. **90**, 451–477 (1987)

[7] Bieri, R. and Renz, B.: *Valuations on free resolutions and higher geometric invariants of groups.* Comment. Math. Helv. **63**, 464–497 (1988)

[8] Bieri, R. and Strebel, R.: *Valuations and finitely presented metabelian groups.* Proc. London Math. Soc. (3) **41**, 439–464 (1980)

[9] Bieri, R. and Strebel, R.: *A geometric invariant for nilpotent-by-abelian-by-finite groups.* J. Pure and Appl. Algebra **25**, 1–20 (1982)

[10] Bieri, R. and Strebel, R.: *Geometric invariants for discrete groups.* Preprint of a book (to be published by W. de Gruyter), Universität Frankfurt a.M. (1992)

[11] Hall, P.: *Finiteness properties for soluble groups.* Proc. London Math. Soc. (3) **4**, 419–436 (1954)

[12] Kropholler, P.: *On groups of type* (FP)$_\infty$. J. Pure and Appl. Algebra. **90**, 55-67 (1993)

[13] Meinert, H.: *Die höheren geometrischen Invarianten Σ^m von Gruppen via Operationen auf CW-Komplexen und der Beweis der Σ^m-Vermutung für metabelsche Gruppen endlichen Prüfer-Ranges.* Dissertation, Universität Frankfurt a.M. (1993)

[14] Meinert, H.: *The higher geometric invariants of direct products of virtually free groups.* Comment. Math. Helv. **69**, 39-48 (1994)

[15] Noskov, G.A.: *Bieri-Strebel invariant and homological finiteness properties of metabelian groups.* SFB-Preprint 93–028, Universität Bielefeld (1993)

[16] Passman, D.S.: *The algebraic structure of group rings.* John Wiley & Sons, New York–London–Sydney–Toronto (1977)

[17] Robinson, D.J.S.: *Finiteness conditions for soluble groups* (2 Vols.). Springer-Verlag, Berlin–Heidelberg–New York (1972)

[18] Robinson, D.J.S.: *A course in the theory of groups.* Grad. Texts in Math. **80**, Springer-Verlag, New York (1982)

Fachbereich Mathematik der Johann Wolfgang Goethe – Universität
Robert-Mayer-Str. 6–10
60054 Frankfurt a.M., Germany
E-mail: meinert@math.uni-frankfurt.de

On calculation of width in free groups

A. YU. OL'SHANSKII [1]

0. Introduction

The present note is motivated by the article [1] of Grigorchuk and Kurchanov, in which a notion of *width* of elements in free groups was considered.

Let F be the free group on generators $x_1, \ldots, x_n$, and let $r_1, \ldots, r_m$ be words in the alphabet $X = \{x_1^{\pm 1}, \ldots, x_n^{\pm 1}\}$. For any element w of the normal closure $N = \langle N \rangle^F$ of $R = \{r_1, \ldots, r_m\}$ in F there is an identity

$$w = \prod_{i=1}^{k} s_i r_{j_i}^{n_i} s_i^{-1} \quad (r_{j_i} \in R, s_i \in F, n_i \in \mathbb{Z}) \tag{1}$$

in F (which may, if desired, be chosen with all exponents $n_i = \pm 1$).

Define the R-*width* of $w \in N$ to be the least integer k for which there is an expression (1) for w. If $w \notin N$ then we define the width of w to be infinite. In the case where $m = n$ and $r_i = x_i$ for all $i = 1, \ldots, n$, we recover the notion of width with respect to the free basis. It is proved in [1] that this width can be effectively calculated. In the present paper we make the following observation.

Theorem. *Let R ba a finite subset of a free group F of finite rank. Then there exists an algorithm to calculate the R-width of elements of F if and only if the group $G = F/N = \langle x_1, \ldots, x_n \mid r_1, \ldots, r_m \rangle$ has soluble word problem.*

Since solubility of the word problem for G is equivalent to the the existence of an algorithm to recognise when the width is finite, this is clearly a necessary condition. Moreover, to prove it is sufficient, it is enough to produce a method for calculating the R-width of an element w known to belong to N. In what follows, we will assume that we are given an expression (1) for w in which all exponents n_i are ± 1, and the number of terms $s_i r_{j_i}^{\pm 1} s_i^{-1}$ is ℓ, say.

1. Weight of a word

Clearly, we may assume that every word $r_i \in R$ is cyclically reduced, and has the form $r_i = q_i^{d_i}$, where q_i is not a proper power in F. Moreover, we will assume that q_i and q_j are equal if they are conjugate in F; in this case we will say that r_i and r_j are *commensurable*.

[1] Partially supported by the Russian Foundation of Fundamental Investigations grant N° 011 1541.

Clearly commensurability is an equivalence relation on R. If $C = \{r_1 = q^{d_1}, \ldots, r_s = q^{d_s}\}$ is a commensurability class, we introduce the word $r_C = q^d$, where $d = \gcd(d_1, \ldots, d_s)$. Let $\bar{R}$ denote the set of words r_C for all commensurability classes C. Clearly $N = \langle R \rangle^F = \langle \bar{R} \rangle^F$.

The *weight* of an expression

$$w = \prod_{i=1}^{t} p_i \bar{r}_i^{u_i} p_i^{-1} \quad \bar{r}_i \in \bar{R}, p_i \in F, u_i \in \mathbb{Z} \tag{2}$$

is defined to be the sum $\sum_{i=1}^{t} \nu_i$, where ν_i is defined to be the minimal number of factors $g_1, \ldots, g_{\nu_i}$ such that $\bar{r}_i^{u_i} = g_1 \cdot \ldots \cdot g_{\nu_i}$, and each g_j is a power of some $r_j \in C$, where C is the commensurability class such that $\bar{r}_i = r_C$. The *weight* of w is defined to be the minimum weight of all expressions (2) for w.

Lemma 1. *The weight of a word $w \in N$ coincides with its R-width.*

Proof. Any expression (1) can clearly be rewritten as an expression (2) by substituting a power of the appropriate $r_C \in \bar{R}$ for each $r \in R$. The numbers ν_i for this expression are all equal to 1, and it follows that the weight of w is not greater than its R-width. Conversely, it is clear from the definition of the ν_i that any expression (2) of weight $k = \sum \nu_i$ can be transformed to an expression (1) with k terms in the sum. It follows that the width of w is no greater than its weight.

2. Weight of a van Kampen diagram

For any expression of the form (2) we can construct a van Kampen diagram Δ (see [2], chapter 5) with t faces, such that each oriented edge e of Δ is labelled by a reduced word $\phi(e)$ in X, the boundary label of the i-th face Π (that is, the clockwise product of the edge labels of $\partial\Pi$) is (letter-for-letter) equal to a cyclic permutation of $\bar{r}_i^{u_i}$, and the label of $\partial\Delta$ is (letter-for-letter) equal to a cyclic permutation of w. Conversely, any such van Kampen diagram with t faces effectively gives rise to an expression of the form (2). We define the *weight* of a van Kampen diagram of this kind to be the weight of the corresponding expression (2). Hence our task is to construct effectively a van Kampen diagram of minimal weight for a given word $w \in N$.

We shall say that a pair of faces Π_1, Π_2 in Δ is *reducible* if there is an edge e in $\partial\Pi_1 \cap \partial\Pi_2$ starting with which one reads $\tilde{r}^{u_1}$ and $\tilde{r}^{u_2}$ as labels for $\partial\Pi_1$ and $\partial\Pi_2$ respectively, where $\tilde{r}$ is a cyclic permutation of some $\bar{r} \in \bar{R}$. A diagram Δ is *reduced* if it contains no reducible pair of faces. If Δ does contain a reducible pair Π_1, Π_2, then these can be combined to form a single face Π with boundary label (a cyclic permutation of) $\tilde{r}^{u_1 - u_2}$. (Some care is required here - in the process of identifying edges of Π so that its boundary label becomes cyclically reduced, it can happen that part of the resulting diagram forms a sphere, which can then be removed - see [2], chapter 5 for details.) In any case, Δ is replaced by a diagram Δ' with fewer faces, and of weight no greater than that of Δ. Hence we have the following result.

Lemma 2. *The weight of a word w coincides with the least weight of all reduced diagrams with boundary label w.*

3. Proof of Theorem.

Consider a reduced diagram Δ that is 'minimal' in the sense of Lemma 2. Recall from section 0 that w has an expression of the form (1) involving ℓ terms $s_i r_{j_i}^{\pm 1} s_i^{-1}$, where ℓ is not less than the R-width of w. By Lemmas 1 and 2, Δ has at most ℓ faces.

The diagram Δ has no vertices of degree 2, with the possible exception of the vertex on $\partial \Delta$ from which one reads the boundary label w. A simple corollary of the Euler formula for planar graphs yields the inequality

$$E < 3\ell, \tag{3}$$

where E is the number of (unoriented) edges of Δ. Let M denote the maximum length of the relators r_i, so that

$$M \geq |q_i| \tag{4}$$

for all i, where $|\cdot|$ denotes the length of a word.

Assume that faces Π_1 and Π_2 of Δ have a common edge e with label w_0 such that

$$|w_0| \geq 2M. \tag{5}$$

Suppose $q_1^{v_1} = \bar{r}_1^{v_1/d_1}$ and $q_2^{v_2} = \bar{r}_2^{v_2/d_2}$ be the labels of $\partial \Pi_1$ (clockwise) and $\partial \Pi_2$ (counterclockwise) respectively. Then w_0 is simultaneously a subword of a power of q_1 and of a power of q_2. The inequalities (4) and (5) imply that q_1 and q_2 are commensurable in F (see, e.g. [3]), and hence equal, by assumption. Moreover, the boundary labels of Π_1 and Π_2 read from e are powers of the same cyclic permutation of q_1 (and hence also of $\bar{r}_1$). In particular $\Pi_1 \neq \Pi_2$, and Δ is not reduced, a contradiction.

Thus $|w_0| < 2M$ and the some of the lengths of the label of edges of $\partial \Delta \cap \partial \Pi$ is greater than $|\partial \Pi| - 6\ell M$ for any face Π by (3). But the sum of the lengths of labels on edges of $\partial \Delta$ is not greater than $|w|$, so we have an inequality $|\partial \Pi| < |w| + 6\ell M$ for all Π.

Finally, we observe that the number of van Kampen diagrams satisfying the above conditions is effectively bounded, for the perimeter is at most $|w|$, the number of faces is at most ℓ, and the perimeter of each face is at most $|w| + 6\ell M$.

The weight of each face (and hence of each diagram) can be easily and effectively calculated, as this calculation reduces to a problem involving divisibility of integers. The algorithm claimed in the theorem consists in examining an effectively constructed finite list of planar diagrams with boundedly many edges, labelling edges by words of bounded length, and calculating the weight of each. Of all reduced diagrams in this list with boundary label w, we chose the one of least weight, and this is the R-width of w by Lemmas 1 and 2.

Remark. In the case where $m = n$ and $r_i = x_i$ for all i, the reduced diagram Δ has no interior edges at all, as no pair of faces with labels $x_i^{u_1}$ and $x_i^{u_2}$ could have a common edge in a reduced diagram. So the exponent sum $\sum |n_i|$ in the corresponding expression (1) is bounded above by $|w|$, as asserted by the main lemma of [1].

Acknowledgments. I wish to thank Jim Howie and the referee for their constructive criticisms.

References

[1] Р. И. Григорчук и П. Ф. Курчанов, О ширине елементов в свободных группах, Укр. Матем. Журнал **43** (1991), no. 7-8, 911-918. English translation: R. I. Grigorchuk and P. F. Kurchanov, On the width of elements in free groups, *Ukrainian Math. J.* **43** (1991), no. 7-8, 850-856 (1992).

[2] R. C. Lyndon and P. E. Schupp, *Combinatorial Group Theory*, Springer-Verlag, Heidelberg, 1977.

[3] N. J. Fine and H. S. Wilf, Uniqueness theorems for periodic functions, *Proc. Amer. Math. Soc.* **16** (1965), 109-114.

Department of Mechanics and Mathematics,
Moscow State University,
Moscow 119899
Russia.

Hilbert modular groups and isoperimetric inequalities

CHRISTOPHE PITTET

Abstract

Let Γ_K be a Hilbert modular group. It acts by isometries on the product X of $n = [K:\mathbb{Q}]$ hyperbolic planes. There is a Γ_K-invariant core X_0 obtained by removing from X a disjoint union of horoballs such that the quotient $\Gamma_K \backslash X_0$ is compact. In the case $[K:\mathbb{Q}] = 2$ the boundary components of X_0 admit *Sol* geometry. This implies exponential lower bounds for the Dehn functions of Hilbert modular groups when $[K:\mathbb{Q}] = 2$.

1 Introduction

If $\langle S; R \rangle$ is a finite presentation for a group Γ, the Dehn function of this presentation is defined by

$$n \rightarrow \sup_{|w| \leq n} A(w)$$

where w is a reduced word on the alphabet S, containing at most n letters, which belongs to the normal closure of the symmetrical set R. The area $A(w)$ of w is the minimal number of relations $r \in R$, counted with multiplicity, required to write w as a product of conjugates of relations.

Among the innumerable asymptotic invariants defined by Gromov in [Gro93], the Dehn function $\delta(n)$ is one of the most studied.

In this paper we detail some aspects of point 3.H of "Asymptotic invariants of infinite groups" [Gro93]. We give details and references for elementary proofs of the facts mentioned in the abstract.

Let us say few words about how the behaviour of Dehn functions of Hilbert modular groups should appear as a special case in a more general setting. The Dehn function of $SL(3, \mathbb{Z})$ is exponential (see [CEH+92] 10 and [Gro93] 3.I) and, according to Thurston, $SL(n, \mathbb{Z})$ has quadratic Dehn functions if $n > 3$. If $[K:\mathbb{Q}] > 2$ one expects polynomial Dehn functions for Γ_K (see [Gro93] 5.A9). The group $SL(n, \mathbb{Z})$ is an irreducible lattice in the semi-simple Lie group $SL(n, \mathbb{R})$ and the group Γ_K is an irreducible lattice in the semi-simple Lie group $SO(2, 1) \times .. \times SO(2, 1)$ where the number of factors equals $[K:\mathbb{Q}]$. In the above two families the exponential lower bounds hold when the real rank of the Lie group is 2. One expects that this is true whenever Γ is an irreducible lattice in a semi-simple Lie group of real rank 2 ([LP]). Dehn functions for irreducible lattices in semi-simple Lie groups always have exponential upper bounds (see [Gro93] 3.I'2).

2   The discrete embedding of Γ_K in $PSL(2,\mathbb{R})^n$

Let K be a totally real number field of degree $n = [K : \mathbb{Q}]$. Let $\sigma_i : K \to \mathbb{R}$, $1 \leq i \leq n$ be the n distinct embeddings of K in $\mathbb{R}$ (see [Sam67] 2.8 and 4.2). Let O_K be the ring of integers of K (see [Sam67] 2.1 and 2.5). For example, if $K = \mathbb{Q}[\sqrt{2}]$, then $O_K = \{x + y\sqrt{2} : x, y \in \mathbb{Z}\}$.

Let $SL(2, O_K)$ be the group of two by two matrices with coefficients in O_K and determinant 1. The Hilbert modular group $\Gamma_K = PSL(2, O_K)$ associated to K is the quotient of $SL(2, O_K)$ by its center $\{\pm 1\}$. Let $\sigma : K \to \mathbb{R}^n$ be the map which sends x to $(\sigma_1(x), ..., \sigma_n(x))$. The image of O_K under σ in $\mathbb{R}^n$ is a lattice of rank n (see [Sam67] 2.1 and 4.2). Since the only discrete subgroups of $\mathbb{R}$ are cyclic, the projection of this lattice on each factor is dense provided $n > 1$. If $M \in PSL(2, O_K)$ is represented by the matrix $\begin{pmatrix} \alpha & \beta \\ \gamma & \delta \end{pmatrix} \in SL(2, O_K)$, define $\sigma_i(M)$ as the class of $\begin{pmatrix} \sigma_i(\alpha) & \sigma_i(\beta) \\ \sigma_i(\gamma) & \sigma_i(\delta) \end{pmatrix}$ in $PSL(2, O_K)$. The embedding $PSL(2, O_K) \to PSL(2, \mathbb{R})^n$ sending M to $(\sigma_1(M), ...\sigma_n(M))$ is discrete (this follows from the fact that $\sigma(O_K)$ is discrete in $\mathbb{R}^n$). The image is irreducible in the sense that if $\sigma_i(M) = id$ for some i then this holds for all i (this is because the σ_i are field homomorphisms). Since $PSL(2, \mathbb{R})$ is the isometry group of the hyperbolic plane, this implies that Γ_K acts properly discontinuously by isometries on the product X of n copies of the hyperbolic plane.

3   The Γ_K-invariant core X_0 of X

The purpose of this section and of the next one is to prove the following fact.

Proposition 3.1 *There exists a Γ_K-invariant submanifold X_0, obtained by removing from X a disjoint union of horoballs such that the quotient $\Gamma_K \backslash X_0$ is compact.*

This implies in particular, that Γ_K is finitely presented and is quasi-isometric to X_0 with its induced Riemannian metric (see [Tro90] 19).

Let us recall some geometric facts to introduce the notion of a cusp. The group Γ_K acts on the ideal boundary S^{2n-1} of X (see [BGS85] I.3). The ideal boundary of each factor $\mathbb{H}^2$ in X is a circle S^1 in S^{2n-1}. The product $S^1 \times ... \times S^1 = T^n$ is a Γ_K-invariant torus in S^{2n-1}. Consider the natural action of $PSL(2, O_K)$ on $P^1(K)$. The embeddings $\sigma_i : K \to \mathbb{R}$ induce embeddings $\sigma_i : P^1(K) \to P^1(\mathbb{R})$ of projective lines. Together they define an embedding of $P^1(K)$ into the product of n copies of $P^1(\mathbb{R})$. Identifying $P^1(\mathbb{R})$ with the ideal boundary of $\mathbb{H}^2$ and observing that the natural action of $PSL(2, \mathbb{R})$ on $P^1(\mathbb{R})$ is also the action of the isometries of $\mathbb{H}^2$ on its ideal boundary, we can view $P^1(K)$ as a Γ_K-invariant subspace of T^n in S^{2n-1}.

Definition 3.1 *A cusp for $\Gamma_K = PSL(2, O_K)$ is an orbit of $PSL(2, O_K)$ in $P^1(K)$.*

For example if $n = 1$, the group $PSL(2,\mathbb{Z})$ has only one cusp in $P^1(\mathbb{Q})$. The following lemma is a translation of Dirichlet's theorem (see [Sam67] 4.3) on the finiteness of the ideal class group of a number field.

Lemma 3.1 *The set of cusps of Γ_K in $P^1(K)$ is finite.*

For the proof see [VdG88] I.1.1.

We now define functions which measure "the (inverse) distance from a cusp". If a is a non-zero ideal of O_K let $N(a)$ be its norm (see [Sam67] 3.5). If $\kappa = (\alpha : \beta) \in P^1(K)$ we are free to assume $\alpha, \beta \in O_K$ (just multiply α and β by some $m \in \mathbb{Z}$). Let $a = \langle \alpha; \beta \rangle$ be the ideal generated by α and β. Let $\mu_\kappa : X \to \mathbb{R}_+^*$ defined by

$$\mu_\kappa(z) = N(a)^2 \prod_{i=1}^{n} \frac{Im(z_i)}{|\sigma_i(\alpha) - \sigma_i(\beta)z_i|^2}.$$

If $x \in K^*$ and if $x\alpha$ and $x\beta$ are in O_K then $N(xa) = |N(x)||N(a)|$. If $x \in K^*$ then $\prod_{i=1}^{n} |\sigma_i(x)| = |N(x)|$. Hence μ_κ does not depend on the choice of the representatives α and β for κ. Furthermore, if $\gamma \in \Gamma_K$ then

$$\mu_{\gamma\kappa}(\gamma z) = \mu_\kappa(z), \forall z \in X.$$

This follows from two things. First if $\begin{pmatrix} a & b \\ c & d \end{pmatrix}$ is in $SL(2, O_K)$ and represents γ then $\langle a\alpha + b\beta; c\alpha + d\beta \rangle = \langle \alpha; \beta \rangle$. Secondly if $M \in SL(2,\mathbb{R})$, $w \in \mathbb{C}$ with $Im(w) > 0$ and $x, y \in \mathbb{R}$ then

$$\frac{Im(Mw)}{|x - yMw|^2} = \frac{Im(w)}{|x - yw|^2}.$$

Lemma 3.2 *There is a constant $C > 0$ such that for all $t = (t_1, ..., t_n) \in \mathbb{R}^n$ with $t_i \neq 0, 1 \leq i \leq n$ there is a unit $u \in O_K$ such that*

$$|\sigma_i(u)t_i| \leq C \prod_{j=1}^{n} |t_j|^{1/n}, \ 1 \leq i \leq n.$$

Proof. Consider the hyperplane of $\mathbb{R}^n$ defined by

$$W = \{x \in \mathbb{R}^n : \sum_{i=1}^{n} x_i = 0\}.$$

Given $p \in \mathbb{R}^n$, there is an $x \in W$ such that

$$p_i - x_i \leq \frac{1}{n} \sum_{j=1}^{n} p_j, \ 1 \leq i \leq n;$$

for instance, we can take $x_i = \frac{n-1}{n}p_i - \frac{1}{n}\sum_{j\neq i}p_j$. Hence if Λ is a lattice in W there is a constant C such that, given $p \in \mathbb{R}^n$, there is an $x \in \Lambda$ such that

$$p_i - x_i \leq C + \frac{1}{n}\sum_{j=1}^{n}p_j, \; 1 \leq i \leq n.$$

Let $L : K^* \to \mathbb{R}^n$ be defined by $L(x) = (log|\sigma_1(x)|, ..., log|\sigma_n(x)|)$. Let $U \subset O_K$ be the group of units. According to Dirichlet's unit theorem (see [BS] 2.4 or [Sam67] 4.4), $L(U)$ is a lattice in W. Applying the exponential map to the inequalities above proves the lemma.

If $r > 0$ and $\kappa \in P^1(K)$, let $X_\kappa(r) = \{z \in X : \mu_\kappa(z) > r\}$. This is a horoball associated to κ. See Lemma 4.1.

Lemma 3.3 *There exists $r > 0$ such that, if $\sigma, \tau \in P^1(K)$ are distinct then $X_\sigma(r) \cap X_\tau(r) = \emptyset$.*

Proof. See [VdG88] I.2.1 where the previous lemma is used (but not proved).
Let $x \in \mathbb{R}^n$. We denote by

$$||x|| = \sup_{1\leq i\leq n} |x_i|$$

the sup norm. If $u \in \mathbb{R}^n$ we denote by ux the vector $(u_1x_1, ..., u_nx_n) \in \mathbb{R}^n$. The following lemma follows from an application of the pigeon hole principle (see [Fre90] I.3.6.1 for a proof).

Lemma 3.4 *Let Λ be a lattice in $\mathbb{R}^n$. There exists a constant $C > 0$ such that, for each $x \in \mathbb{R}^n$ and each $\epsilon > 0$, there are $u, v \in \Lambda$ with $u \neq 0$ such that $||ux + v|| < \epsilon$ and $||u|| < C/\epsilon$.*

Lemma 3.5 *There exists $s > 0$ such that $X = \bigcup_{\kappa\in P^1(K)} X_\kappa(s)$.*

Proof. The formula $\mu_{\gamma\kappa}(\gamma z) = \mu_\kappa(z)$ for $\gamma \in \Gamma_K$ implies, for any $r > 0$, that the set $\bigcup_{\kappa\in P^1(K)} X_\kappa(r)$ is Γ_K-invariant. Therefore it is enough to find $s > 0$ such that given $z \in X$ there is a $\gamma \in \Gamma_K$ such that $\gamma z \in \bigcup_{\kappa\in P^1(K)} X_\kappa(s)$. According to 3.2 there is a constant $C_1 > 0$ such that if $z \in X$ and if y_i is the imaginary part of z_i, then there is a unit $u \in O_K$ such that

$$|\sigma_i(u)y_i^{1/2}| \leq C_1 \prod_{j=1}^{n} |y_j|^{1/2n}, \; 1 \leq i \leq n.$$

Replacing z by $\begin{pmatrix} u & 0 \\ 0 & u^{-1} \end{pmatrix} z$ and denoting by x_i and y_i the real and imaginary parts of the ith coordinate of this new point, we have, since $\prod_{i=1}^{n} \sigma_i(u)^2 = 1$, that

$$y_i \leq C_1^2 \prod_{j=1}^{n} y_j^{1/n}, \; 1 \leq i \leq n.$$

We now apply Lemma 3.4 to the lattice $\sigma(O_K)$ of $\mathbb{R}^n$ with $x = (x_1, ..., x_n)$ and with $\epsilon = \prod_{j=1}^{n} y_j^{1/2n}$, to obtain $\alpha, \beta \in O_K$ with $\beta \neq 0$ such that

$$|\sigma_i(\alpha) - \sigma_i(\beta)x_i|^2 \leq \prod_{j=1}^{n} y_j^{1/n}, \ 1 \leq i \leq n$$

and

$$|\sigma_i(\beta)|^2 \leq C_2 \prod_{j=1}^{n} y_j^{-1/n}, \ 1 \leq i \leq n.$$

where C_2 is a constant depending only on K. Separating the real and the imaginary part and using the three systems of inequalities above gives

$$\prod_{i=1}^{n} |\sigma_i(\alpha) - \sigma_i(\beta)z_i|^2 \leq (1 + C_2 C_1^4)^n \prod_{i=1}^{n} y_i.$$

Let $s = (1 + C_2 C_1^4)^{-n}$. Let $\kappa = (\alpha : \beta)$. As the norm of an ideal is bigger or equal to 1 it follows that $\mu_\kappa(z) \geq s$.

Let $r > 0$ be big enough so that $\bigcup_{\kappa \in P^1(K)} X_\kappa(r)$ is a disjoint union. Let

$$X_0 = X \setminus \bigcup_{\kappa \in P^1(K)} X_\kappa(r).$$

This set is Γ_K-invariant. We will now study the subgroups of Γ_K stabilizing a boundary component of X_0.

4   Cusp groups and horospheres

Let $\kappa \in P^1(K)$. Define $\Gamma_\kappa = \{\gamma \in \Gamma_K : \gamma\kappa = \kappa\}$. Let $r > 0$. Denote by

$$S_\kappa(r) = \{z \in X : \mu_\kappa(z) = r\}$$

the border of $X_\kappa(r)$. It follows from the formula $\mu_{\gamma\kappa}(\gamma z) = \mu_\kappa(z)$ where $\gamma \in \Gamma_K$ that $\gamma \in \Gamma_\kappa$ if and only if $\gamma S_\kappa(r) = S_\kappa(r)$. Up to conjugation by $g \in PSL(2, K)$ sending κ to $(1 : 0) = \infty$ it is enough to study Γ_∞ acting on $S_\infty(r)$ (see [VdG88] I.3). We fix $r = 1$ and denote $S_\infty(1)$ by S. Let

$$H = \left\{ \begin{pmatrix} e^t & x \\ 0 & 1 \end{pmatrix} : t, x \in \mathbb{R} \right\}.$$

We identify X with the Lie group G defined as a direct product of n copies of H. A left-invariant metric on G is given by $ds^2 = \sum_{i=1}^{n} e^{-2t_i} dx_i^2 + dt_i^2$. With this identification S is the semi-direct product defined by the exact sequence

$$0 \rightarrow \mathbb{R}^n \rightarrow S \rightarrow V \rightarrow 0$$

where the projection sends $(x_1, ..., x_n, e^{t_1}, ..., e^{t_n})$ to $(t_1, ..., t_n)$ and where

$$V = \{(t_1, ..., t_n) \in \mathbb{R}^n : \sum_{i=1}^{n} t_i = 0\}.$$

This is because $\mu_\infty(z) = \prod_{i=1}^{n} Im(z_i) = \prod_{i=1}^{n} e^{t_i}$.

Lemma 4.1 *The hypersurface S is a horosphere of X.*

Proof. The one-parameter subgroup of G given in coordinates by $c(t) = (0, ..., 0, e^t, ..., e^t)$ is a geodesic which is orthogonal to S. As S is invariant in G, for every $s \in S$ the left-translated geodesic $sc(t)$ is at finite Hausdorff distance from $c(t)$. This shows that S is an hypersurface in G orthogonal to an asymptotic family of geodesics.

Each $\gamma \in \Gamma_\infty$ is represented by a matrix of the form

$$\begin{pmatrix} \epsilon^2 & a \\ 0 & 1 \end{pmatrix}$$

where ϵ is a unit of O_K and $a \in O_K$. The map $\Gamma_\infty \to S$ sending such a matrix to

$$\left(\begin{pmatrix} \sigma_1(\epsilon^2) & \sigma_1(a) \\ 0 & 1 \end{pmatrix}, ..., \begin{pmatrix} \sigma_n(\epsilon^2) & \sigma_n(a) \\ 0 & 1 \end{pmatrix} \right)$$

makes Γ_∞ appear as a lattice in S. More precisely,

Lemma 4.2 *The quotient $\Gamma_\infty \backslash S$ is a T^n-bundle over T^{n-1}.*

Proof. The exact sequence

$$0 \to \mathbb{R}^n \to S \to V \to 0$$

restricts to an exact sequence

$$0 \to \sigma(O_K) \to \Gamma_\infty \to \mathbb{Z}^{n-1} \to 0$$

where the group $\sigma(O_K)$ is a lattice in $\mathbb{R}^n$ and, according to Dirichlet's unit theorem, where the group

$$\mathbb{Z}^{n-1} = \{2(log|\sigma_1(\epsilon)|, ..., log|\sigma_n(\epsilon)|) : \epsilon \in U\}$$

is a lattice in V. It follows that the solvmanifold $\Gamma_\infty \backslash S$ is a T^n-bundle over T^{n-1}.

In the case $n = 2$ the torus bundle over the circle so obtained has a *Sol* geometry (see [Thu82]). The same things are true for $\kappa \in P^1(\mathbb{R})$. As those manifolds are compact it follows from Lemma 3.1 and Lemma 3.5 that $\Gamma_K \backslash X_0$ is compact.

5   Dehn functions

Let M be a simply connected Riemannian manifold. Let $c : S^1 \to M$ be a piecewise smooth loop of length $l(c)$. Consider all continuous and piecewise smooth maps $h : D^2 \to M$ from the Euclidian unit disk to M such that $h|_{\partial D^2} = c$. The area of the loop c is defined as

$$A(c) = \inf_h \int_{D^2} J_h d\mu$$

where J_h is the Jacobian of h.

Recall that Sol is the semi-direct product of $\mathbb{R}^2$ with $\mathbb{R}$ where $t \in \mathbb{R}$ acts on $\mathbb{R}^2$ by the matrix $\begin{pmatrix} e^t & 0 \\ 0 & e^{-t} \end{pmatrix}$. We choose the coordinates (x, y, t) on Sol where x, y are the canonical coordinates on $\mathbb{R}^2$. A left-invariant metric on Sol is given by $e^{-2t}dx^2 + e^{2t}dy^2 + 2dt^2$.

Lemma 5.1 *There is a constant $C > 0$ and a family of loops*

$$c_k : S^1 \to Sol, \ k \in \mathbb{N}$$

such that $l(c_k) < 4k + 4$ and $A(c_k) > e^k / C$.

Proof. Let $k \in \mathbb{N}$. Consider the rectangle in $\mathbb{R}^2$ whose corners are

$$(0, 0), (1, 0), (1, e^k), (0, e^k).$$

Denote by h_1, v_1, h_2, v_2 the sequence of oriented sides of the rectangle read conterclockwise. Let $p : Sol \to \mathbb{R}^2$ be the projection $(x, y, t) \to (x, y)$. The path $\tilde{v}_1$ in Sol between $(1, 0, 0)$ and $(1, e^k, 0)$ obtained by successively following the paths

$$(1, 0, s), \ 0 \geq s \geq -k$$

$$(1, s, -k), \ 0 \leq s \leq e^k$$

$$(1, e^k, -k + s), \ 0 \leq s \leq k$$

projects onto v_1. As $\int_0^{e^k} e^{-k} dy = 1$ we deduce that $l(\tilde{v}_1) \leq 2k + 1$. In a similar way we define a path $\tilde{v}_2$ between $(0, e^k, 0)$ and $(0, 0, 0)$ which projects on v_2 and such that $l(\tilde{v}_2) \leq 2k + 1$. Let $c_k : S^1 \to Sol$ be the loop obtained as the composition $h_1 \tilde{v}_1 h_2 \tilde{v}_2$. Hence $p \circ c_k$ is the rectangle and

$$l(c_k) \leq 4k + 4.$$

Let $h_L : D^2 \to p^{-1}(v_2)$ be a map such that $h_L|_{\partial D^2}$ is the loop $v_2^{-1} \tilde{v}_2$. (If c is a path we denote by c^{-1} the same path with opposite orientation.) Let $h_M : D^2 \to \mathbb{R}^2$ be a map such that $h_M|_{\partial D^2}$ is the rectangle. Let $h_R : D^2 \to p^{-1}(v_1)$ be a map such that $h_R|_{\partial D^2}$ is the loop $\tilde{v}_1 v_1^{-1}$. Let $h_0 : D^2 \to Sol$ be obtained by gluing h_L with h_M along v_2 and h_M with h_R along v_1. Then $h_0|_{\partial D^2} = c_k$. Let ω be the pull-back by p of the volume form on $\mathbb{R}^2$. We have

$$\int_{D^2} h_0^*\omega = \int_{D^2} h_L^*\omega + \int_{D^2} h_M^*\omega + \int_{D^2} h_R^*\omega = \int_{D^2} h_M^*\omega = e^k.$$

Let $h : D^2 \to Sol$ be a map such that $h|_{\partial D^2} = c_k$. Let α be a 1-form such that $d\alpha = \omega$; for example $\alpha = x\,dy$. By Stokes formula

$$\int_{D^2} h^*\omega = \int_{S^1} c_k^*\alpha = \int_{D^2} h_0^*\omega = e^k.$$

As ω is left-invariant, ω is bounded. Hence there is a constant $C > 0$ depending only on the metric of Sol such that

$$\int_{D^2} J_h d\mu \geq 1/C \int_{D^2} h^*\omega.$$

Hence $A(c_k) \geq e^k/C$.

We refer the reader to [Alo90], [Ger91] and [LS77] for definitions and properties of Dehn functions, Dehn or Van-Kampen diagrams and Cayley graphs.

Proposition 5.1 *Let K such that $[K\!:\!\mathbb{Q}] = 2$. The Dehn function of Γ_K is at least exponential.*

As explained in [Gro93] 3.H4, exponential upper bounds are obtained by first finding a quadratic filling for a loop in X and then by projecting the filling disk onto X_0 using the normal projection of X to the boundary horospheres of X_0. (Using this argument in the combinatorial setting of Dehn functions seems to require technical results about minimal surfaces.)

In the course of the proof of Prop. 5.1 we will use the following fact.

Let $r > 0$. The projection

$$p_\infty : \{z \in X : \mu_\infty(z) \leq r\} \to \{z \in X : \mu_\infty(z) = r\}$$

along geodesics normal to the horosphere $S_\infty(r)$ does not increase distances and hence does not increase areas.

Now we prove Prop. 5.1.

Proof. Let Y be the 2-complex associated to a finite presentation $\langle S; R \rangle$ of Γ_K. Let $\tilde{Y}$ be the universal covering of Y. The 1-skeleton of $\tilde{Y}$ is the Cayley graph $\mathcal{C}(\Gamma_K, S)$. Let $r > 0$ as in the definition of X_0. Let $x_0 \in S_\infty(r)$. The map $\Gamma_K \to X_0$ sending γ to γx_0 can be extended to a map $\rho : \tilde{Y} \to X_0$ which is piecewise smooth and Γ_K-equivariant. Each 2-cell in $\tilde{Y}$ corresponds to an $r \in R$. Let

$$C_1 = \sup_{r \in R} A(\rho(r)).$$

Let $c_k : S^1 \to S_\infty(r)$ be a family of loops as in Lemma 5.1 (notice that $S_\infty(r)$ with its induced Riemannian metric is isometric to *Sol* with a left-invariant metric). We reparametrise those loops by arc-length.

$$c_k : [0, l(c_k)] \to S_\infty(r).$$

As $\Gamma_\infty \backslash S_\infty(r)$ is compact, there is a constant $C_2 > 0$ such that for all $k \in \mathbb{N}$ and all $i \in \{0, 1, .., [l(c_k)]\}$ there is a $w_k(i) \in \Gamma_\infty$ such that

$$d(\rho(w_k(i)), c_k(i)) < C_2$$

where the distance is considered in $S_\infty(r)$. We choose geodesic segments

$$[w_k(0), w_k(1)], ..., [w_k(i), w_k(i+1)], ..., [w_k([l(c_k)]), w_k(0)]$$

in $\mathcal{C}(\Gamma_K, S)$ and define w_k as the loop in $\mathcal{C}(\Gamma_K, S)$ obtained by successively following those segments. As the action of Γ_K on X is proper there is a constant $C_3 > 0$ such that the word norm of w_k satisfies

$$|w_k| < C_3 l(c_k) \ \forall k \in \mathbb{N}.$$

Let D_k be a Dehn diagram for w_k. Let $\iota : D_k \to \tilde{Y}$ be the cellular natural map sending the base point to the identity. Let $A(D_k)$ be the number of 2-cells of D_k. We have

$$C_1 A(D_k) \geq A(\rho(\iota(D_k))) \geq A(p_\infty(\rho(\iota(D_k)))).$$

Now if M is a simply connected Riemannian manifold diffeomorphic to $\mathbb{R}^n$ which admits a cocompact group of isometries, then for any $L > 0$ there is a constant $C > 0$ such that if $c : S^1 \to M$ is a piecewise smooth loop with $l(c) < L$ then $A(c) < C$. We apply this fact to the loops in $S_\infty(r)$ obtained, for each k and each i, by successively following the paths

$$c_k(s), \ i \leq s \leq i+1$$
$$[c_k(i+1), \rho(w_k(i+1))]$$
$$p_\infty \rho[w_k(i+1), w_k(i)]$$
$$[\rho(w_k(i)), c_k(i)]$$

(by $[a, b]$ we mean a geodesic between two points a and b). Hence there is a constant C_4 such that

$$A(p_\infty(\rho(\iota(D_k)))) \geq A(c_k) - C_4 l(c_k).$$

Acknowledgements Thanks to C. Drutu, K. Vogtmann, M. Burger, E. Leuzinger, A. Papasoglu and P. Pansu for discussions and motivations.

References

[Alo90]   J.M. Alonso.   Inégalités isopérimétriques et quasi-isométries.
          C.R.A.S. Série 1, 311:761–764, 1990.

[BGS85]   W. Ballmann, M. Gromov, and V. Schroeder. *Manifolds of non-
          positive curvature*, volume 61. Birkhäuser, 1985.

[BS]   Z.I. Borevich and I.R. Shafarevich. *Number theory*. Academic
          Press.

[CEH+92] J. Cannon, D. Epstein, D. Holt, S. Levy, M. Paterson and
          W. Thurston. *Word processing in groups*. Bartlett and Jones,
          1992.

[Fre90]   E. Freitag. *Hilbert Modular Forms*. Springer-Verlag, 1990.

[Ger91]   S.M. Gersten. Dehn functions and l_1-norms of finite presentations.
          In G. Baumslag, F.B. Cannonito, and C.F. Miller, editors, *Proceed-
          ings of the Workshop on Algorithms, Words Problems and Classi-
          fication in Combinatorial Group Theory*. Springer-Verlag, 1991.

[Gro93]   M. Gromov. Asymptotic invariants of infinite groups. In G. A.
          Niblo and M. A. Roller, editors, *Geometric Group Theory, Vol-
          ume 2*, London Mathematical Society Lecture Note Series 182,
          Cambridge University Press, 1993.

[LP]   E. Leuzinger and Ch. Pittet. Isoperimetric inequalities for lattices
          in rank 2. In preparation.

[LS77]   R.C. Lyndon and P.E. Schupp. *Combinatorial Group Theory*.
          Springer-Verlag, 1977.

[Sam67]   P. Samuel. *Théorie algébrique des nombres*. Hermann, 1967.

[Thu82]   W. Thurston. Three dimensional manifolds, kleinien groups and
          hyperbolic geometry. *Bull. Amer. Math. Soc.*, 3(6):357–381, 1982.

[Tro90]   M. Troyanov. Espaces à courbure négative et groupes hyper-
          boliques. In E. Ghys and P. De la Harpe, editors, *Sur les groupes
          hyperboliques d'après M. Gromov*, pages 47–68. Birkhäuser, 1990.

[VdG88]   G. Van der Geer. *Hilbert Modular Surfaces, Ergebnisse der Math-
          ematik und ihrer Grenzgebiete* 16. Springer-Verlag, 1988.

Laboratoire de Topologie
Bât. 425
Université de Paris-Sud
ORSAY 91405 CEDEX, France.

On systems of equations
in free groups

ALEXANDER A. RAZBOROV

Abstract

The main purpose of this presentation is to introduce non-specialists into recent developments in the study of systems of equations in free groups and, at a very intuitive level, into the related techniques.

1 Introduction

The most important class of algebraic systems for which the question of the algorithmic decidability of their elementary theories was left open after the remarkable breakthrough in the area achieved in the 60's, is the class of free groups. It is also open whether the elementary theories of non-abelian free groups with different number of generators coincide or not; both these questions are usually attributed to Tarski.

The natural approach to these questions is to begin with first-order formulae from a restricted class. The simplest nontrivial class is made by existential formulae i.e. those of the form

$$\exists x_1 \ldots \exists x_n \, A(x_1, \ldots, x_n), \tag{1}$$

where A is open. It turns out that even for this class Tarski's first problem becomes very hard, and the second becomes non-trivial.

In fact, if we additionally require the formula A in (1) to be positive (that is to contain no negations) then already this case bears all the difficulties of the case of arbitrary existential formulae (see e.g. [20]). With this additional requirement, the formula (1) is equivalent to a finite disjunction of systems of equations in free groups. The latter objects are the main target of this presentation.

The two natural questions about systems of equations in free groups are:

1. *Is it algorithmically decidable whether a given system has at least one solution?*

2. *How to describe the solution set?*

Both these questions have been answered in affirmative by now. The machinery involved in settling them is, however, extremely complicated. In this presentation we try to explain it at a very rough and intuitive level.

The paper is organized as follows. Section 2 contains necessary definitions. In Section 3 we address Question 1, and in Section 4 we address Question 2.

The reader who has got that far will undoubtedly have been convinced of the extreme importance of so-called quadratic equations for the whole theory. In Section 5 we present a result which indicates that those equations make a proper hierarchy. An informal implication of this is that one can not hope to get a drastic simplification of the description from Section 4.

The paper is concluded with a brief overview in Section 6.

2   Notations and Definitions

We will denote the free group with basis $x_1, \ldots, x_n$ by $\langle x_1, \ldots, x_n \rangle$. The symbol $=$ will stand for the group equality, and $A_1 \ldots A_r \overset{\text{x}}{=} B_1 \ldots B_s$, will mean, by definition, that both left- and right-hand sides are reduced and equal to each other. $\partial(A)$ is the length of the word A.

Vectors of letters like $\phi_1, \ldots, \phi_m$ or $X_1, \ldots, X_n$ etc. will be denoted by $\bar{\phi}, \bar{X}, \ldots$ When we need to consider simultaneously several vectors of the same type, we use superscripts: $\bar{X}^{(1)}, \ldots, \bar{X}^{(h)}, \ldots$

We fix the following group alphabets:

$$\Sigma_0 \rightleftharpoons \{x_1, \ldots, x_n, \ldots, x_1^{-1}, \ldots, x_n^{-1}, \ldots\} \quad - \quad \text{the alphabet of}$$
$$\textit{variables},$$
$$\Sigma_1 \rightleftharpoons \{a_1, \ldots, a_\omega, \ldots, a_1^{-1}, \ldots, a_\omega^{-1}, \ldots\} \quad - \quad \text{the alphabet of}$$
$$\textit{coefficients}.$$

Let F_0 and F_1 be free groups with the set of generators Σ_0, Σ_1 respectively. A *system of equations in a free group* is a system of equalities of the form

$$\begin{cases} \phi_1(\bar{x}, \bar{a}) = 1, \\ \cdots \\ \phi_m(\bar{x}, \bar{a}) = 1, \end{cases} \tag{2}$$

where $\phi_1, \ldots, \phi_m$ are words over $\Sigma_0 \cup \Sigma_1$. A system is *trivial* if all its equations have the form $1 = 1$. A system is *coefficient-free* if its equations do not contain coefficients. A *solution* of the system (2) is a vector $\bar{X} \in F_1$ such that $\phi_i(\bar{X}, \bar{a}) = 1$ for all $1 \leq i \leq m$. We denote by $\mathrm{Sol}(\bar{\phi})$ the set of all solutions of the system (2).

The *length* of a solution $\bar{X} = (X_1, \ldots, X_n)$ [of a system $\bar{\phi} = (\phi_1, \ldots, \phi_m)$] is $\partial(\bar{X}) \rightleftharpoons \sum_{i=1}^n \partial(X_i)$ [$\partial(\bar{\phi}) \rightleftharpoons \sum_{i=1}^m \partial(\phi_i)$, respectively]. A solution $\bar{X}$ of a system $\bar{\phi}$ is *minimal* if $\partial(\bar{X}) \leq \partial(\bar{X}')$ for every $\bar{X}' \in \mathrm{Sol}(\bar{\phi})$. Obviously, any system which has at least one solution also has a minimal solution.

A system is *quadratic* if every variable which occurs in the system, occurs there exactly twice. It is very easy to see that quadratic systems can always be reduced to single quadratic equations. However, for the sake of uniformity, we often will prefer to talk of arbitrary quadratic systems rather than of single equations.

3   Algorithmic Decidability

The algorithm for deciding whether $\mathrm{Sol}(\bar{\phi}) = \emptyset$ was given by Makanin [19, 20][1]. It is appropriate, however, to give a short survey of preceding results before elucidating some ideas of that work.

The case of quadratic equations is easy. All the necessary ideas for solving this case can be extracted from [8, 22] and [10, Sections 1.6, 1.7].

Lyndon [9] constructed an algorithm for solving equations in one variable, and Lorents [17] noticed that his algorithm can also be applied to *systems* of equations in one variable.

Khmelevskiĭ [28, 29] established the algorithmic decidability for systems in which every equation contains at most two variables and, moreover, has either the form $\phi(x, \bar{a}) = \psi(y, \bar{a})$ or $W(x, y) = A(\bar{a})$, where all letters which may appear in ϕ, ψ, W, A are explicitly displayed. Ozhigov [24] was able to handle equations in two variables.

As we mentioned above, the algorithm for solving arbitrary systems in free groups was given by Makanin [19, 20].

In fact, all these results (with the important exception [9, 17]) use the same idea of simplifying the pair $\left(\bar{\phi}, \bar{X}\right)$, where $\bar{X} \in \mathrm{Sol}(\bar{\phi})$, by applying to $\bar{\phi}$ some elementary transforms. It is the kind of allowed transforms and the requirements on the arrangement of equations in the systems which make the difference between different approaches. In this paper, however, we consider things only at a very intuitive level, and the easiest way to explain the basic idea is to start with the toy-like case of quadratic systems.

So, assume that we have a non-trivial quadratic system

$$\bar{\phi}(x_1, \ldots, x_n, a_1, \ldots, a_\omega),$$

and imagine for a moment that $\bar{X}$ is its solution. As $\bar{\phi}\left(\bar{X}, \bar{a}\right) = 1$, the vector $\left(\bar{X}, \bar{a}\right)$ is *not* Nielsen reduced and hence can be simplified by an elementary Nielsen transformation. This transformation may have one of the following forms:

$$X_i \text{ is crossed out since } X_i = 1, \tag{3}$$

$$X_i^\varepsilon \longrightarrow X_i^\varepsilon U, \tag{4}$$

where $U \in \left\{X_1^{\pm 1}, \ldots, X_n^{\pm 1}, a_1^{\pm 1}, \ldots a_\omega^{\pm 1}\right\}$. It is important that in case (4) the transformation is *attached to* the system $\bar{\phi}$ (cf. [10, Section 1.6]) that is $x_i^\varepsilon u$ is a subword of one of its equations. Here $u = U$ if $U \in \Sigma_1$ and $u = x_j^\delta$ if $U = X_j^\delta$. Denote by $\left(\bar{X}', \bar{a}\right)$ the vector that results from this transformation.

Let $\bar{\phi}'(\bar{x}', \bar{a})$ be the system obtained from $\bar{\phi}$ by the substitution $x_i \longrightarrow 1$ in case (3), and by the substitution $x_i^\varepsilon \longrightarrow x_i^\varepsilon u^{-1}$ in case (4). Then $\bar{X}'$ is its solution.

---

[1]the original proof in [19] contained a gap shortly after fixed in [20]

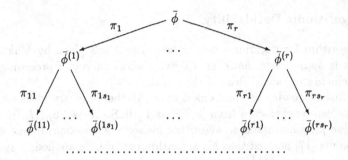

Figure 1: The tree made of elementary transforms

Let's now forget about the imaginary solution $\bar{X}$ and see what we can do without it. We still can apply to $\bar{\phi}$ all elementary transformations (3) and those transformations (4) which are attached to it. This results in a list of systems

$$\bar{\phi}^{(1)}(\bar{x}^{(1)}, \bar{a}) = 1, \ldots, \bar{\phi}^{(r)}(\bar{x}^{(r)}, \bar{a}) = 1$$

and group homomorphisms

$$\pi_1 : \langle \bar{x}, \bar{a} \rangle \longrightarrow \langle \bar{x}^{(1)}, \bar{a} \rangle, \ldots, \pi_r : \langle \bar{x}, \bar{a} \rangle \longrightarrow \langle \bar{x}^{(r)}, \bar{a} \rangle$$

such that $\pi_i(a_j) = a_j$ and

$$\text{Sol}(\bar{\phi}) = \bigcup_{i=1}^{r} \left\{ \pi_i(\bar{X}^{(i)}) \,\middle|\, \bar{X}^{(i)} \in \text{Sol}(\bar{\phi}^{(i)}) \right\}. \tag{5}$$

Here $\pi_i\left(\bar{X}^{(i)}\right)$ means that we interpret $\bar{X}^{(i)}$ as a homomorphism

$$\langle \bar{x}^{(i)}, \bar{a} \rangle \longrightarrow F_1$$

and compose this with π_i to get a vector $\bar{X}$.

The effective transforms of one system $\bar{\phi}$ into a finite set of systems $\bar{\phi}^{(1)}, \ldots, \bar{\phi}^{(r)}$ with the characteristic property (5) is the core of almost all known approaches. This is what we called above *elementary transforms*; (5) basically means that we have reduced the system $\bar{\phi}$ to $\bar{\phi}^{(1)}, \ldots, \bar{\phi}^{(r)}$.

We can iterate the elementary transforms and get a tree (see Figure 1).

Now we once again bring in consideration a solution $\bar{X}$ and note that the elementary Nielsen transformations can not be applied to $\left(\bar{X}, \bar{a}\right)$ forever. Instead, they terminate at some point, and the only way it may happen is when we arrive at a trivial system. This means that $\text{Sol}(\bar{\phi}) \neq \emptyset$ *if and only if the tree in Figure 1 contains a trivial system*. The main problem (which, once again, is characteristic) is that this tree is in general infinite.

The remedy in the case of quadratic systems is very simple. Namely, $\partial\left(\bar{\phi}^{(i)}\right) \leq \partial\left(\bar{\phi}\right)$. This is obvious for singular transformations $x_i \longrightarrow 1$, and if we apply a regular transformation $x_i^\varsigma \longrightarrow x_i^\varsigma u^{-1}$ attached to $\bar{\phi}$ then the length indeed increases by two but then it is restored to at least its original value as $u^{-1}u$ gets cancelled. Hence the tree in Figure 1 in fact contains only finitely many different systems, and the algorithm for solving $\bar{\phi}$ looks as simple as that.

Start constructing the tree crossing out all repetitions of the same system (the subtrees rooted at such systems are isomorphic so we may consider only one of them). When the process is over, look at the result. The original system $\bar{\phi}$ has at least one solution if and only if the tree contains at least one trivial system.

For the general case this naive approach immediately breaks apart. We can not guarantee any longer that the tree contains only finitely many different systems, hence we do not have a weapon against the potential infinity. In later works this is circumvented by a much more careful choice of the class of systems, their solutions, elementary transforms etc. allowed in the tree. Let us briefly sketch some ideas along these lines from the culminating work of Makanin [19, 20].

The first step (which is very easy) is to reduce our original system $\bar{\phi}\left(\bar{X},\bar{a}\right)$ to finitely many systems of the form

$$
\begin{cases}
\psi_1(\bar{x},\bar{a}) \doteq \theta_1(\bar{x},\bar{a}), \\
\quad \cdots \\
\psi_u(\bar{x},\bar{a}) \doteq \theta_u(\bar{x},\bar{a}).
\end{cases}
\tag{6}
$$

This is done simply by writing down all possible ways in which the words $\phi_i\left(\bar{X},\bar{a}\right)$ may get canceled.

The next step is to arrange the resulted systems (6) into a special form which Makanin, somewhat ironically, called *generalized equations*. We omit the exact definition.

Then we (similarly to the case of quadratic systems but in a much more sophisticated way) define elementary transforms for generalized equations so that the analogue of the crucial property (5) still holds. These elementary transforms are designed to control as many parameters of generalized equations as possible, just as Nielsen elementary transformations controlled the length of the quadratic systems.

Next we construct the analogue of the tree in Figure 1. Makanin's elementary transforms never increase the length of the solution, and those transforms which preserve it can be applied along any branch only finitely often. Which means that, once again, $\mathrm{Sol}(\bar{\phi}) \neq \emptyset$ iff the tree contains the trivial generalized equation.

The problem with the potential infinity of the tree is this time much more elaborate. We still can cross out the repetitions in the tree but what remains still may contain infinite branches along which generalized equations get more and more complicated. Makanin showed, however, that if the original solution $\bar{X}$ was *minimal* then it is possible to find a *recursive* bound on the number of elementary transforms leading from $\left(\bar{\phi}, \bar{X}\right)$ to the trivial generalized equation. One of the crucial properties of minimal solutions necessary (but not sufficient) for this purpose is Bulitko's Lemma [15]: *if $(X_1, \ldots, X_n)$ is a minimal solution of a system $\bar{\phi}$ then X_i may not contain $P^{\gamma(\bar{\phi})}$ as a subword, where P is non-empty and $\gamma(\bar{\phi})$ is an explicit integer-valued recursive function*. If we are interested only in the *solvability* of the system $\bar{\phi}$, it is good enough, and we can terminate the construction of the tree at a point which is specified by a *recursive* function given in advance.

We conclude this section with a brief survey of known partial results related to Tarski's problems. For the second problem (that is whether the elementary theories of non-abelian free groups with different number of generators coincide or not) the strongest results were obtained by Merzlyakov [23] and Sacerdote [13]. Namely, Merzlyakov showed that the set of positive formulae which are true on a non-abelian free group is indeed independent of its rank. Sacerdote established the same thing for the class of $\forall \ldots \forall \exists \ldots \exists$-formulae, that is for formulae of the form

$$\forall x_1 \ldots \forall x_n \exists y_1 \ldots \exists y_m \ A(x_1, \ldots, x_n, y_1, \ldots, y_m)$$

with A open. These results do *not* use Makanin's technique.

Makanin's algorithm immediately gives rise to an algorithm recognizing the set of true positive existential formulae. Based upon ideas of Merzlyakov, Makanin [20] noticed that it can be also used for recognizing true formulae *either* from the class of positive formulae *or* from the class of existential formulae. The latter result was strengthened by him in [21], where the decidability for the class of $\forall \exists \ldots \exists$-formulae was established (notice that we have here only *one* universal quantifier, not a block).

4 Description of the Solution Set

There are three known languages for describing $\mathrm{Sol}(\bar{\phi})$.

The first, logical language is based upon the notion of *parametric functions*. Very good examples are provided by Lyndon's parametric words [9] and Nielsen-Khmelevskiĭ function [28, 29]. Lyndon used his parametric words for describing the solution set of equations in one variable, and this description was extended to the case of one-variable systems and considerably simplified by Lorents [17]. The final result proved independently by Appel [1] and Lorents [18] says that for any such system $\bar{\phi}$,

$$\mathrm{Sol}(\bar{\phi}) = \bigcup_{i=1}^{h} \left\{ A_i B_i^{\mu} C_i \mid \mu \in \mathbb{Z} \right\}, \qquad (7)$$

and this representation can be constructed effectively. Khmelevckiĭ [28, 29] proved that the solution set of those systems for which he established the algorithmic decidability (see Section 3) can always be represented in terms of Lyndon's parametric words and Nielsen-Khmelevskiĭ functions.

The second approach is that of *fundamental sequences* which was suggested in [25]. It has more algebraic flavour and is based upon the well-known observation that $\mathrm{Sol}(\bar{\phi})$ is actually isomorphic to the set of coefficient-preserving homomorphisms from $\mathrm{Hom}\left(H(\bar{\phi}), F_1\right)$. Here $H(\bar{\phi})$ is the finitely presented group $\langle x_1, \ldots, x_n, a_1, \ldots, a_\omega \mid \phi_1(\bar{x}, \bar{a}) = 1, \ldots, \phi_m(\bar{x}, \bar{a}) = 1 \rangle$. One advantage of this approach is that it allows us to combine into one scheme both the representation of $\mathrm{Sol}(\bar{\phi})$ and *the proof* that all vectors $\bar{X}$ obtained from this representation are indeed in $\mathrm{Sol}(\bar{\phi})$.

At this conference, however, the most appropriate language is geometric. Before we continue, let me stress that the differences between all three languages are actually very minor.

Once again, we begin with the case of quadratic equations. Let us once more consider the tree in Figure 1 but instead of removing repetitions in that tree, we factor it by the corresponding equivalence relation. Then we delete from the resulting finite graph all systems from which no paths lead to trivial systems (as follows from the discussion in the previous section, they do not have solutions anyway).

We end up with a finite directed graph whose nodes are marked with systems of equations in a free group, and every edge of the form

$$e : \bar{\phi}^{(1)}\left(\bar{x}^{(1)}, \bar{a}\right) \longrightarrow \bar{\phi}^{(2)}\left(\bar{x}^{(2)}, \bar{a}\right)$$

is marked by a coefficient-preserving homomorphism $\pi_e : \langle \bar{x}^{(1)}, \bar{a} \rangle \longrightarrow \langle \bar{x}^{(2)}, \bar{a} \rangle$. This graph has one input node marked by the original system $\bar{\phi}$, and all sink nodes are marked by trivial systems. The main property of this graph (compare with (5)) is that

$$\mathrm{Sol}(\bar{\phi}) \;=\; \left\{ \pi_p\left(\bar{X}'\right) \mid p \text{ is a path from the input node to a sink} \right.$$
$$\left. \text{node, and } \bar{X}' \text{ is an arbitrary vector} \right\}.$$

In connection with this, note that trivial systems usually have "dummy" variables, but their solution set is also trivial and consists of all possible vectors whatsoever. π_p is defined in the natural way.

A graph with this properties is called the *solution graph* for the system $\bar{\phi}$. In order to better understand its structure, we recall that two nodes u and v lie in the same *strong component* iff there exists a loop containing both of them. This is an equivalence relation and, factoring by it the solution graph, we get an *acyclic* graph. Now, we can separate all I/O paths in the factor graph and come up with a tree. All this means that we can represent the solution graph as in Figure 2. Here $\bar{\phi}$ is the original system, $\bar{\psi}^{(1)}, \ldots, \bar{\psi}^{(h)}$ are all trivial, and circles correspond to strong components. Although it is not clear a priori, in all known constructions the edges lying within a strong component can be

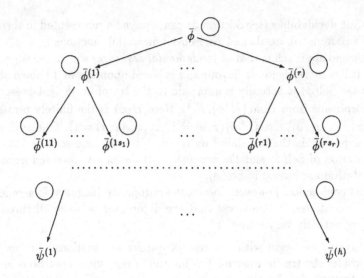

Figure 2: The solution graph

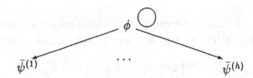

Figure 3: The solution graph for quadratic equations

inverted. Hence we may assume that every strong component consists of a single node, and that the set of all loops within that component possesses the group structure. Moreover, this group actually consists of automorphisms of a certain group $G(\bar{\phi})$ associated with the system $\bar{\phi}$ sitting at that node. This observation lies in the heart of the algebraic approach to describing $\mathrm{Sol}(\bar{\phi})$ mentioned at the beginning of this section.

Coming back to the case of quadratic equations, I want to mention a remarkable result proved independently by Comerford and Edmonds [3] and by Grigorchuk and Kurchanov [6] (some partial results in that direction were obtained in [8, 22, 14, 28, 29, 4, 11]). Namely, for that case the general picture in Figure 2 can be simplified as shown in Figure 3. Another way of stating their result is that, while transforming an arbitrary solution $\bar{X}$ along the tree in Figure 1, we can apply first a sequence of regular transforms (4) alone (this give us the circle on Figure 3) followed by a sequence of singular transforms (3).

Now we turn to the general case. Makanin's algorithm alone does not suffice since it essentially uses the minimality of the solution $\bar{X}$. This difficulty

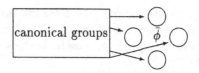

Figure 4: The structure of strong components

was overcome in two steps by Razborov [25, 27]. The key idea is to bring into consideration transforms which are inverse to Makanin's elementary transforms. When we meet a situation which is not liked by Makanin's algorithm then, instead of the minimality assumption, we use a combination of original and inverse transforms in order to "shrink" our solution while retaining the generalized equation. A number of new difficulties arises on this way, e.g. the fact that inverse transforms may in general increase the length of the solution. All these difficulties are, however, tractable. The final result [27] says that for every system $\bar{\phi}$ we can effectively construct a finite solution graph describing $\mathrm{Sol}(\bar{\phi})$.

It is clear already from Figure 2 that it is the strong components which make the most essential part of this description (as this is the only place where the potential infinity comes in). Something can be additionally said about their structure. Namely, every strong component can be assumed to have the form as shown in Figure 4, where the petals correspond to so-called *canonical groups of automorphisms* [27] closely related to quadratic equations. We address the reader to [27, 16] for the precise definition and confine ourselves here to one typical example.

Assume that the system $\bar{\phi}$ in Figure 4 has the special form

$$\left\{ \begin{array}{l} \bar{\psi}(\bar{z}, \bar{a}) = 1 \\ \prod_{i=1}^{n} w_i^{-1} U_i(\bar{z}, \bar{a}) w_i \cdot \prod_{j=1}^{g} [v_{2j-1}, v_{2j}] = U_0(\bar{z}, \bar{a}). \end{array} \right. \tag{8}$$

Let P be the pointwise stabilizer of the set

$$\left\{ u_1, \ldots, u_n, \prod_{i=1}^{n} w_i^{-1} u_i w_i \cdot \prod_{j=1}^{g} [v_{2j-1}, v_{2j}] \right\}$$

in $\mathrm{Aut}(\bar{u}, \bar{v}, \bar{w})$. Then every automorphism $\sigma \in P$ gives rise, in a natural way, to an automorphism $\sigma' \in \mathrm{Aut}(\bar{a}, \bar{z}, \bar{v}, \bar{w})$ preserving $\bar{a}, \bar{z}$ and the equations of the system (8). We take a finite system of generators $\sigma_1, \ldots, \sigma_\ell$ of P (it can be found effectively e.g. by the general result of McCool, see [10, Chapter 1.6]) and form one petal in Figure 4 by including to it ℓ edges marked $\sigma_1', \ldots, \sigma_\ell'$.

Very loosely speaking, the result from [25, 27] reduces arbitrary systems of equations in a free group to a finite number of auxiliary *quadratic* equations. One application of this result was given in [25], namely the solution of the so-called rank problem: the rank of a coefficient-free system of equations in a free group is algorithmically computable. Here the *rank* of a system is, by definition, the maximal possible rank of subgroups generated by its solutions. Some of the techniques from [27] were also useful in the study of group actions on $\mathbb{R}$-trees (see [2, 5]).

5   The Hierarchy of Quadratic Equations

In view of the description given in the previous section, it is important to study quadratic equations in a free group. Another good reason for this is their close connection to the Poincaré Conjecture (see e.g. [10, Section 4.4] or [16, §2.2]). One natural question is whether their hierarchy is proper or, more generally, isn't it possible to start with a *finite* number of *arbitrary* systems $\bar{\phi}^{(1)}, \ldots, \bar{\phi}^{(\ell)}$ so that for any other system $\bar{\phi}$, Sol($\bar{\phi}$) can be "decomposed" on the base of Sol $\left(\bar{\phi}^{(1)}\right), \ldots,$ Sol $\left(\bar{\phi}^{(\ell)}\right)$. The intuitive answer is "no". The proof, however, is a more complicated story, and it was given in [12] (this result was originally announced in [26]).

For the purposes of this section the geometric language becomes somewhat awkward and does not let us to formulate the result in its full generality. So, we switch to the more general language of parametric functions. Fix one more group alphabet

$$\Sigma_2 \rightleftharpoons \{u_1, \ldots, u_n, \ldots, u_1^{-1}, \ldots, u_n^{-1}, \ldots\}$$

of *free word variables*. Let $P^{(1)}, P^{(2)}, \ldots, P^{(m)}, \ldots$ be some sets of absolutely arbitrary nature. We associate with each $P^{(i)}$ countably many *parameter variables* $p_1^{(i)}, \ldots, p_k^{(i)}, \ldots$

A *parametric function of rank r* is an arbitrary mapping

$$F(p_1^{(i_1)}, \ldots, p_l^{(i_l)}) : P^{(i_1)} \times \cdots \times P^{(i_l)} \longrightarrow \{u_1^{\pm 1}, \ldots, u_r^{\pm 1}\}^*.$$

We can think alternatively of F as of the mapping

$$F : P^{(i_1)} \times \cdots \times P^{(i_l)} \times \underbrace{\Sigma_1^* \times \cdots \times \Sigma_1^*}_{r} \longrightarrow \Sigma_1^* \qquad (9)$$

which takes $p_1^{(i_1)} \in P^{(i_1)}, \ldots, p_l^{(i_l)} \in P^{(i_l)}$ and $U_1, \ldots, U_r \in \Sigma_1^*$ to the result of substituting into $F(p_1^{(i_1)}, \ldots, p_l^{(i_l)})$ the words $U_1, \ldots, U_r$ for the free word variables $u_1, \ldots, u_r$. We consider F as a function symbol

$$F(p_1^{(i_1)}, \ldots, p_l^{(i_l)}, u_1, \ldots, u_r)$$

in the first order many-sorted logic with the interpretation provided by (9).

Given a family $\mathcal{L}$ of parametric functions, we define *parametric terms over* $\mathcal{L}$ merely as those terms in the many-sorted signature $\langle \Sigma_1, {}^{-1}, \cdot, \mathcal{L} \rangle$ which take values in Σ_1^*. See [12] for more details.

Let now $\bar{\phi}(x_1, \ldots, x_n, \bar{a})$ be a system of equations, and $T_i^{(j)}(\bar{p}, \bar{u})$ $(1 \le i \le n, \ 1 \le j \le h)$ be a matrix of parametric terms over some $\mathcal{L}$. We say that this matrix *represents* $\mathrm{Sol}(\bar{\phi})$ *over* $\mathcal{L}$ if

$$\left\{ \bar{T}^{(j)}\left(\bar{p}, \bar{U}\right) \,\middle|\, 1 \le j \le h, \ \boldsymbol{p}_k^{(\nu)} \in P^{(\nu)}, \ \bar{U} \in \Sigma_1^* \right\} = \mathrm{Sol}(\bar{\phi}).$$

The connection with the geometric language from the previous section is clear. If we have a solution graph represented as in Figure 2, then for every internal node marked by $\bar{\phi}'(x_1, \ldots, x_r, a_1, \ldots, a_\omega) = 1$ we introduce the parametric functions

$$F_1(p, u_1, \ldots, u_{r+\omega}), \ldots, F_r(p, u_1, \ldots, u_{r+\omega}),$$

where p runs over the fundamental group of the corresponding strong component. Namely, $F_i(p, u_1, \ldots, u_{r+\omega})$ is the result of substituting into $\pi_p(x_i)$ letters $u_1, \ldots, u_{r+\omega}$ for $x_1, \ldots, x_r, a_1, \ldots, a_\omega$. Then for every leaf $\bar{\psi}^{(j)}$ in the tree we can define parametric terms $\bar{T}^{(j)}(\bar{p}, u_1, \ldots, u_{r_j})$ representing exactly those solutions which terminate at $\bar{\psi}^{(j)}$. The free word variables $u_1, \ldots, u_{r_j}$ are in one-to-one correspondence with the set of (dummy) variables of $\bar{\psi}^{(j)}$.

Of course, the definition of parametric functions is quite general. For example, if we have the canonical coefficient-free equation

$$\prod_{i=1}^{g}[x_i, y_i] = 1 \tag{10}$$

then it is very easy to represent its solution set by parametric functions of rank g. We merely enumerate all its solutions in $\langle u_1, \ldots, u_g \rangle$.

However, while not quite adequate for positive results, this language is perfect to state the following theorem:

Theorem 5.1 *Let $g \ge 3$. Then the solution set of the equation* (10) *cannot be represented over the language consisting of* **all** *parametric functions of rank at most* $(g-1)$.

As a corollary, we obtain that no *finite* language $\mathcal{L}$ suffices to represent the solution set of all canonical quadratic equations (10). As we noticed in the beginning of this section, this result is not unexpected. In fact, the main motivation for pursuing the proof was to see which kind of techniques we do need to prove things like that. We conclude this section with mentioning one concept discovered on this way as it may be of independent interest.

The following hierarchy $\mathcal{H}(d, r)$ of words over Σ_2 lies in the heart of the proof:

$$\mathcal{H}(0, r) \rightleftharpoons \Sigma_2,$$
$$\mathcal{H}(d+1, r) \rightleftharpoons \cup \left\{ \mathrm{Gp}(A_1, \ldots, A_r) \,\middle|\, A_1, \ldots, A_r \in \mathcal{H}(d, r) \right\}.$$

The members of this hierarchy majorize the sets of values of parametric terms. And what we basically need for the proof is to establish "lower bounds" for it: that is to show that some explicit words do not belong to certain members of this hierarchy. Such a machinery is indeed developed in [12].

6 Discussion and Open Problems

As in the previous sections, we begin with the case of quadratic equations. Here the situation is rather well understood, and the results from [3, 6] mentioned in Section 4 show that studying quadratic equations (or systems) is in fact more or less equivalent to studying mapping class groups of 2-manifolds. Of course, the latter area is still rich in important open problems, but this is already another story.

Turning to the general case, it should be noticed first that Makanin's algorithm is of purely existential nature. Indeed, Kościelski and Pacholsky [7] noticed that it is not even primitively recursive. It would be extremely interesting to reduce the complexity of the algorithm or (which, to my feeling, is the problem of the same sort) to be able to apply it to at least some concrete interesting systems.

Now, disregarding the complexity questions, the problem of describing $\mathrm{Sol}(\bar{\phi})$ for arbitrary $\bar{\phi}$ is reduced to solving finitely many auxiliary quadratic equations. I feel, however, that this reduction is incomplete since every strong component of the solution graph still may contain several "canonical petals" (see Figure 4). It would be very nice to show that actually only one such petal per node is sufficient. This would immediately imply that every system $\bar{\phi}$ can be reduced to a "triangular" system of quadratic equations. I feel that this question might be solved by refining the technique from [27].

Turning to Tarski's first problem, the next interesting class of formulae to be considered is made by $\forall \ldots \forall \exists \ldots \exists$-formulae. As we noticed in Section 3, Makanin solved the partial case when there is only one universal quantifier. Attempts to generalize his result to a sequence of universal quantifiers immediately lead to several natural questions about the description of $\mathrm{Sol}(\bar{\phi})$. One of them was already mentioned in the previous paragraph. Another natural thing to ask along these lines is the following.

Can we assume that the number of dummy (= free word) variables in every trivial system appearing in the solution graph is strictly less than the number of variables in the original system (unless the latter is itself trivial)? Note for comparison that it is true for systems in one variable, that is the Lyndon-Lorents description (7) does not involve free word variables. It is this circumstance which allowed Makanin to solve the case with one universal quantifier. If the conjecture above is true then for the case of arbitrary $\forall \ldots \forall \exists \ldots \exists$-formulae we might try to apply induction on the number of universal quantifiers.

An upper bound on the number of free word variables was noticed in [27] but it is only quadratic and, which is worse, is given in terms of the *length* of $\bar{\phi}$, not in terms of the number of variables. Somewhat curiously, the main problem with resolving this conjecture lies in the obvious transition from the original system to (6) since we apparently must introduce new variables for pieces resulting from the process of cancellation.

7 Acknowledgement

This work was supported by the grant # 93-011-16015 of the Russian Foundation for Fundamental Research.

References

[1] K. I. Appel, One variable equations in free groups. *Proc. Amer. Math. Soc.*, 19 (1968) 912–918.

[2] M. Bestvina and M. Feighn, Stable actions of groups on real trees. Manuscript, 1992.

[3] L. P. Comerford and C. C. Edmunds, Solutions of equations in free groups. In *Group Theory, Proceedings of the Singapore Group Theory Conference held at the National University of Singapore, 1987*, 347-56. Walter de Gruyter, 1989.

[4] M. Culler, Using surfaces to solve equations in free groups. *Topology*, 20 (1981) 133-145.

[5] D. Gaboriau, G. Levitt and F. Paulin, Pseudogroups of isometries and Rips' theorem on free actions on $\mathbb{R}$-trees. *Israel Journal of Mathematics*, to appear (1994).

[6] R. I. Grigorchuk and P. F. Kurchanov, On quadratic equations in free groups. *Contemporary Mathematics*, 131 (1992) 159–171, (Part 1).

[7] A. Kościelski and L. Pacholski, Complexity of unification in free groups and free semi-groups. In *Proceedings of the 31st IEEE FOCS*, 824-829, 1990.

[8] R. C. Lyndon, The equation $a^2b^2 = c^2$ in free groups. *Michigan Math. J.*, 6 (1959) 89–95.

[9] R. C. Lyndon. Equations in free groups. *Trans. Amer. Math. Soc.*, 96 (1960) 445–457.

[10] R. C. Lyndon and P. E. Schupp, *Combinatorial Group Theory*. Springer-Verlag, New York/Berlin, 1977. рус. пер.: Р. Линдон, П. Шупп, Комбинаторная теория групп, М.: Мир, 1980.

[11] D. Piollet, Solutions d'une equation quadratique dans le groupe libre. *Discrete Math.*, 59 (1986) 115–123.

[12] A. Razborov, On the parameterization of solutions for equations in free groups. *International Journal of Algebra and Computation*, 3 (1993) 251-273.

[13] G. S. Sacerdote, Elementary properties of free groups. *Trans. Amer. Math. Soc.*, 178 (1972) 127–138.

[14] H. Zieschang, Alternierende Produkte in freien Gruppen. *Abh. Math. Sem. Univ. Hamburg*, 27 (1964) 13–31.

[15] В. К. Булитко. Об уравнениях и неравенствах в свободной группе и свободной полугруппе. *Ученые записки мат. кафедр. Тульский гос. пед. ин-т*, 2:242–253, 1970. V. K. Bulitko, On equations and inequalities in a free group and in a free semigroup, *Proceedings of the Math. Dep. Tula State Pedagogic Institute*, 2 (1970), 242-253.

[16] Р. И. Григорчук и П. Ф. Курчанов. Некоторые вопросы теории групп, связанные с геометрией. In *Итоги науки и техники. Современные проблемы математики, фундаментальные направления*, 58. ВИНИТИ, 1990. R. I. Grigorchuk and P. F. Kurchanov, Some Questions of Group Theory Related to Geometry, *Encycl. Math. Sci.* 58 (2) Springer-Verlag (1993).

[17] А. А. Лоренц. Решение систем уравнений с одним неизвестным в свободных группах. *Докл. АН СССР*, 148(6):1253–1256, 1963. A. A. Lorenc, The solution of systems of equations in one unknown in free groups, *Soviet Math. Dokl.*, 4(1963), 262-266.

[18] А. А. Лоренц. О представлении множеств решений систем уравнений с одним неизвестным в свободных группах. *Докл. АН СССР*, 178(2):290–292, 1968. A. A. Lorenc, On the representation of solution sets of systems of equations with one unknown in a free group, *Soviet Math. Dokl.*, 9(1968).

[19] Г.С. Маканин. Уравнения в свободной группе. *Изв. АН СССР, сер. матем.*, 46(6):1199–1273, 1982. G. S. Makanin, Equations in a free group, *Math. USSR Izv.*, 21(1983), 483-546.

[20] Г. С. Маканин. Разрешимость универсальной и позитивной теорий свободной группы. *Изв. АН СССР, сер. матем.*, 48(4):735–749, 1984. G. S. Makanin, Decidability of universal and positive theories of a free group, *Math. USSR Izv.*, 1984.

[21] Г. С. Маканин. Об одном разрешимом фрагменте элементарной теории свободной группы. In С. И. Адян, editor, Вопросы кибернетики. Сложность вычислений и прикладная математическая логика, pages 103–114. ВИНИТИ, Москва, 1988. G. S. Makanin, On one decidable fragment of the elementary theory of a free group. In *Problems of Cybernetics. Complexity Theory and Applied Mathematical Logic*, ed.: S. I. Adian. Moscow, 1988, pages 103-114.

[22] А. И. Мальцев. Об уравнении $zxyx^{-1}y^{-1}z^{-1} = aba^{-1}b^{-1}$ в свободной группе. *Алгебра и логика*, 1(5):45–50, 1962. A. I. Malcev, On the equation $zxyx^{-1}y^{-1}z^{-1} = aba^{-1}b^{-1}$ in a free group, *Algebra i Logika*, 1, 45-50 (1962).

[23] Ю. И. Мерзляков. Позитивные формулы на свободных группах. *Алгебра и логика*, 5(4):25–42, 1966. Yu. I. Merzlyakov, Positive formulae over free groups, *Algebra i Logika*, 5(4):25-42 (1966).

[24] Ю. И. Ожигов. Уравнения с двумя неизвестными в свободной группе. *Докл. АН СССР*, 268(4):808–814, 1983. Yu. I. Ozhigov, Equations with two unknowns in a free group, *Soviet Math. Dokl.*, 27(1983).

[25] А. А. Разборов. О системах уравнений в свободной группе. *Изв. АН СССР, сер. матем.*, 48(4):779–832, 1984. A. A. Razborov, On systems of equations in a free group, *Math. USSR Izvestiya*, 25(1):115-162, 1985.

[26] А. А. Разборов. Об уравнениях в свободной группе, общие решения которых не представимы в виде суперпозиции конечного числа параметрических функций. In *Тезисы 9 всесоюзного симпозиума по теории групп*, page 54, Москва, 1984. A. A. Razborov, An equation in a free group whose set of solutions does not allow a representation as a superposition of a finite number of parametric functions. In *Proceedings of the 9th All-Union Symposium on the Group Theory*, Moscow, 1984, p. 54.

[27] А. А. Разборов. О системах уравнений в свободной группе. PhD thesis, МГУ, 1987. A. A. Razborov, On systems of equations in a free group, Moscow State University, 1987.

[28] Ю. И. Хмелевский. Системы уравнений в свободной группе, I. *Изв. АН СССР, сер. матем.*, 35(6):1237–1268, 1971. Yu. I. Khmelevskiĭ, Systems of equations in a free group I,*Math. USSR Izv.*, 5(1971).

[29] Ю. И. Хмелевский. Системы уравнений в свободной группе, II. *Изв. АН СССР, сер. матем.*, 36(1):110–179, 1972. Yu. I. Khmelevskiĭ, Systems of equations in a free group II, *Math. USSR Izv.*, 6(1972).

Steklov Mathematical Institute
Vavilova 42,
117966, GSP–1, Moscow,
RUSSIA

Cogrowth and essentiality in groups and algebras

AMNON ROSENMANN [1]

Abstract

The cogrowth of a subgroup is defined as the growth of a set of coset representatives which are of minimal length. A subgroup is essential if it intersects non-trivially every non-trivial subgroup. The main result of this paper is that every function $f : \mathbb{N} \cup \{0\} \longrightarrow \mathbb{N}$ which is strictly increasing, but at most exponential, is equivalent to a cogrowth function of an essential subgroup of infinite index of the free group of rank two. This class of functions properly contains the class of growth functions of groups.

The notions of growth and cogrowth of right ideals in algebras are introduced. We show that when the algebra is without zero divisors then every right ideal, whose cogrowth is less than that of the algebra, is essential.

1 Growth, Cogrowth and Essentiality in Groups

1.1 Growth and Cogrowth of Subgroups

A *growth function* $\Gamma_S(n)$ on a set S with a length function l on it is defined by

$$\Gamma_S(n) := \operatorname{card}\{s \in S \mid l(s) \le n\}, \tag{1}$$

assuming that $\Gamma_S(n)$ is finite for each n. A preorder is given on the growth functions by

$$\Gamma_1(n) \preceq \Gamma_2(n) \iff \exists C \, [\Gamma_1(n) \le C\Gamma_2(Cn)]. \tag{2}$$

The notion of growth when applied to finitely generated groups (see [6] for an overview) has been investigated mainly after Milnor's paper ([12]). A geometric interpretation can be given, for example, when computing the growth function of the fundamental group of a Riemannian manifold. In order to avoid the dependence of the length function upon the generating set of the group, an equivalence relation is used

$$\Gamma_1(n) \sim \Gamma_2(n) \iff \exists C \, [\Gamma_1(n) \le C\Gamma_2(Cn) \,\&\, \Gamma_2(n) \le C\Gamma_1(Cn)]. \tag{3}$$

[1]Supported by the Minerva Fellowship

We will also use the following notation

$$\Gamma_1(n) \prec \Gamma_2(n) \iff \Gamma_1(n) \preceq \Gamma_2(n) \ \& \ \Gamma_1(n) \nsucc \Gamma_2(n). \tag{4}$$

The growth function of the group G will be denoted by $\Gamma_G(n)$, when referring to its equivalence class and also when some fixed generating set is assumed (but omitted in the notation).

When H is a subgroup of G we may speak of the *cogrowth* of H in G, denoted $\Gamma_{G/H}(n)$. This is defined to be the growth of a (complete) set of coset representatives for H in G which is "minimal" in the sense that every representative is of minimal length in its coset relative to the given group-generating set of G. Clearly, any other set of coset representatives will grow at most as fast as a minimal set. We also notice that cogrowth functions of H relative to different generating sets of G are equivalent. (Remark: it does not matter if we take right or left cosets because the inverses of the right representatives can be used as left representatives). Constructing the set of right coset representatives by induction on length, we see that it can always be chosen so that its elements are initially closed, that is arranged in a form of a tree. Such a set is called a Schreier transversal (see [11]). Presenting an order on the generating set of G induces a "ShortLex" total order on G (comparing elements first by length and then by the lexicographic order). Relative to this order each subgroup has a unique minimal transversal, which is also a Schreier transversal.

The growth of a subgroup $H < G$ with respect to the generators of G was studied by Grigorchuk (see [4]). We denote this growth function by $\Gamma_H^{(G)}(n)$ (to distinguish it from the growth function of H when considered a *group*). A connection between the different growth functions can be given by (assuming a fixed generating set)

$$\sum_{i=0}^{n} \gamma_{G/H}(i)\Gamma_H^{(G)}(n-i) \le \Gamma_G(n) \le \sum_{i=0}^{n} \gamma_{G/H}(i)\Gamma_H^{(G)}(n+i), \tag{5}$$

where $\gamma_{G/H}(i) := \Gamma_{G/H}(i) - \Gamma_{G/H}(i-1)$, that is the coset representatives of length exactly i. The left inequality comes from the fact that different cosets are disjoint subsets. The right inequality is by the definition of the cogrowth function through the coset representatives of minimal length.

When H is a normal subgroup of G then the cogrowth of H describes the growth of the group G/H. This imposes restrictions on the cogrowth of H:

$$\Gamma_{G/H}(n_1 + n_2) \le \Gamma_{G/H}(n_1 - 1) + \gamma_{G/H}(n_1)\Gamma_{G/H}(n_2). \tag{6}$$

Since each right coset is also a left one when H is normal, the minimal Schreier transversal tree T, relative to a ShortLex order, is suffix as well as prefix-closed. In other words, each subtree T' of T is "covered" by T when putting the root of T over the root of T'. (This is also the reason why a finitely generated subgroup of a free group which contains a non-trivial normal subgroup is of finite index.) Thus if $g = g_1 g_2 \in T$ when written in reduced form then the subtree $T(g)$ with root in g is "contained" (in the above sense) in the subtree $T(g_2)$. Therefore, there exists a descending chain of subtrees of T, which is of length $l(g)$ and terminates with $T(g)$. This tendency of T to "close" itself raised the question if there are groups of non-exponential but also non-polynomial growth. Milnor and Wolf showed that a f.g. solvable group has polynomial growth if it is virtually-nilpotent and otherwise has exponential growth ([13], [16]). Gromov ([7]) showed that virtually-nilpotent groups are the only ones with polynomial growth. Then Grigorchuk succeeded to obtain remarkable examples of groups of "intermediate growth" (between n^d and e^n), and to show that the set of growth degrees of finitely generated groups is of the continuum cardinality (see [5], [6] and also [3]).

If H_1, H_2 are subgroups of G then $\Gamma_{G/H_1 \cap H_2}(n) \geq max\{\Gamma_{G/H_1}(n), \Gamma_{G/H_2}(n)\}$ for every n (as usual, we assume here that the generating set of G is fixed). We also know that the intersection of two subgroups of finite index can be at most of the product of the indices. But in fact $\Gamma_{G/H_1 \cap H_2}$ behaves in this manner all along the way.

Proposition 1.1 *If $H_1, H_2 < G$ then $\Gamma_{G/H_1 \cap H_2}(n) \leq \Gamma_{G/H_1}(n)\Gamma_{G/H_2}(n)$ for every n.*

Proof. Let T_1, T_2 and T be minimal right Schreier transversals for H_1, H_2 and $H_1 \cap H_2$ respectively. Each element $g \in T$ can be represented by a pair (g_1, g_2), $g_i \in T_i$, where g_i is the representative of the coset $H_i g$ of H_i. By the minimality of the lengths of the coset representatives, $l(g_i) \leq l(g)$ for each i. The result then follows since the pairs are distinct.

The proposition gives a sufficient condition for the intersection of two subgroups to be non-trivial:

$$\Gamma_{G/H_1}(n)\Gamma_{G/H_2}(n) \prec \Gamma_G(n) \implies \Gamma_G(n) \not\preceq \Gamma_{G/H_1}(n)\Gamma_{G/H_2}(n)$$
$$\implies \exists n \,[\, \Gamma_{G/H_1}(n)\Gamma_{G/H_2}(n) < \Gamma_G(n) \,]$$
$$\implies H_1 \cap H_2 \text{ is non} - \text{trivial}.$$

For example, if $\Gamma_G(n) \sim d^{n^e}$, where $d > 1, 0 < e \leq 1$, and $\Gamma_{G/H_1}(n), \Gamma_{G/H_2}(n) \prec \Gamma_G(n)$ then $H_1 \cap H_2$ is non-trivial.

If $H_2 < H_1 < G$ then an element $g \in G$ of minimal length in $H_1 g$ is also of minimal length in the coset $H_2 g$, thus a minimal Schreier transversal for H_1 can be chosen to be a subset of a minimal transversal for H_2.

Let $1 \longrightarrow H \xrightarrow{i} F \xrightarrow{\pi} G \longrightarrow 1$, where F is free and finitely generated on X. Then there is a 1-1 correspondence between a minimal Schreier transversal T for $i(H)$ in F and the set of elements of the group G with generating set $\{\pi(x) \mid x \in X\}$, given by $g = x_{i_1} \cdots x_{i_n} \longleftrightarrow \pi(g) = \pi(x_{i_1}) \cdots \pi(x_{i_n})$, where g and $\pi(g)$ here are already in reduced form (note that we do not require the extension to split). Then a (minimal) Schreier transversal for a subgroup G' of G can be represented (with the above correspondence) by a (minimal) Schreier transversal for $\pi^{-1}(G')$ in F which can be taken to be a subset of T. This means that the cogrowth functions of subgroups of finitely generated groups are all cogrowth functions of subgroups of finitely generated free groups (and as will be seen in Theorem 1.2, it suffices to consider the free group on 2 generators for the equivalence classes of the cogrowth functions).

1.2 Essential Subgroups and Their Cogrowth Functions

We come now to essential subgroups. We call a subgroup $H < G$ *essential* if it intersects non-trivially every non-trivial subgroup of G. Clearly the family $\mathcal{E}$ of essential subgroups of a given group is a filter: if $H_1 \in \mathcal{E}$ and $H_1 < H_2$ then $H_2 \in \mathcal{E}$, and if $H_1, H_2 \in \mathcal{E}$ then $H_1 \cap H_2 \in \mathcal{E}$. Also a conjugate of an essential subgroup is essential. For example, if G is torsion-free then clearly every subgroup of finite index is essential. In finitely generated free groups, a subgroup is of finite index if and only if it is finitely generated and essential.

In contrast to the situation in algebras (as will be shown in the next section), one cannot tell whether a subgroup of infinite index of a free group is essential or not just by knowing the cogrowth of the subgroup. This is due to the "one-dimensionality" of a subgroup generated by a single element. For example, let G be the free group on the two generators x, y and let H be the normal closure of the subgroup of G generated by x. Then a minimal Schreier transversal for H consists of all powers of y, hence $\Gamma_{G/H}(n) \sim n$. But H is not essential since $H \cap < y > = 1$. Thus, when H is of infinite index it can be of minimal cogrowth and still lack essentiality. On the other hand, H can be essential although it has exponential cogrowth. If H is a normal subgroup of a torsion-free group G then by definition H is essential if and only if G/H is periodic (torsion). The well known examples of essential subgroups $H \lhd G$ of exponential cogrowth are when G is free of rank $m \geq 2$ and G/H is the Burnside group $B(m, n)$ with $n \geq 665$ and odd, as shown by Adyan ([1]). Essential normal subgroups of intermediate growth were constructed by Grigorchuk.

Let $\mathcal{CG}$ be the class of functions $\alpha(n)$ of the following type. $\alpha : \mathbb{N} \cup \{0\} \longrightarrow \mathbb{N}$ is the sequence of partial sums $\sum_{i=0}^{n} f_i$ of the series $\sum_{i=0}^{\infty} f_i$, such that (i) the f_i-s are zero on (r, ∞), where $0 \leq r$ and can be ∞, and positive integers otherwise, with $f_0 = 1$; (ii) there exists $0 < d$ such that $f_{i+1} \leq df_i$ for every i. Clearly $\mathcal{CG}$ includes the cogrowth functions of subgroups, but as seen from (6), the set of equivalence classes of the growth functions of groups is properly contained in the set of equivalence classes of the members of $\mathcal{CG}$.

Theorem 1.2 *Let G be the free group on $X = \{x_1, x_2\}$. Then for every $\alpha(n) \in \mathcal{CG}$ there exists an essential subgroup H of G such that $\Gamma_{G/H}(n) \sim \alpha(n)$.*

Proof. If $\alpha(n)$ is eventually-constant then any subgroup of finite index can be taken. So let us assume that $\alpha(n)$ is not bounded. We will construct in

an inductive way a right Schreier transversal T for H. T will contain two
types of sections constructed alternately: those responsible for the desired
growth (g-sections), and those to ensure the essentiality (e-sections). The
growth function of T, $\Gamma_T(n)$, will be $\preceq \Gamma_{G/H}(n)$, which is the growth function
of a minimal Schreier transversal tree, but by an appropriate definition of the
coset function, and by letting the g-sections be of sufficient depth (length)
compared to the e-sections, T can be constructed such that $\Gamma_T(n) \sim \Gamma_{G/H}(n)$
(if each g-section will be of depth equal to that of the next e-section, the
growth of T will be at least as half the growth of $\Gamma_{G/H}$).

We start with a g-section. Here we prevent the occurrence of the same
generator (or the same inverse of a generator) in adjacent edges. Hence, each
non-root vertex will have either 1 or 2 out-going edges. As we will see later,
an essential section can be constructed to be of growth $2n$, so no problem will
be to achieve growth equivalent to the minimal possible growth of $\alpha(n)$. As
for the maximal growth, $\alpha(n)$ can grow at most as d^n, for some positive d.
So if c is such that $2^c \geq d$ then $2^{cn} \geq d^n$, and we can construct T such that
$\alpha(n) \leq \Gamma_T(cn)$. Then this inequality can surely be reached when $\alpha(n)$ grows
slower than d^n, and thus $\Gamma_T(n)$ can be bounded by

$$\alpha(c^{-1}n) \leq \Gamma_T(n) \leq \alpha(2n). \tag{7}$$

(In general, $\Gamma_T(n)$ can be constructed so that it grows at the fastest possible
rate, but not exceeding $\alpha(n)$, resulting in some averaging of $\alpha(n)$, making it
"smoother" in places of great jumps.)

Suppose we constructed the tree T up to depth p_1, being the first g-section.
We label also each edge of T by some $x \in X \cup X^{-1}$, such that the labels on
the vertices are the elements one obtains by reading off the edge labels in
a path that starts at the root and terminates at the given vertex. We then
partially define the coset function π. If an edge labelled by $x \in X \cup X^{-1}$ goes
from the vertex g to the vertex h then $\pi(gx) = h$ and $\pi(hx^{-1}) = g$, that is
$\pi(g) = g$ for every $g \in T$. Otherwise, we define π on the set

$$\{\, gx \mid g \in T,\ l(g) < p_1 - 1 \text{ and } x \in X \cup X^{-1} \,\} \tag{8}$$

in the following way. We go as far as possible on a path in the *opposite*
direction. That is, we start at g, and go from a vertex h' to a vertex h''
whenever $\pi(h'x^{-1})$ is defined and $h'' = \pi(h'x^{-1})$. If the vertex h is the
endpoint of this path (in general, h can be g itself) then we define $\pi(gx) = h$.
We notice that by the restriction of not labeling two adjacent edges with the
same letter, such a path as described above will be of length 1 (except near
the root where it can be of length 2). The same process of defining π will
take place at the following stages, but then, due to the e-sections, paths as
above could be of greater length. The definition of π on other elements of G
is then according to the inductive rule $\pi(gx) = \pi(\pi(g)x)$.

Next we construct an e-section. Assume an ordered list of the elements of G
is given. We take the first element g of this list, and "travel" along T as long

as possible with powers of g, starting from the root 1 and using the function π. If we happen to get back to the root after some k-th power of g then we are done: $g^k \in H$, and we can take the next element in the list. Otherwise, we will extend T and π so that $< g >$ will intersect H non-trivially. (One may look at T as representing an automaton, which has to be extended so that g will be accepted by it. Here the states of the automaton are the vertices, with 1 the accepting state, the input alphabet is $X \cup X^{-1}$ and the function is π.) The idea is to form two paths starting from the root, one of a positive power of g and the other of a negative power, and to "tie" these paths using the function π. Assume that when written in reduced form we have $g = h_1 h_2 h_1^{-1}$, where h_2 is of minimal length. We start a travel from the root of T, this time with h_1 followed by powers of h_2. It may still happen that we will return to $\pi(h_1)$ after some powers of h_2, and in this case too we are done. (We note that it is impossible to return to the same vertex after some powers of h_2 before first visiting $\pi(h_1)$. This is because by the very definition of π, if $\pi(h_1 h_2^r) = \pi(h_1 h_2^s)$ then $\pi(h_1 h_2^{s-r}) = \pi(h_1)$.) Otherwise, we reach a vertex $h \in T$ on which $\pi(hx)$ is not yet defined, where x is the next "input" letter of $h_1 h_2^k$ for some $k \geq 0$. We then increase T by adding a path, starting with the vertex hx, according to the rest of $h_1 h_2^k$. The same process of increasing the tree is then done with a path that goes from the root with h_1 followed by powers of h_2^{-1}. Since no prefix of h_2^{-1} equals a suffix of h_2, the two added paths must become separated one from another, and after that happens we need not increase the tree anymore. The endpoints of the two added paths are vertices u_1 and u_2 such that $u_1 = \pi(h_1 h_2^r)$ and $u_2 = \pi(h_1 h_2^{-s})$, for some $r, s \geq 0$. Assume now that $h_2 = xw$, where $x \in X \cup X^{-1}$, $w \in G$ and h_2 is written in reduced form. Then we further extend the tree by adding a w^{-1}-segment at the vertex u_2 and define $\pi(u_1 x) = u_2 w^{-1}$, and necessarily $\pi(u_2 w^{-1} x^{-1}) = u_1$. Later, this construction will result (after a sufficient extension of π) in

$$\pi(g^{r+s+1}) = 1, \tag{9}$$

i.e. $< g >$ will intersect H non-trivially. Let us call T_1 the current tree we have.

Next we construct a g-section, extending T_1 and the coset function π, according to the growth function $\alpha(n)$, the same as before. If needed, we first widen the tree in the part of the e-section to reach the desired growth, but we do not change π where it is already defined. Again, we do not define π on the boundary of the current tree, to ensure further increasing of the tree. Following this stage comes an e-section with the next element in the list. The result is a tree T_2. We continue in this way indefinitely and define $T = \bigcup_i T_i$. Clearly

$$\pi(\pi(gx)x^{-1}) = g \tag{10}$$

for every $g \in T$, $x \in X \cup X^{-1}$. This makes T a right Schreier transversal for a unique subgroup H of G, for which π is the function giving the coset representatives (see [8], [9]). H is freely generated by the non-trivial elements of the form $gx(\pi(gx))^{-1}$, where $g \in T$, $x \in X$ (Nielsen-Schreier theorem).

As said before, $\Gamma_T(n)$ can be made equivalent to $\alpha(n)$. On the other hand, if g is a vertex in a g-section (and assume it is not on the boundary of the section) then $l(\pi(gx)) \leq l(g) + 1$, for any $x \in X \cup X^{-1}$. If g is in an e-section then $l(\pi(gx)) \leq l(g) + r$, where r is the length of the e-section. Therefore, if each g-section has at least the depth of the next e-section, we get for any $g \in G$

$$l(\pi(g)) \leq 2l(g). \tag{11}$$

Thus, the length of each element in T is at most twice the length of the minimal element in its coset, and so $\Gamma_{G/H}(n) \leq \Gamma_T(2n)$ and they are equivalent (because $\Gamma_T(n) \leq \Gamma_{G/H}(n)$). Combining it with (7) gives

$$\Gamma_{G/H}(n) \sim \Gamma_T(n) \sim \alpha(n). \tag{12}$$

Finally, the e-sections make sure that for every element of G some positive power of it lies in H, that is H is essential.

We remark that with a little more effort (by adding segments corresponding to the different group elements), the subgroups constructed in the theorem above can have the additional property of not containing any subgroup which is normal in G.

As we have seen, even when the group is torsion-free a normal subgroup of infinite index need not be essential although it can be of minimal cogrowth. However, the following simple observation expresses the "largeness" of normal subgroups in special cases. If G is a torsion-free group which does not contain a non-cyclic abelian subgroup then every two non-trivial normal subgroups of G have non-trivial intersection. To see it, let $1 \neq x \in H_1$, $1 \neq y \in H_2$, where H_1 and H_2 are normal subgroups of G. Then $xyx^{-1}y^{-1} = x(yx^{-1}y^{-1}) \in H_1$ and also $xyx^{-1}y^{-1} = (xyx^{-1})y^{-1} \in H_2$ and the result follows.

2 Growth, Cogrowth and Essentiality in Algebras

Let R be an associative algebra with a unit generated on a finite set X over a field K. Having the length function on the free semigroup X^* generated by X and the grading of R by the subspaces $R^{(n)} = \sum_{i=0}^n KX^i$, the growth function on R is defined by

$$\Gamma_R(n) := dim\ R^{(n)} \tag{13}$$

(see [15], and also [2] for a generalization). The length of an element $r \in R$ is the smallest n such that $r \in R^{(n)}$. As in groups, the equivalence class of a growth function does not depend upon the set of generators.

If I is a right ideal of R then its growth function is $\Gamma_I(n) := dim\ (I \cap R^{(n)})$, and its cogrowth $\Gamma_{R/I}(n) := \Gamma_R(n) - \Gamma_I(n)$. The definition can then be extended to subspaces of R by $\Gamma_V(n) := dim\ (V \cap R^{(n)})$, and if $V \supseteq U$ then $\Gamma_{V/U}(n) := \Gamma_V(n) - \Gamma_U(n)$.

Let V be a complementary subspace to I, that is $R = I + V$ and $I \cap V = 0$. Then a basis T for V can consist of a set of (monic) monomials, which moreover is initially closed, that is forms a tree. Introducing a ShortLex order on X^* and extending it in the usual manner to a partial order on R, such a basis can be formed from all monomials which are minimal in their cosets (see [10]), resulting in a unique minimal Schreier transversal T (similar to the group case). Then we get

$$\begin{aligned} \Gamma_I(n) &= card\{g \mid g \text{ is a leading monomial of some } r \in I,\ l(r) \leq n\}, \\ \Gamma_{R/I}(n) &= card\{g \in T \mid l(g) \leq n\}. \end{aligned}$$

In fact, when R is a group algebra KG and H a subgroup of G then the Schreier transversals for the right ideal I of R generated by the elements $h - 1$, where $h \in H$, coincide with the Schreier transversals for H in G, and thus $\Gamma_{G/H}(n) = \Gamma_{R/I}(n)$ (keeping the generating set for G fixed).

The situation concerning intersection of right ideals is simpler than intersection of subgroups because in each $R^{(n)}$ we have intersection of finite dimensional subspaces.

If I is a right ideal and $r \in R$ then $(I : r) := \{s \in R \mid rs \in I\}$.

Proposition 2.1 *If $\Gamma_I(n) \npreceq \Gamma_{R/I}(n)$ then for every $0 \neq r \in R$, $(I : r) \neq 0$.*

Proof. We may assume that $l(r) \geq 1$. Since $\Gamma_I(n) \npreceq \Gamma_{R/I}(n)$ then

$$\forall C\ \exists n\ [\ \Gamma_R(n) \geq \Gamma_I(n) > C\Gamma_{R/I}(Cn) \geq C\Gamma_{rR+I/I}(Cn)\ ]. \tag{14}$$

Hence $\Gamma_{rR+I/I}(n) \prec \Gamma_R(n)$. Taking $C = 2l(r)$ we get $Cn \geq n + l(r)$ and by (14)

$$\exists n\ [\ \Gamma_R(n) > \Gamma_{rR+I/I}(n + l(r))\ ]. \tag{15}$$

We look now at $\{rg \mid g \in G,\ l(g) \leq n_0\}$, where G is the set of (monic) monomials in R, and n_0 is such that the inequality in (15) holds. We have here $\Gamma_R(n_0)$ elements of length $\leq n_0 + l(r)$. By (15) these elements are linearly dependent modulo I, hence there exists some $0 \neq s = \sum_{l(g) \leq n_0} a_g g$, $a_g \in K$, $g \in G$, such that $rs \in I$, i.e. $(I : r) \neq 0$.

We now come to essentiality of right ideals. Essential right ideals are more common than essential subgroups. From a geometrical point of view, the difference is that a right ideal generated by a single element grows in a cone-like manner, whereas cyclic subgroups are "1-dimensional".

Corollary 2.2 *Let R be without zero divisors and let I be a right ideal of R. If $\Gamma_{R/I}(n) \prec \Gamma_R(n)$ then I is essential.*

Proof. Since I is not empty it contains a right regular element. Therefore its growth function is equivalent to the growth function of R. Thus $\Gamma_{R/I}(n) \prec \Gamma_I(n)$, and by Proposition 2.1 I is essential.

Remark: The converse of the above does not hold.

When R is the group algebra KG, where G is free of rank 2 and K is a field, then for every $\alpha(n) \in C\mathcal{G}$ (as defined in the previous section) there exists an essential right ideal I of R with $\Gamma_{R/I}(n) \sim \alpha(n)$. For the functions which are $\prec 2^n$ we can take the right ideal generated by the "right augmentation ideal" of the subgroups constructed in Theorem 1.2. For exponential growth we can take, for example, the fractal ideals defined in [14].

References

[1] Adyan, S.I. (1975). *The Burnside problem and identities in groups.* Nauka, Moscow, 1975 (Russian) [(1979) Ergebnisse der Mathematik und ihrer Grenzgebiete, **95**. Springer-Verlag].

[2] Aljadeff, E., Rosset, S. (1988). *Growth and uniqueness of rank.* Israel J. Math., Vol. **64**, No. 2, 251-256.

[3] Fabrykowski, J., Gupta, N. (1985). *On groups with sub-exponential growth functions.* J. Indian Math. Soc., Vol **49**, 249-256.

[4] Grigorchuk, R.I. (1978). *Symmetrical random walks on discrete groups.* In (Ed. Dobrushin, R.L., Sinai, Ya.G.): Multicomponent random systems. Nauka, Moscow, 132-152 [English transl. (1980) *Advances in probability and related topics*, Vol. **6**, 285-325. Marcel Dekker].

[5] Grigorchuk, R.I. (1983). *On Milnor's problem of group growth.* Dokl. Ak. Nauk SSSR, **271**, 31-33 (Russian) [English transl. (1983): Soviet Math. Dokl., **28**, 23-26].

[6] Grigorchuk, R.I. (1990). *On growth in group theory.* Proc. of the International Congress of Mathematicians, Kyoto, 1990, 325-338.

[7] Gromov, M. (1981). *Groups of polynomial growth and expanding maps.* Publ. Math. IHES, **53**, 53-73.

[8] Hall, M., Radó, T. (1948). *On Schreier systems in free groups.* Trans. AMS, **64**, 386-408.

[9] Hall, M. (1949). *Coset representations in free groups.* Trans. AMS, **67**, 421-432.

[10] Lewin, J. (1969). *Free modules over free algebras and free group algebras: the Schreier technique.* Trans. AMS, **145**, 455-465.

[11] Lyndon, R.C., Schupp, P.E. (1977). *Combinatorial group theory.* Springer-Verlag.

[12] Milnor, J. (1968). *A note on curvature and fundamental group.* J. Diff. Geom., **2**, 1-7.

[13] Milnor, J. (1968). *Growth of finitely generated solvable groups.* J. Diff. Geom., **2**, 447-449.

[14] Rosenmann, A. (1993). *Essentiality of fractal Ideals.* Inter. J. Algebra and Computation **3**, 425-445.

[15] Rowen, L.H. (1988) *Ring theory.* Academic Press.

[16] Wolf, J.A. (1968). *Growth of finitely generated solvable groups and curvature of Riemannian manifolds.* J. Diff. Geom., **2**, 421-446.

Institute for Experimental Mathematics
Universität GH Essen
45326 Essen
Germany
email: amnon@exp-math.uni-essen.de

Regular geodesic languages for 2-step nilpotent groups

MICHAEL STOLL

1 Introduction

One possibility to show that a group G with finite semigroup generating system S has rational growth series is to exhibit a regular language $L \subset S^*$ consisting of geodesic words and mapped bijectively onto G. Machi and Schupp have even conjectured that the existence of such an L is equivalent to the rationality of the growth series ([2], conjecture 8.7).

The aim of this paper is to show that this approach does not work for nilpotent groups. More precisely, we show that if G is 2-step nilpotent with maximal free abelian quotient $\overline{G}$ and S is any finite set of semigroup generators for G, then for every regular and geodesic (with respect to G) language $L \subset S^*$, the natural map $L \to \overline{G}$ has finite fibers. We conjecture that this holds for all nilpotent groups. If G is not virtually abelian, this implies in particular that L cannot be mapped bijectively onto G.

This also gives a counterexample to the conjecture of Machi and Schupp, for it is known that the discrete Heisenberg group with its standard generating set,

$$ H = \langle a, b \mid [a, b] \text{ central} \rangle \, ; $$

has rational growth series [3], but our theorem implies that no regular and geodesic language can be mapped bijectively onto H.

2 Notations and Definitions

Let G denote an arbitrary finitely generated group. Let S be some set (whose elements will be called *letters*) and $\varphi : S \to G$ some map such that the extension of φ to a semigroup homomorphism $\varphi : S^* \to G$ is onto. For $w \in S^*$, the *length* $l(w)$ is the number of letters in w. For $g \in G$, define $l_S(g)$, the *length of g with respect to S*, to be $\min\{l(w) \mid w \in S^*, \varphi(w) = g\}$. A word $w \in S^*$ is called *geodesic*, if $l(w) = l_S(\varphi(w))$. Every subword of a geodesic word is also geodesic.

For an element g of G, define the *stable length* to be

$$ \lambda_S(g) = \lim_{n \to \infty} \frac{1}{n} l_S(g^n) = \inf_n \frac{1}{n} l_S(g^n) $$

(the limit exists and equals the inf, because $l_S(g^{m+n}) \leq l_S(g^m) + l_S(g^n)$). Obviously, $\lambda_S(g) \leq l_S(g)$, $\lambda_S(g^n) = n\lambda_S(g)$ for $n \geq 0$ and $\lambda_S(gg') \leq \lambda_S(g) + \lambda_S(g')$ if g and g' commute.

A subset $L \subset S^*$ is called a *language*. If L is a language, let $S_L \subset S$ denote the set of letters occurring in some element of L. We write S_w for $S_{\{w\}}$. L is called *geodesic*, if all its elements are. L is called *almost geodesic*, if $l_S(\varphi(w)) = l(w) - o(l(w))$ for $w \in L$, i.e. if for every $\varepsilon > 0$ there is some n_0 such that for all $w \in L$ with $l(w) \geq n_0$ we have $l_S(\varphi(w)) \geq (1 - \varepsilon)l(w)$.

An element $g \in G$ is called *straight*, if $\lambda_S(g) = l_S(g)$, i.e. if $l_S(g^n) = n\, l_S(g)$ for all $n \geq 0$. A word $w \in S^*$ is called *straight*, if all powers w^n are geodesic, or equivalently, if $l(w) = \lambda_S(\varphi(w))$, i.e. if w is a shortest representative of a straight element. A language $L \subset S^*$ is called *straight*, if all its elements are. A subset $T \subset S$ is called *straight*, if T^* is (or equivalently, if T^* is geodesic).

We will regard S also as giving generators for any quotient of G, so that the above definitions make sense for elements g in a quotient group.

3 Straight elements and straight words

Lemma 1 *Let $U \subset Z(G)$ be a central subgroup of G and assume $\lambda_S(z) = 0$ for all $z \in U$. Let $G \ni g \mapsto \tilde{g} \in \tilde{G} = G/U$ be the canonical epimorphism. Then $\lambda_S(g) = \lambda_S(\tilde{g})$ for all $g \in G$.*

Proof. To begin with, we note that $\lambda_S(gz) = \lambda_S(g)$ for $g \in G$ and $z \in U$. For $\lambda_S(gz) \leq \lambda_S(g) + \lambda_S(z) = \lambda_S(g)$, since g and z commute. Furthermore, we have obviously $l_S(\tilde{g}) \leq l_S(g)$ for $g \in G$, since a word representing g also represents $\tilde{g}$. This gives $\lambda_S(\tilde{g}) \leq \lambda_S(g)$.

To show the reverse inequality, take an $\varepsilon > 0$. Then there is some $n > 0$ such that $l_S(\tilde{g}^n) \leq n\,\lambda_S(\tilde{g}) + n\varepsilon$. Take a word w of length $l_S(\tilde{g}^n)$ representing $\tilde{g}^n$, and let $h = \varphi(w)$. Then $l_S(h) = l_S(\tilde{g}^n)$ and $\tilde{h} = \tilde{g}^n$, i.e. $h = g^n z$ with some $z \in U$. Now, $\lambda_S(g) = \frac{1}{n}\lambda_S(g^n) = \frac{1}{n}\lambda_S(h) \leq \frac{1}{n}l_S(h) = \frac{1}{n}l_S(\tilde{g}^n) \leq \lambda_S(\tilde{g}) + \varepsilon$. Letting $\varepsilon \to 0$, we get $\lambda_S(g) \leq \lambda_S(\tilde{g})$.

From now on, let G be nilpotent and $\overline{G}$ the maximal free abelian quotient of G. Write $g \mapsto \overline{g}$ for the canonical map $G \to \overline{G}$.

Proposition 1 *For any $g \in G$, $\lambda_S(g) = \lambda_S(\overline{g})$.*

Proof. We proceed by induction on the nilpotency class $c(G)$. If $c(G) = 1$, then G is abelian, and we take for U in Lemma 1 the torsion subgroup of G, which is finite and therefore satisfies the condition $\lambda_S|_U = 0$. The claim then follows from the Lemma, since here $\overline{G} = G/U$.

Now, let $c(G) = c > 1$ and let U be the last nontrivial term in the lower central series of G. Then U is central, and it is well known that $l_S(z^n) = O(n^{1/c})$ for $z \in U$ (cf. [1, 4]). Hence, we can apply Lemma 1 and see that $\lambda_S(g) = \lambda_S(\tilde{g})$. Now, $c(G/U) = c(G) - 1$ and $\overline{G/U} = \overline{G}$, so the claim follows by induction.

Corollary 1 *A word $w \in S^*$ is straight (with respect to G), if and only if it is straight with respect to $\overline{G}$.*

Proof. w is straight with respect to $G \iff l(w) = \lambda_S(\varphi(w)) \iff$ (by Proposition 1) $l(w) = \lambda_S(\overline{\varphi(w)}) \iff w$ is straight with respect to $\overline{G}$.

Corollary 2 *If w is straight, then so is S_w.*

Proof. By Corollary 1, we can assume G to be abelian. Let $v \in S_w^*$. Then some power w^n of w can be permuted such as to have the form uv. Since G is abelian and since w^n is geodesic by assumption, uv and hence v have to be geodesic. Hence S_w^* is geodesic, i.e. S_w is straight.

4   Regular geodesic languages for 2-step nilpotent groups

Within this section, G is a finitely generated 2-step nilpotent group. This means that every commutator $[g, g']$ is central in G. Denote by $\overline{G}$ the maximal free abelian quotient group of G; the canonical map $G \to \overline{G}$ will be written $g \mapsto \overline{g}$, its kernel will be called H.

We begin with some simple observations.

Lemma 2 (Commutators)
a) $(g, h) \mapsto [g, h]$ *is bilinear, i.e.*
   $[gg', h] = [g, h][g', h]$ *and* $[g, hh'] = [g, h][g, h']$.
b) $l_S([g, h]^n) = O(\sqrt{n})$.
c) *For* $h \in H$, $l_S(h^n) = O(\sqrt{n})$.
d) *If $g \in G$ and $h \in H$, then $[g, h]$ has finite order.*
e) *If $g \in G$ is fixed and $h \in G$ varies, then $l_S([g, h]) = O(\sqrt{l_S(h)})$.*

Proof. a) $[gg', h] = g'^{-1}g^{-1}h^{-1}gg'h = g'^{-1}[g, h]h^{-1}g'h = [g, h]g'^{-1}h^{-1}g'h = [g, h][g', h]$, and likewise for $[g, hh']$.
b) Let $n = m^2 + r$ with $0 \le r \le 2m$. Then by part a), $[g, h]^n = [g^m, h^m][g, h]^r$, hence $l_S([g, h]^n) \le 2(m + r)(l_S(g) + l_S(h)) \le 6(l_S(g) + l_S(h))\sqrt{n}$.
c) Some power h^k is in G', i.e. product of commutators. The claim then follows from part b) using that G' is abelian.
d) If $h \in H$, then some power h^m is in G', hence central. Therefore, by part a), $[g, h]^m = [g, h^m] = 1$.
e) Let $x_1, \ldots, x_k$ be the images in G of the elements in S. We can write h as a product of $n = l_S(h)$ of the x_j's. By part a), $[g, h] = [g, x_1]^{n_1} \cdots [g, x_k]^{n_k}$ with $n_1 + \cdots + n_k = n$. Then, by part b), $l_S([g, h]) = O(\sqrt{n_1}) + \cdots + O(\sqrt{n_k}) = O(\sqrt{n})$.

Lemma 3 (Almost geodesic languages) *Let $L, L' \subset S^*$ be languages and $w \in S^*$ a word. Then, if LwL' is almost geodesic, so is LL'.*

Proof. Let $u \in L$ and $u' \in L'$. Denote $g = \varphi(u)$, $g' = \varphi(u')$ and $h = \varphi(w)$. We have $l_S(gg') = l_S(ghg'[g', h]h^{-1}) \geq l_S(ghg') - l_S([g', h]^{-1}) - l_S(h)$. By Lemma 2, $l_S([g', h]^{-1}) \leq C\sqrt{l_S(g')} + C_1$. Hence, with $C_2 = C_1 + l_S(h)$,

$$
\begin{aligned}
l(uu') - l_S(gg') &\leq l(uu') - l_S(ghg') + C\sqrt{l_S(g')} + C_2 \\
&\leq l(uwu') - l_S(ghg') + C\sqrt{l(u')} + C_2 \\
&= o(l(uwu')) \\
&= o(l(uu'))
\end{aligned}
$$

for $l(uu') \to \infty$ by assumption.

Lemma 4 (Almost geodesic languages and straightness)
Let $L_1, L_2, \ldots, L_k$ be languages. If $L_1^* L_2^* \cdots L_k^*$ is almost geodesic, then $S_{L_1 L_2 \cdots L_k}$ is straight.

Proof. Let $w_j \in L_j^*$ with $S_{w_j} = S_{L_j}$. We can assume that for some l, $l(w_j) = l$ for all j. By Corollary 2, it is enough to show that $w = w_1 w_2 \cdots w_k$ is straight. Denote $g_j = \varphi(w_j)$. If w is not straight, then by Proposition 1, $\lambda_S(\overline{g_1 g_2 \cdots g_k}) < l(w) = kl$, hence for some $n > 0$, $l_S(\overline{g_1^n g_2^n \cdots g_k^n}) = nkl(1 - \varepsilon)$ with some $\varepsilon > 0$. By taking w_j^n instead of w_j, we can assume $n = 1$. We then have some $g' \in G$ with $l_S(g') = kl(1 - \varepsilon)$ and $\overline{g'} = \overline{g_1 g_2 \cdots g_k}$, so there is some $h \in H$ such that $g'h = g_1 g_2 \cdots g_k$.

We now take an integer $a \geq 2$ such that $k\varepsilon a \geq 4k + 3$ and set $r_j = a^{k-j}$ for $j = 1, 2, \ldots, k$ and $m = 2\mu a^k$, where μ is a positive integer. Then, writing $m_j = \frac{m}{2r_j} = \mu a^j$, we have (using Lemma 2)

$$
\begin{aligned}
&g_1^{m(1+r_1)} g_2^{m(1+r_2)} \cdots g_k^{m(1+r_k)} \\
&= g_1^{mr_1} g_2^{mr_2} \cdots g_k^{mr_k} g_1^m g_2^m \cdots g_k^m \prod_{i<j} [g_i, g_j]^{m^2 r_j} \\
&= g_1^{mr_1} g_2^{mr_2} \cdots g_k^{mr_k} g'^m h^m [h, g']^{\binom{m}{2}} \prod_{i<j} [g_i, g_j]^{m^2 r_j + \binom{m}{2}} \\
&= g_k^{-(2r_k+1)m_1} \cdots g_2^{-(2r_2+1)m_1} g_1^{mr_1} g_2^{(2r_2+1)m_1} \cdots g_k^{(2r_k+1)m_1} \cdot \\
&\quad g_k^{-(2r_k+1)m_2} \cdots g_3^{-(2r_3+1)m_2} g_2^{mr_2} g_3^{(2r_3+1)m_2} \cdots g_k^{(2r_k+1)m_2} \cdot \\
&\quad \cdots \cdots g_k^{-(2r_k+1)m_{k-1}} g_{k-1}^{mr_{k-1}} g_k^{(2r_k+1)m_{k-1}} \cdot \\
&\quad g'^m h^m [h, g']^{\binom{m}{2}} \prod_{i<j} [g_i, g_j]^{-m/2}.
\end{aligned}
$$

So, by Lemma 2 again,

$$
\begin{aligned}
&l_S(g_1^{m(1+r_1)} g_2^{m(1+r_2)} \cdots g_k^{m(1+r_k)}) \\
&\leq 2l \sum_{i<j} (2r_j + 1)m_i + l \sum_j mr_j + mlk(1 - \varepsilon) + O(\sqrt{m})
\end{aligned}
$$

$$= l(w_1^{m(1+r_1)} w_2^{m(1+r_2)} \cdots w_k^{m(1+r_k)})$$
$$+ 2l\mu a^k (\sum_{i<j} (2a^{i-j} + a^{i-k}) - k\varepsilon) + O(\sqrt{m})$$

$$\le l(w_1^{m(1+r_1)} w_2^{m(1+r_2)} \cdots w_k^{m(1+r_k)})$$
$$+ 2l\mu a^{k-1}(4k+2 - ak\varepsilon) + O(\sqrt{m})$$

$$\le \left(1 - \frac{1}{2ka^{k+1}}\right) l(w_1^{m(1+r_1)} w_2^{m(1+r_2)} \cdots w_k^{m(1+r_k)}) + O(\sqrt{m}).$$

(We have used $\sum_{i<j}(2a^{i-j} + a^{i-k}) \le (2k+1)\sum_{j=1}^{\infty} a^{-j} = (2k+1)/(a-1) \le (4k+2)/a$, $4k+2 - ak\varepsilon \le -1$ and $\sum_j lm(1+r_j) \le klma^k$.) So, for μ large, the words considered are too long by a factor bounded below by some constant > 1, and $L_1^* L_2^* \cdots L_k^*$ cannot have been almost geodesic. Whence a contradiction, and $w_1 w_2 \cdots w_k$ has to be straight.

We are now ready to prove the main result of this paper.

Theorem 1 (Main Theorem) *Let G be 2-step nilpotent and S be any finite set of semigroup generators for G. Then, if $L \subset S^*$ is a regular and geodesic language, the map $L \to G \to \overline{G}$ is finite-to-one, i.e. for each coset of H in G there are only finitely many words in L representing an element in that coset.*

Proof. As a regular language, L is a finite union of languages of the form $w_0 L_1^* w_1 L_2^* w_2 \cdots L_n^* w_n$ (where the w_j are words and the L_j are some languages). Therefore, it suffices to prove the claim when L is a language of this type. L geodesic implies that $L_1^* L_2^* \cdots L_n^*$ is almost geodesic by repeated application of Lemma 3. Then, by Lemma 4, $S_{L_1 L_2 \cdots L_n} = S_{L_1} \cup S_{L_2} \cup \cdots \cup S_{L_n}$ is straight. Thus, given an element $\overline{g}$ of $\overline{G}$, the length of any word $w \in L_1^* L_2^* \cdots L_n^*$ such that $\varphi(w) = \overline{g}$ has to be $l(w) = l_S(\overline{g})$. This gives a bound on the length of words in L representing an element in the coset $\varphi(w_0 w_1 \cdots w_n)gH$ of H in G, whence the Theorem.

It is natural to conjecture that this result remains true for arbitrary nilpotent groups.

Corollary 3 *If G and S are as in the Theorem and furthermore H is infinite (i.e. G is not virtually abelian), then no regular and geodesic language is mapped onto G by φ.*

In particular:

- There is no regular and geodesic language mapped bijectively onto G.

- The language of all geodesics of G (with respect to S) is not regular, i.e. G has infinitely many cone types with respect to S.

References

[1] H. BASS: *The degree of polynomial growth of finitely generated nilpotent groups*, Proc. London Math. Soc. (3) **25** (1972), 603–614

[2] ROSTISLAV I. GRIGORCHUK: *On growth in group theory*, Proceedings of the International Congress of Mathematicians Kyoto 1990, (1991), 325–338

[3] MICHAEL SHAPIRO: *A geometric approach to the almost convexity and growth of some nilpotent groups*, Math. Ann. **285** (1989), 601–624

[4] BERNHARD WEBER: *Zur Rationalität polynomialer Wachstumsfunktionen*, Bonner Mathematische Schriften **197**, 1989

Mathematisches Institut der Universität
Beringstr. 4
D–53115 Bonn
Germany
email: stoll@rhein.iam.uni-bonn.de

Finding indivisible Nielsen paths for a train track map

EDWARD C. TURNER

Abstract

An algorithm is described for determining the indivisible Nielsen paths for a train track map and therefore the subgroup of elements of the fundamental group fixed by the induced automorphism.

In my talk at the Edinburgh Conference I described the "procedure" for finding fixed points of an automorphism of a free group that is implicit in [6] and explicit in [2] and [7]. I made the point that this is a procedure and not an algorithm since there is no way in general of knowing how long to persist before being sure that all fixed elements have been found—although it has been shown to be effective for positive automorphisms [2]. The example of Stallings [8], p99, figure 3 (Example 1 below) was presented to show that one may need more persistence than expected. I also discussed the notion of an *indivisible Nielsen path (INP)* which was introduced in [1] and used as a fundamental tool in [4]. After the talk, Bestvina asked me whether the procedure could be adapted to determine the INPs for a train track map, the determination of which is an essential part of the algorithm introduced in [1] for finding the fixed words of the induced automorphism. This paper shows how to do this. All irreducible automorphisms have train track representatives so this provides a straightforward means (given the train track map!) of computing the generator of the fixed subgroup. These methods also apply to automorphisms with fixed subgroup of maximal rank since each such automorphism has an easily determined train track representative on a bouquet of circles [4]. It seems likely that these ideas can be generalized to make the procedure effective for a relative train track map as well, and thus to apply to general automorphisms.

We begin in section 1 with a review of the ideas introduced in [6], in particular the definition of D_ϕ, and their use in determining $\text{Fix}(\phi)$. In section 2, we show how to generalize the construction of D_ϕ and in section 3 the generalized procedure is shown to be effective in the case of a train track map and to give all the INPs. Our notational conventions are as follows: Γ is a finite graph with a specified basepoint v_*, $f : \Gamma \to \Gamma$ is a graph map (sending edges to paths) which is a homotopy equivalence, $f(v_*) = v_*$ and $\phi : \pi_1(\Gamma, v_*) \to \pi_1(\Gamma, v_*)$ is the automorphism of the free group F_r induced by f. A *path* is understood to join vertices and to be reduced. The inverse of a group element a is a^{-1} and of an edge E is $\overline{E}$ (and $\overline{E_1\,E_2} = \overline{E_1} \cdot \overline{E_2}$).

1 The graph D_ϕ for $\phi \in \mathrm{Aut}\,(F_r)$.

Throughout this section we assume that $\Gamma = W = \bigvee_{i=1}^r S^1$, the wedge of r circles, v_* is the wedge point, and that $f : W \to W$ is the standard map associated with ϕ. Then $\mathrm{Fix}(\phi) = \{w \in F_r \mid \phi(w) = w\}$ is a subgroup of $\pi_1(W, v_*)$ and so determines a covering space $\widetilde{W}_{Fix(\phi)}$ of W. $\widetilde{W}_{Fix(\phi)}$ is a graph with a base point $\tilde{v}_*$ covering v_* whose edges are labeled with the generators of F_r so that the labels on closed loops at $\tilde{v}_*$ are precisely the elements of $\mathrm{Fix}(\phi)$. The vertices of $\widetilde{W}_{Fix(\phi)}$ are labeled as follows.

Vertex labeling in $\widetilde{W}_{Fix(\phi)}$: The vertex $\tilde{v}$ has label $w = \bar{\ell}\phi(\ell)$ where ℓ is the label on some path in $\widetilde{W}_{Fix(\phi)}$ from $\tilde{v}_*$ to $\tilde{v}$. Each label is a word in F_r.

It is easily checked that this does not depend on the path chosen. (*Note:* This labeling convention agrees with that of [2] and is the inverse of that of [6] and [7].) The labels on adjacent points are related as indicated below.

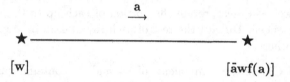

Definition. D_ϕ is the graph whose vertices are labeled by words $w \in F_r$, whose directed edges are labeled by the generators a_i of F_r where labeled edges join labeled vertices as indicated in the diagram above. The vertex with label w will be denoted by $[\mathbf{w}]$. The covering projection is $p_\phi : D_\phi \to W$.

The component of D_ϕ containing the vertex $[\mathbf{1}]$ is $\widetilde{W}_{Fix(\phi)}$, and it's not hard to see that the component of D_ϕ containing the vertex $[\mathbf{w}]$ is the covering space of W corresponding to the subgroup $\mathrm{Fix}(i_w \circ \phi)$, where i_w is the inner automorphism $i_w(v) = wvw^{-1}$. This observation can be exploited (as in [5]) to obtain information about the outer automorphism class of ϕ.

Relative to this labeling, there is a preferred direction out of each vertex $[\mathbf{w}]$ (denoted ▶ —), namely in the direction of the edge labeled by the first letter of w. The key observation of [6] was that there are finitely many *special edges* for which neither end is the preferred out direction (called *repulsive edges* in [2]) and each such edge corresponds to the occurrence of a generator in its image as follows:

$$\phi(a_i) = u a_i v \qquad \rightsquigarrow \qquad \underset{[\mathbf{u^{-1}}]}{\underline{\quad\quad}\blacktriangleleft\bigstar} \overset{\overset{\textstyle a_i}{\longrightarrow}}{\text{------------}} \underset{[\mathbf{v}]}{\bigstar\blacktriangleright\underline{\quad\quad}}$$

There are also finitely many edges that are the preferred directions out of both of their ends and these correspond to occurrences of the inverse of a generator in the image of the generator. All other edges have consistent orientations at both ends. Thus if the special edges are removed from D_ϕ, the remaining graph has edge orientations so that there is at most one out edge at each vertex—and so each component has rank 0 or 1 ([6], Proposition 2). A core C_ϕ of D_ϕ (a finite subgraph which carries the full fundamental group) can be constructed as follows:

Construction of C_ϕ .

1. Start with the *special edges* and their endpoints, the *special vertices*, denoted ★. Include [1] (always a sink) as a special vertex.

2. "Flow" from each special vertex other than [1], following the preferred direction out of each newly encountered vertex until another special vertex is encountered, a vertex is encountered a second time (forming a loop) or a ray is produced.

3. If two rays meet, retain the portion of each up to the vertex at which they meet. Discard the rest of each ray and all of any ray that meets no others.

If D_ϕ^v and C_ϕ^v are the components of D_ϕ and C_ϕ containing v, then $D_\phi^v \setminus C_\phi^v$ is the disjoint union of trees: those that contain rays discarded in step (3) above flow toward the ray and then to ∞, the others flow toward C_ϕ^v. C_ϕ^v is a finite subgraph of D_ϕ^v that has the same fundamental group, so if C_ϕ^v can be constructed algorithmically, Fix(ϕ) has been effectively computed. Step (1) is trivial to do since it is just a matter of identifying the occurrences of generators in their own images, and in fact the number of such occurrences is an upper bound for the rank of Fix(ϕ) (see [2], [7]). The flow is also very easy to compute: the vertex $[a_{i_1} a_{i_2} \ldots a_{i_k}]$ flows to $[a_{i_2} a_{i_3} \ldots a_{i_k} \phi(a_{i_1})]$.

$$\overset{\displaystyle a_{i_1}}{\longrightarrow}$$

★ ▶——————————————————— ★

$[a_{i_1} a_{i_2} \cdots a_{i_k}]$ $\qquad\qquad\qquad\qquad\qquad$ $[a_{i_2} a_{i_3} \cdots a_{i_k} \phi(a_{i_1})]$

The problem in using this procedure is in knowing whether a particular ray has been followed long enough to be sure that all encounters with special vertices, other rays or repeated vertices have already occurred. The following example, discovered by John Stallings [8], shows that more persistence than expected may be needed.

Example 1. Let $\sigma : F_4 \to F_4$ be the following automorphism, where the boxed letters are the occurrences of generators in their images.

$$\sigma(a) = d\;\boxed{a}\;c \qquad\qquad \sigma(c) = c^{-1}a^{-1}b^{-1}a\;\boxed{c}$$
$$\sigma(b) = c^{-1}a^{-1}d^{-1}ac \qquad \sigma(d) = c^{-1}a^{-1}bc$$

There are two special edges—labeled **a** and **c**—joining two pairs of special vertices—$\{[d^{-1}],[c]\}$ and $\{[a^{-1}bac],[1]\}$. The vertex $[c]$ flows to $[c^{-1}a^{-1}b^{-1}ac]$ and then back to $[c]$: this is what happens in general for an edge that is the preferred direction out of both of its vertices. The vertex $[d^{-1}]$ flows to $[c]$ in five steps as indicated below.

$\overset{d^{-1}}{\longrightarrow}$	$\overset{c^{-1}}{\longrightarrow}$	$\overset{a}{\longrightarrow}$	$\overset{c}{\longrightarrow}$	$\overset{d}{\longrightarrow}$

★▶────── ★▶────── ★▶────── ★▶────── ★▶────── ★

$[d^{-1}]$ $[c^{-1}b^{-1}ac]$ $[ac]$ $[cdac]$ $[db^{-1}ac]$ $[c]$

The vertex $[c^{-1}a^{-1}b^{-1}a]$ flows to $[1]$ in **49 steps!** as follows with a path labeled by the word **w** where

$$w = \; a^{-1}bd^{-1}c^{-1}a^{-1}d^{-1}ad^{-1}c^{-1}b^{-1}acdadacdcdbcda^{-1}a^{-1}d^{-1}$$
$$a^{-1}d^{-1}c^{-1}a^{-1}d^{-1}c^{-1}b^{-1}d^{-1}c^{-1}d^{-1}c^{-1}daabcdaccdb^{-1}a^{-1}$$

$\overset{a^{-1}}{\longrightarrow}$	$\overset{b}{\longrightarrow}$	$\overset{b^{-1}}{\longrightarrow}$	$\overset{a^{-1}}{\longrightarrow}$

★▶────── ★▶────── ★▶ · · · · · ·★▶────── ★▶────── ★

$[a^{-1}bac]$ $[bd^{-1}]$ $[a^{-1}c^{-1}a^{-1}d^{-1}]$ $[1]$

49 steps

The graph C_ϕ is shown below. As each flow line has encountered a special vertex, there are no algorithmic problems in this example. $\mathrm{Fix}(\phi)$ is generated by $c^{-1}w$ and $\mathrm{Fix}(i_c \circ \phi)$ is generated by $a^{-1}d^{-1}c^{-1}acd$. All the remaining components of D_ϕ contain no special edges and so have rank at most 1. What is interesting about this example is the unexpectedly large numbers of steps required relative to the lengths of the images of the generators.

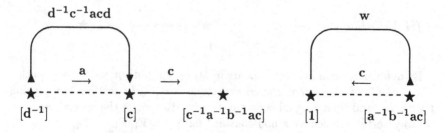

2 The graph D_f for $f : \Gamma \to \Gamma$.

We now return to the notation conventions described in the introduction. As in section **1**, there is a covering graph $\tilde{\Gamma}_{\mathrm{Fix}(\phi)}$ of Γ with base point $\tilde{v}_*$ whose edges are labeled by the edges of Γ so that the labels on closed loops based at $\tilde{v}_*$ are precisely the elements of $\mathrm{Fix}(\phi)$. The vertices of $\tilde{\Gamma}_{\mathrm{Fix}(\phi)}$ are labeled as follows.

Vertex labeling in $\tilde{\Gamma}_{\mathrm{Fix}(\phi)}$: The vertex $\tilde{v}$ has label $\mu = \bar{\ell} f(\ell)$ where ℓ is the label on some path from $\tilde{v}_*$ to $\tilde{v}$. Each label μ is a path in Γ with the property that $f(\mu(0)) = \mu(1)$: such paths will be called *f-paths*.

Again it is easy to check that this labeling is well-defined and that edge and vertex labels correspond as indicated below. Note that μ is an f-path with the same initial vertex as E if and only if $\overline{E}\mu f(E)$ is an f-path.

$$\overset{E}{\longrightarrow}$$

$$\bigstar \text{\textemdash\textemdash\textemdash\textemdash\textemdash\textemdash\textemdash} \bigstar$$

$$[\mu] \qquad\qquad\qquad\qquad [\overline{E}\mu f(E)]$$

The constructions of section 1 then generalize, leading to the following.

Definition. D_f is the graph whose vertices are labeled by f-paths $\mu \in \Gamma$, whose directed edges are labeled by the edges E_i of Γ with labeled edges joining labeled vertices as indicated in the diagram above. The vertex with label μ will be denoted by $[\mu]$. The constant path at vertex v (an f-path only if v is fixed) is denoted by $\mathbf{1_v}$.

The component of D_f containing the vertex $[\mathbf{1_{v_*}}]$ is $\tilde{\Gamma}_{Fix(\phi)}$. If μ is not a constant path, then the preferred direction out of $[\mu]$ is in the direction of the edge of D_f labeled by the first edge of μ. There are finitely many *special edges* of D_f for which neither end is the preferred out direction and each such edge corresponds to the occurrence of an edge E of Γ in its own image as follows:

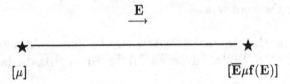

$$f(E_i) = \mu E_i \nu \qquad \rightsquigarrow$$

$$\overset{E_i}{\longrightarrow}$$

$$\bigstar \text{-----------} \bigstar$$

$$[\mu^{-1}] \qquad\qquad\qquad [\nu]$$

By definition, such an edge appears in D_f only if both μ and ν are f-paths. The core C_f of D_f is constructed exactly as was C_ϕ in section **1**, replacing **1** with $\mathbf{1_{v_*}}$ and including all constant f-paths $\mathbf{1_v}$ among the special vertices. For any vertex labeled by a non-constant path $\mu = \mathbf{E_{i_1} E_{i_2} \dots E_{i_k}}$,

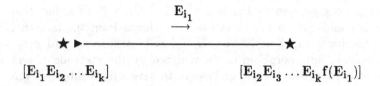

Example 2: Let $f : \Gamma \to \Gamma$ be as shown below.

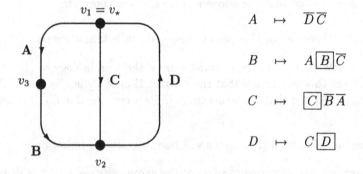

$$A \mapsto \overline{D}\,\overline{C}$$

$$B \mapsto A\,\boxed{B}\,\overline{C}$$

$$C \mapsto \boxed{C}\,\overline{B}\,\overline{A}$$

$$D \mapsto C\,\boxed{D}$$

The map $f : \Gamma \to \Gamma$

The three boxed letters determine special edges, all of which join f-paths. The graph C_f is the rectangle shown below.

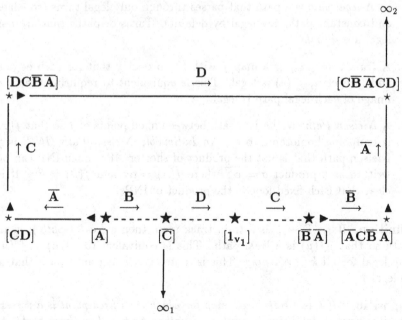

The core C_f

It follows that the induced automorphism $\phi : \pi_1(\Gamma, v_*) \to \pi_1(\Gamma, v_*)$ has fixed subgroup generated by the loop $C\overline{B}\,\overline{A}\,\overline{D}\,\overline{C}ABD$. (The fact that the rays to ∞_1 and ∞_2 do not meet can be deduced from the fact that, relative to the basis $\{a, b\} = \{AB\overline{C},\ \overline{D}\,\overline{C}AB\overline{C}\}$, $\phi(a) = ba$ and $\phi(b) = b^2a$, is a positive automorphism easily analyzed by the methods of section 1 (or directly by ad hoc means) and shown to have a fixed subgroup generated by $[b, a^{-1}] = ba^{-1}b^{-1}a$.) In section 3, we show that for a train track map, such special information is not necessary. Since f is a train track map, this will provide a direct proof that the above picture of C_f is complete.

3   The effectiveness of the procedure for train track maps.

We assume now that f is a *train track map* in the terminology of [1]. The main results of this section are that the construction of C_f is effective in this case (Theorem A) and that the structure of INPs is contained in C_f (Theorem B).

The basic notions from [1] that we will be using are the following.

1. A *turn* is a pair of germs of edges at the same vertex— a turn is degenerate if the germs are the same and non-degenerate otherwise. Turns are mapped by f in the obvious way.

2. A *legal turn* is a turn whose image under f^n is non-degenerate for all n. A *legal path* is a path that passes through only legal turns (so edges and constant paths are legal by default). Turns or paths that are not legal are *illegal*.

3. A *train track map* is a map f with the property that for each edge e and integer n, $f^n(e)$ is legal. This is equivalent to requiring that the image of each legal path is legal.

4. A *Nielsen Path (NP)* σ is a path between fixed points of f so that $f(\sigma)$ is endpoint homotopic to σ. An *Indivisible Nielsen Path (INP)* is a Nielsen path that is not the product of shorter NPs. Each INP can be written as a product $\sigma = \alpha\overline{\beta}$ where $f(\alpha) = \alpha\tau$ and $f(\beta) = \beta\tau$. It is clear that each fixed loop is the product of INPs.

Definition. If $f : \Gamma \to \Gamma$ is a train track map, then a *legal f-path* μ is an f-path so that $\mu f(\mu)$ is a legal path. This is equivalent to $\mu f(\mu)\dots f^k(\mu)$ being legal for all k. (*Warning:* This is more than being an f-path that is also legal.)

Proposition. *If f is a train track map for which each fixed point is a vertex, then all the vertex labels in C_f are legal f-paths. As one flows from vertex to vertex in C_f, the labeling paths have non-decreasing length.*

Proof. Since each fixed point is a vertex, the only way an edge can appear in its own image is at the beginning or the end and no edge is fixed: thus there are the following three possibilities.

$$f: \quad \begin{array}{ccccc} A & \mapsto & A\alpha & \mapsto & A\alpha f(\alpha) \\ B & \mapsto & \beta B & \mapsto & \beta B f(\beta)\beta B \\ C & \mapsto & C\gamma C & \mapsto & C\gamma C f(\gamma)C\gamma C \end{array}$$

The corresponding special paths μ, in addition to the constant paths at the fixed points, are α, $\overline{\beta}$, $\overline{\gamma}\overline{C}$ and $C\gamma$ for which the paths $\mu f(\mu)$ are parts of the legal paths $f^2(A), f^2(\overline{B}), f^2(\overline{C})$, and $f^2(C)$ respectively: thus each is a legal f-path. If $\mu = E_1 E_2 \ldots E_k$ is a legal f-path, then

$$(E_2 \ldots E_k f(E_1))f(E_2 \ldots E_k f(E_1))$$

is part of $\mu f(\mu)f^2(\mu)$ and is therefore legal—thus all labels encountered in the construction of C_f are legal f-paths. In particular, there are no cancellations encountered. Note also that if μ is a legal f-path all of whose forward images under the flow have the same length, then μ must be composed entirely of periodic edges.

It is now clear that in the construction of C_f, one can tell how long to follow a ray to be sure whether or not it meets another special vertex—just persist until the vertex label is longer than that for any special vertex or all short labels are exhausted. Another idea is needed to be sure that a ray will not meet another or close on itself if followed. In [2], the idea was that D_ϕ and $D_{\phi^{-1}}$ are the same graph with different edge directions and that including the information for ϕ^{-1} provided the extra information to solve the problem. We use the same idea, but since the homotopy equivalence f may not have an inverse (just a homotopy inverse), we need the following somewhat more complicated construction.

Definition. Suppose that $g, h : \Gamma \to \Gamma$ are graph maps. Then the graph $D_{g,h}$ has vertices and edges as follows.

1. For each vertex $v \in \Gamma$ and path μ from $h(v)$ to $g(v)$ there is a vertex of $D_{g,h}$ labeled $[v, \mu]$.

2. For each edge $E \in \Gamma$ from v to v' and μ as above, there is an edge labeled $E \in D_{g,h}$ joining vertices as indicated below.

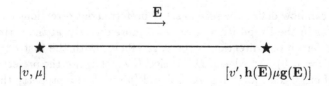

The graph $D_{g,h}$ covers Γ by $p_{g,h} : D_{g,h} \to \Gamma$, $p_{g,h}([v,\mu]) = v$ and $p_{g,h}$ carries each edge labeled E to the edge E. This corresponds to the following picture in Γ.

$$
\begin{array}{ccccccc}
\overset{E}{\longrightarrow} & & \overset{h(E)}{\longleftarrow} & \overset{\mu}{\longrightarrow} & & \overset{g(E)}{\longrightarrow} & \\
\star \rule{2cm}{0.4pt} \star & & \star \rule{2cm}{0.4pt} \star \rule{2cm}{0.4pt} \star \rule{2cm}{0.4pt} \star & & & & \\
v \qquad\quad v' & & h(v') \qquad h(v) \qquad g(v) \qquad g(v') & & & &
\end{array}
$$

Proposition. *Suppose that $f, g, h : \Gamma \to \Gamma$ are graph maps, that $gf = h$ and that $H : \Gamma \times I \to \Gamma$ is a homotopy from the identity id to h. Then there are maps $\widetilde{id}$ and $\widetilde{H}$*

$$ D_f \xrightarrow{\widetilde{id}} D_{g,h} \xrightarrow{\widetilde{H}} D_g $$

which are locally injective. Thus a set of preferred out directions for the vertices of D_f can be defined by pulling back those of D_g by $\widetilde{H} \circ \widetilde{id}$.

Proof. The map $\widetilde{id}$ is given by

$$
\begin{array}{ccc}
\overset{E}{\longrightarrow} & \overset{\widetilde{id}}{\longrightarrow} & \overset{E}{\longrightarrow} \\
\star \rule{2cm}{0.4pt} \star & & \star \rule{2cm}{0.4pt} \star \\
{[\mu]} \qquad\quad [\mu'] & & [\mu(0), \mathbf{g}(\overline{\mu})] \qquad [\mu'(0), \mathbf{g}(\overline{\mu'})]
\end{array}
$$

Since $\mu' = \overline{E}\mu f(E) \Rightarrow g(\overline{\mu'}) = gf(\overline{E})g(\overline{\mu})g(E) = h(\overline{E})g(\overline{\mu})g(E)$, this makes sense. For each vertex $v \in \Gamma$, let ρ_v be the path from v to $h(v)$ determined by the homotopy H: namely $\rho_v(t) = H(v,t)$. Then map $\widetilde{H}$ is given by

$$
\begin{array}{ccc}
\overset{E}{\longrightarrow} & \overset{\widetilde{H}}{\longrightarrow} & \overset{E}{\longrightarrow} \\
\star \rule{2cm}{0.4pt} \star & & \star \rule{2cm}{0.4pt} \star \\
{[\nu,\mu]} \qquad\quad [\nu',\mu'] & & [\rho_\nu\mu] \qquad\quad [\rho_{\nu'}\mu']
\end{array}
$$

which makes sense since $\mu' = h(\overline{E})\mu g(E) \Rightarrow \rho_{v'}h(\overline{E})\mu g(E) = \overline{E}\rho_v\mu g(E)$. Note that since all labels are understood to be reduced and the paths $\rho_{v'}h(\overline{E})$ and $\overline{E}\rho_v$ are homotopic, they determine the same label.

We can now define the second set of preferred out directions for the vertices of D_f as in the Proposition. Described more directly, at the vertex $[\mu]$ in D_f, the preferred out direction is the edge labeled with the first edge in $\rho_v g(\overline{\mu})$, where $v = \mu(0)$. An edge of D_f labeled E which is not the preferred out direction of either end joins vertices $[\mu]$ and $[\mu']$ with the property that $\rho_v g(\overline{\mu}) =$

$E_1 E_2 \ldots E_k$, $E_1 \neq E$ and $\rho_{v'} g(\overline{\mu'}) = E_1' E_2' \ldots E_\ell', E_1' \neq \overline{E}$. But $\rho_{v'} g(\overline{\mu'}) = \rho_{v'} g(\overline{E}\mu f(E)) = \rho_{v'} g f(\overline{E}) g(\overline{\mu}) g(E) = \overline{E} \rho_v g(\overline{\mu}) g(E) = \overline{E} E_1 E_2 \ldots E_k g(E)$. The only way this can fail to begin with $\overline{E}$ is that $g(E) = g(\mu)\overline{\rho_v} E \sigma$ for some σ. So for each edge E of Γ we need to consider the corresponding special vertices of D_g and determine which can be expressed as $g(\overline{\mu})\rho_v$, and since g is a homotopy equivalence, if there is a solution, there is only one. Thus we again have finitely many such edges and they are trivial to locate.

Example 2 revisited. The map $f : \Gamma \to \Gamma$ of Example 2 is in fact a train track map: there are two illegal turns—$\{A, D\}$ at v_1 and $\{B, D\}$ at v_2—neither of which occurs in the image of any edge. The maps g and h and the paths ρ_v are the following, where $\lambda = \overline{B}\,\overline{A}\,\overline{D}\,\overline{C}$.

f	g	h	
$A \mapsto \overline{D}\,\overline{C}$	$A \mapsto CD$	$A \mapsto \overline{D}\,\overline{B}\,\overline{A}\,\overline{D}\,\overline{C}$	$= \overline{D}\lambda$
$B \mapsto A\boxed{B}\overline{C}$	$B \mapsto A\boxed{B}D$	$B \mapsto CDAB\overline{C}$	$= \overline{\lambda}\overline{C}$
$C \mapsto \boxed{C}\overline{B}\,\overline{A}$	$C \mapsto \boxed{C}D$	$C \mapsto C\overline{B}\,\overline{A}\,\overline{D}\,\overline{C}$	$= C\lambda$
$D \mapsto C\boxed{D}$	$D \mapsto AB\boxed{D}$	$D \mapsto CDABD$	$= \overline{\lambda}D$

$$\rho_{v_1} = 1_{v_1} \qquad \rho_{v_2} = \overline{B}\,\overline{A}\,\overline{D}\,\overline{C} = \lambda, \qquad \rho_{v_3} = \overline{A}\,\overline{D}\,\overline{B}\,\overline{A}\,\overline{D}\,\overline{C} = \overline{A}\,\overline{D}\lambda$$

There are six edges E which appear in their own images under g and for each there is a path μ from the initial point v of E to $f(v)$ so that $g(E) = g(\mu)\overline{\rho_v} E\sigma$. They are the following:

$$E = B, \overline{B}, C, \overline{C}, D, \overline{D} \quad \Rightarrow \quad \mu = \overline{A}C\overline{B}\,\overline{A}, \overline{B}\,\overline{A}, 1_{v_1}, \overline{B}\,\overline{A}, \overline{C}, 1_{v_1}.$$

Thus there are three inverse pairs of special edges relative to the pullback directions and all lie in C_f which is shown below with the pullback labels on the vertices and the three special edges indicated by ♦.

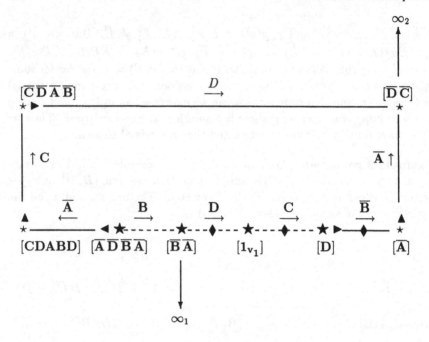

C_f **with pullback labels**

Notice that at the points at the beginnings of the two rays to ∞, the preferred out directions for D_f and the pullback are different.

Theorem A. *The procedure of section 2 is effective for train track maps.*

Proof. We need only show that it can be decided how long to follow each flow line to be sure that all encounters with a) special vertices or b) other rays or c) the ray itself have occurred.

Claim:

  a) If the flow line is followed until its labels are longer than those of all special vertices or all short labels are exhausted, then no more special vertices will be met.

  b) Suppose that R_1 and R_2 are two rays of D_f which contain no special edges for the pullback and that R_1 and R_2 have initial edges which are *not* the preferred out direction for the pullback. Then R_1 and R_2 are either disjoint or one is the terminal segment of the other. Furthermore, on every ray R there is an edge whose pullback direction is not the D_f direction. (In fact, the directions differ for all but finitely many edges in the ray.)

  c) Suppose that the greatest common divisor of the periods of periodic edges is d and that for any edge, if $f^n(E)$ is composed entirely of periodic edges for some n, then it is for $n = e$. Then if a flow line cycles back on itself, it does so within $d + e$ steps.

Proof of claim. Claim a) is clear, as remarked earlier. The first part of Claim b) is precisely 4.8 of [2] rephrased in the language of D_f and their argument applies—if R_1 and R_2 were different and met at some point, then at that point the pullback directions would have to point back towards the initial points of both rays, contradicting the uniqueness of out directions. The second part of Claim b) is effectively 4.10 of [2]. Suppose that the ray R starts at $[\mu]_*$ and that the label from $[\mu]_*$ to the ith vertex of R is X_i. Then for large i, $f(X_i) = X_i \cdot Y_i \cdot Z_i$ in reduced form, $\lim_{i \to \infty} |Y_i| = \infty$ and if $|Y_i| = j$, then $X_{i+j} = X_i Y_i$. Let $C(g)$ be the cancellation bound for g (using bounded cancellation for homotopy equivalences of graphs) and let j be large enough so that $|Y_j| > C(g) + max_v |\rho_v|$. Then

$$\begin{aligned}|g(X_j Y_j)| &\leq |g(X_j Y_j)| + |g(Z_j)| \leq |g(X_j Y_j Z_j)| + C(g) = |h(X_j)| + C(g) \\ &< |h(X_j)| + |Y_j| - max_v |\rho_v|\end{aligned}$$

Suppose Y_j is a path from $[\mu]_j$ to $[\mu]'_j$. Then the pullback direction at $[\mu]'_j$ is given by the first letter of

$$\begin{aligned}\rho_{v'_j} g(\overline{(\overline{X_j Y_j})f(X_j Y_j)}) &= \rho_{v'_j} h(\overline{(\overline{X_j Y_j})}g(X_j Y_j) \\ &= \rho_{v'_j} \overline{h(Y_j)h(X_j)}g(X_j Y_j) = \overline{Y_j}\rho_{v_j} h(\overline{X_j})g(X_j Y_j).\end{aligned}$$

But this must be the first letter of $\overline{Y_j}$, so that the pullback direction is back along R towards $[\mu]_*$.

Finally, any flow cycle must be composed of vertices whose labels are composed entirely of periodic edges and the length of the cycle is a divisor of d. Any vertex that flows into the cycle must do so within e steps.

Theorem B. *Suppose that f is a train track map for which all fixed points are vertices. Then the INPs for f are the images under p_f of the paths in C_f joining vertices labeled by constant paths 1_v and which pass through no other such vertices.*

Proof. This is clear, since every INP lifts to such a path and every such path projects to an INP.

Example 2 re-revisited. To apply this to the map f of Example 2, we need to subdivide the edge $B = B_1 B_2$ and redefine f to $\tilde{f}$ appropriately as

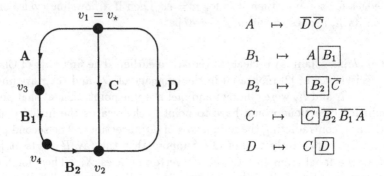

$$A \mapsto \overline{D}\,\overline{C}$$

$$B_1 \mapsto A\boxed{B_1}$$

$$B_2 \mapsto \boxed{B_2}\,\overline{C}$$

$$C \mapsto \boxed{C}\,\overline{B_2}\,\overline{B_1}\,\overline{A}$$

$$D \mapsto C\boxed{D}$$

The subdivided $\tilde{f}$

Then $D_{\tilde{f}}$ is the corresponding subdivision of D_f. In the diagram below with only the new vertex new vertex labeled by a constant path shown: all edges labeled B are subdivided and all labels $B = B_1 B_2$.

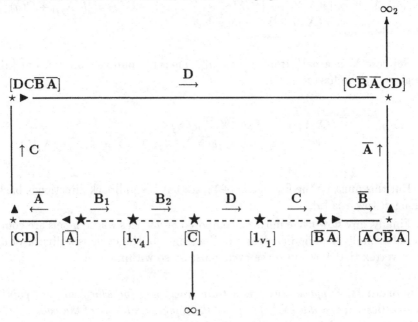

There are two INPs $\sigma_1 = \alpha_1 \overline{\beta_1}$ and $\sigma_2 = \alpha_2 \overline{\beta_2}$ for $\tilde{f}$, both joining v_4 to v_1. The INPs σ_1 and σ_2 and the action of $\tilde{f}$ are as follows.

$$\sigma_1 = \alpha_1 \cdot \overline{\beta_1} = \quad B_2 \cdot D \quad \overset{\tilde{f}}{\mapsto} \quad B_2 \overline{C} \cdot CD \quad \sim \sigma_1$$

$$\sigma_2 = \alpha_2 \cdot \overline{\beta_2} = \overline{B_1}\overline{A}CD \cdot AB\overline{C} \mapsto \overline{B_1}\overline{A}CDC\overline{B}\,\overline{A}CD \cdot \overline{D}\,\overline{C}AB\overline{C}AB\overline{C} \sim \sigma_2$$

The INPs for $\tilde{f}$

In other words, the effect of f is to pull each INP out along the ray to ∞ at its unique illegal turn, and the flow directions show the direction of the stretch. It can be shown that this phenomenon is typical for train track maps. Thus to get the INP structure for a train track map, first subdivide at each fixed point that is interior to an edge and then apply the procedure. In Example 2 the fact that the pullback directions at the initial points of the rays to ∞_1 and ∞_2 differ from the preferred D_f-directions guarantees that the rays do not meet. The fact that there are no periodic edges guarantees that they do not eventually cycle.

References

[1] Bestvina, M. and Handel, M., Train tracks and automorphisms of free groups, *Annals of Math.* **135** (1992) 1–53

[2] Cohen, M.M. and Lustig, M., On the dynamics and the fixed point subgroup of a free group automorphism, *Invent. Math.* **96** (1989) 613–638

[3] Collins, D.J. and Turner, E.C., Efficient representatives for automorphisms of free products, Preprint, 1992

[4] Collins, D.J. and Turner, E.C., All automorphisms of a free group with max rank fixed subgroup, in preparation

[5] Gaboriau, D., Levitt, G., and Lustig, M., A dendrological proof of the Scott conjecture for automorphisms of free groups, preprint

[6] Goldstein, R.Z., and Turner, E.C., Fixed subgroups of homomorphisms of free groups, *Bull. London math. Soc.* **18** (1986) 468–470

[7] Imrich, W., Krstič. S. and Turner, E.C., A bound for the rank of the fixed point subgroup, in: *Cycles and rays* (G. Hahn et al, eds.), Kluwer Academic Publishers Dordrecht (1990) 113–132

[8] Stallings, J.R., Graphical theory of automorphisms of free groups, in: *Combinatorial group theory and topology,* (Gersten, S.M. and Stallings, J.R., eds.), Annals of Math. Studies **111** Princeton Univ. Press Princeton, N.J. 1987 79–106

Department of Mathematics and Statistics,
SUNYA,
Albany,
NY 12222
U.S.A.
email: ET968ALBNYVMS

More on Burnside's problem

In 1902 W. Burnside formulated his famous problem about periodic groups:

> Is it true that every m generated group of exponent n is finite?

> If the answer is "yes" then what is the maximal order of such a group?

This problem has positive solutions for groups of exponents $2, 3, 4, 6$. Let $B(m, n)$ denote the largest m generator group of exponent n ($n = 2, 3, 4, 6$). Then

(i) $|B(m, 2)| = 2^m$, $|B(m, 3)| = 3^{m + \binom{m}{2} + \binom{m}{3}}$ (Levi and van der Waerden [15]),

(ii) $|B(m, 4)| = 2^k$, where $\frac{1}{2} 4^m \leq k < \frac{1}{2}(4 + 2 \cdot \sqrt{2})^m$ (see A. Mann [17])

(iii) $|B(m, 6)| = 2^a \cdot 3^{b + \binom{b}{2} + \binom{b}{3}}$, where $a = 1 + (m - 1)3^{m + \binom{m}{2} + \binom{m}{3}}$, $b = 1 + (m - 1)2^m$ (M. Hall, [8]).

P. S. Novikov and S. I. Adian [20] constructed infinite finitely generated groups of exponent n for each odd $n \geq 4381$. Later S. I. Adian improved the lower bound for odd exponents down to 665 [2]. Recently S. Ivanov and I. Lysenok announced the existence of infinite finitely generated groups of exponent 2^n for sufficiently big n. All that has made the second part of Burnside's question meaningless.

However, in spite of the counterexamples, a substantial part of Burnside's conjecture has proved to be correct. This "positive" development is related to the so-called restricted Burnside problem which is a "finitary" version of the original question. The problem is as follows:

> Is it true that the orders of all finite m generator groups of exponent n are bounded above by a function $f(m, n)$ of m and n?

P. Hall and G. Higman [9] showed that the restricted Burnside problem has a positive solution if: (a) the restricted Burnside problem has a positive solution for prime power exponent; (b) there are only finitely many finite simple groups of a given exponent, (c) the outer automorphism group of any finite simple group is solvable. Thus, modulo the classification of finite simple groups, the problem has been reduced to the case of prime power exponent.

In 1959 A. I. Kostrikin gave a positive solution to the restricted Burnside problem when the exponent is prime (see [13, 14, 15, 24]). In 1989 E. Zelmanov ([26,27]) solved the restructed Burnside problem for groups of prime power exponents.

Now the problem of what might be the maximal order of an m generator finite group of exponent n again makes sense.

For groups of a prime exponent p a primitive recursive upper bound for the orders of finite m generator groups was found by S. I. Adian and A. A. Razborov [3] (see also [15]). Their proof was based on the original solution of A. I. Kostrikin. Even to give an idea of the order of magnitude of the primitive recursive functions involved we will need to define classes of the Grzegorchyk hierarchy (see [1], [7]). For each $n \geq 0$ define a function Gr_n by

$$Gr_0(x,y) = x + y, \ Gr_1(x) = x^2 + 2,$$

$$Gr_{n+2}(0) = 2, \ Gr_{n+2}(x+1) = Gr_{n+1}(Gr_{n+2}(x)).$$

For example,
$$2^{2^x} < Gr_2(x) < 3^{3^x}$$

and

$$2^{2^{\cdot^{\cdot^{\cdot^2}}}} < Gr_3(x) < 3^{3^{\cdot^{\cdot^{\cdot^3}}}}$$

where both towers have height $2x + 1$. The class $Gr^n, n \geq 1$, consists of functions which can be obtained from zero function, the successor function, the projection functions, and the function Gr_{n-1} by composition and limited recursion. Thus polynomials lie in Gr^2 while exponentials lie in Gr^3. The class of all primitive recursive functions is the union $\bigcup_{n \in N} Gr^n$.

Functions from the classes Gr^2, Gr^3, Gr^4, Gr^5 are referred to as functions of polynomial, exponential, toweric or wowzer (!) type respectively (see [7]).

The upper bound found in the work of S. I. Adian and A. A. Razborov was a wowzer as were the bounds given in [15].

The appearance of these unpleasant functions can be related to the very nature of the Burnside problem: it is a problem of Ramsey type (see [7]). By a theorem of Ramsey type we mean an assertion that a sufficiently large system contains a sufficiently large regular subsystem. The most well known example of this kind is the theorem of van der Waerden which states that for any m, n there exists a number $N(m, n)$ such that having the first N numbers $1, 2, \cdots, N$ coloured in m colours we always get a monochromatic arithmetic progression of length n. The problem of Burnside asks if there exists a number $N(m, n)$ such that an arbitrary word of length $N(m, n)$ in the free group on m generators can be "reduced" to a word containing an n periodic subword. On the other hand the van der Waerden theorem can be reformulated as an assertion about finiteness of a certain m generator semigroup (m is the number of colours) with n playing a rôle close to that of an exponent.

It was an open problem for many years whether an upper bound in the van der Waerden theorem (and other major Ramsey-type theorems) can be primitive recursive. This problem was solved in 1988 by S. Shelah [21]. The upper bound of Shelah is a wowzer. It is still an open problem (formulated by Shelah, see also [7]), whether it can be reduced to a toweric function.

Zelmanov's original solution of the restricted Burnside problem for prime power exponent did not yield even a primitive recursive bound. It included A. I. Shirshov's proof of the so-called $N(k,s,n)$-lemma which is a typical Ramsey-type statement ([22]). The proof proceeds by a double induction on n and k and yields a variant of Ackerman's function [1] which is known to grow faster than any primitive recursive function.

The revised proof due to M. Vaughan-Lee and Zelmanov [25] yielded the explicit toweric bound for the order of an m generator finite group G of exponent $q = p^k$:

$$|G| \leq \left. m^{m^{\cdot^{\cdot^{\cdot^m}}}} \right\} q^{q^q}.$$

We believe, however, in a much stronger conjecture. It concerns upper bounds $N(m,q)$ for classes of nilpotency of finite m generator groups of exponent $q = p^k$.

Conjecture *There exists an upper bound $N(m,q)$ that is polynomial in m and exponential in q.*

Conjecture *For groups of prime exponent p there exists an upper bound $N(m,p)$ that is linear in m and exponential in p.*

S. I. Adian and N. N. Repin [4] proved that if the prime p is sufficiently large then there exists a finite 2 generator group of exponent p of class at least $2^{p/15}$.

Note that the known lower bounds for the function $N(m,p)$ in the van der Waerden theorem ($n = p$ being a prime number) are also polynomial in m and exponential in p (see [7]).

The linear growth of classes of nilpotency with respect to m is known for groups of exponent 5. G. Higman [11] proved that $9m$ can be taken for an upper bound. Later Havas, Newman and Vaughan-Lee [10] improved the estimates down to $6m$.

Recently Vaughan-Lee proved the existence of a polynomial upper bound for classes of nilpotency of finite m generator groups of exponent 7 (personal communication).

The probem of a polynomial upper bound $N(m,q)$ is related to the so-called problem of additional laws raised by B. Neumann [18] and H. Neumann [19].

The counterexamples to the Burnside problem due to P. S. Novikov and S. I. Adian show that the law $x^n = 1$ is not sufficient for a group to be locally finite. On the other hand from the positive solution of the restricted Burnside problem it follows that the property of a group of exponent n to be locally finite can be expressed in the language of laws. There exists a system of "additional" laws $w_i = 1$ such that a group of exponent n is locally finite if and only if it satisfies the additional laws $w_i = 1$. Immediately after the work of A. I. Kostrikin [13] had appeared B. Neumann and H. Neumann asked if there exists a finite system of additional laws for groups of prime exponent. In the language of varieties the question sounds as follows: is the Kostrikin variety of locally finite groups of exponent p finitely based?

G. Endimioni [12] noticed that the existence of a polynomial (in m) upper bound $N(m, p)$ for classes of nilpotency of finite m generator groups of exponent p implies the positive solution of the additional laws problem. Indeed, let $B(p)$ be the free Burnside group in the variety of groups of exponent p on the countable set of generators $x_1, x_2, \cdots$. Consider a sequence of elements in the group $B(p)$: $w_1 = x_1$, $w_{n+1} = ((x_{n+1}, w_n), w_n)$, where $(x, y) = x^{-1}y^{-1}xy$ is the group commutator. Each element w_m involves m generators $x_1, \cdots, x_m$ and lies in the $(2^m - 1)$th term of the lower central series. If there exists a polynomial upper bound $N(m, p)$ then we have

$$N(m, p) < 2^m - 1$$

for some m. This implies that $w_m = 1$. Thus, every locally finite group of exponent p satisfies the law $w_m = 1$. On the other hand, let us show that a group G of exponent p that satisfies a law $w_m = 1$ is locally finite. If this is not the case then without loss of generality (see [15]) we will assume that G does not contain a nontrivial locally finite normal subgroup. Let k be the minimal number such that $w_k = 1$ is a law in G, $k > 1$ and let $a = w_{k-1}(a_1, \cdots, a_{k-1}) \neq 1$ for some elements $a_1, \cdots, a_{k-1} \in G$. Then $((G, a), a) = (1)$. It is easy to see that an element a with such a property generates an abelian normal subgroup of G. This normal subgroup is locally finite, a contradiction.

In [13] A. I. Kostrikin mentioned the even stronger conjecture that classes of nilpotency of all finite groups of prime exponent p are bounded from above. It means that there is a constant upper bound for classes of nilpotency of finite groups of exponent p.

To discuss the growth of $N(m, q)$ with respect to m is the same as to discuss the structure of the free locally finite group of exponent q on countably many free generators.

Let us be more precise. As we have mentioned above, the positive solution of the restricted Burnside problem means that all locally finite groups of a given exponent q form a variety in the sense of [19]. Let $\overline{B}(q)$ be the free group in this variety on the free generators $x_1, x_2, \cdots$.

The conjecture about a constant upper bound for classes of nilpotency of finite groups of exponent p is equivalent to the conjecture that the group $\overline{B}(p)$ is nilpotent.

In 1970 this conjecture was proved to be false for groups of exponent 5 by S. Bachmuth, H. Y. Mochizuki, D. Walkup [5] and in 1971 Yu. P. Rasmyslov [21] proved that it is false for groups of all prime exponents $p \geq 5$.

Theorem (Rasmyslov) *For $p \geq 5$ the group $\overline{B}(p)$ is not solvable.*

Thus far, all existing solutions of the restricted Burnside problem (Kostrikin [13,14], Zelmanov [26,27]) are based on the study of associated Lie algebras.

Let G be a finite p-group. Consider the lower central series of G,

$$G = G_1 \geq G_2 \geq \cdots$$

and the abelian group

$$L(G) = \oplus_{i \geq 1} G_i/G_{i+1}.$$

The bracket (Lie) product

$$[a_i G_{i+1}, b_j G_{j+1}] = (a_i, b_j) G_{i+j+1}$$

defines the structure of a Lie ring on $L(G)$.

If G is a group of exponent p then $L(G)$ is an algebra over the field F_p of residues modulo p. This algebra satisfies the Engel identity

$$[\cdots [x, \underbrace{y], y], \cdots, y}_{p-1}] = 0 \quad (E_{p-1})$$

If the exponent of G is $q = p^k$ then the quotient $L(G)/pL(G)$ is an algebra over F_p and $L(G)/pL(G)$ satisfies the linearized Engel identity

$$\sum [\cdots [y, x_{\sigma(1)}], x_{\sigma(2)}], \cdots, x_{\sigma(q-1)}] = 0, \sigma \in S_{q-1}.$$

For groups of exponent p the existence of an upper bound $N(m,p)$ with linear growth in m would follow from the proof of the following conjecture

Conjecture *Let L be a Lie algebra over the field F_p that satisfies the Engel identity (E_{p-1}). Then an arbitrary element $a \in L$ generates a nilpotent ideal in L.*

Indeed, if the Conjecture above is true then there exists a number $N = N(p)$ such that an arbitrary element $a \in L$ generates an ideal that is nilpotent of class $\leq N$.

For a nilpotent group G of exponent p it implies that m arbitrary elements generate a normal subgroup that is nilpotent of class $\leq Nm$.

The conjecture above was proved for $p = 5$ by G. Higman [11]; see also Havas, Newman, Vaughan-Lee ([10]).

For Lie algebras of characteristic p that satisfy the Engel condition $(E_n), n > p$ (this situation is related to groups of exponent $q = p^k$) we conjecture that there exists a number $r = r(n,p)$ such that an arbitrary commutator $[\cdots [a_1, a_2], \cdots, a_r]$ generates a nilpotent ideal. This assertion was proved recently by M. Vaughan-Lee for $p = 7$.

At the present moment the problems of linear and polynomial growth of classes of nilpotency with respect to number of generators seem to be rather difficult. Today our understanding of the restricted Burnside problem and of the group $\overline{B}(q)$ is on the "toweric" level. This proved however sufficient for the following theorem.

Theorem [28] *The group $\overline{B}(p)$ is generalized solvable. More precisely, there exists a normal series of subgroups* $(1) = H_1 < H_2 < \cdots < H_s = \overline{B}(p)$ *such that each factor H_{i+1}/H_i is a product of abelian normal subgroups of the group $\overline{B}(p)/H_i$.*

This theorem means that the group $\overline{B}(p)$ satisfies the law $w_m = 1$ for some $m \geq 1$.

Theorem [28] *For every prime number p there exists $m = m(p)$ such that a group of exponent p is locally finite if and only if it satisfies the law $w_m = 1$.*

This theorem gives a positive solution of the additional laws problem for groups of prime exponent.

Corollary *A group G of exponent p is locally finite if and only if every $m(p)$ generator subgroup of G is finite.*

References

[1] W. Ackermann, Zum hilbertschen Aufbau der reellen Zahlen, Math. Ann. (1938), 118-133.

[2] S. I. Adian, The Burnside Problem and Identities in Groups, Ergebnisse der Mathematik und ihrer Grenzgebiete 95, Springer-Verlag, Berlin, 1979.

[3] S. I. Adian and A. A. Razborov, Periodic groups and Lie algebras. Uspekhi Mat. Nauk 42 (1987), 3-68.

[4] S. I. Adian and N. N. Repin, A lower bound for the order of a maximal finite group of prime exponent, Mat. Zametki 44 (1988), 161-176.

[5] S. Bachmuth, H. Y. Mochizuki and D. Walkup, A nonsolvable group of exponent 5, Bull. AMS, 76 (1970), 638-640.

[6] W. Burnside, On an Unsettled Question in the Theory of Discontinuous Groups, Q. J. Pure Appl. Math. 33 (1902), 230-238.

[7] R. L. Graham, B. L. Rothschild and J. H. Spencer, Ramsey Theory, Wiley-Interscience Ser. in Discrete Math., New York, 1990.

[8] M. Hall, Solution of the Burnside Problem for exponent six, Ill. J. Math., 2 (1958), 764-786.

[9] P. Hall and G. Higman, On the p-length of p-soluble groups and reduction theorems for Burnside's problem, Proc. London Math. Soc., 6
(1956), 1-42.

[10] G. Havas, M. F. Newman and M. R. Vaughan-Lee, A nilpotent algorithm
for graded Lie rings, J. Symbolic Computation 9 (1990), 653-664.

[11] G. Higman, On finite groups of exponent 5, Proc. Camb. Phil. Soc., 52
(1956), 381-390.

[12] G. Endimioni, Conditions de finitude pour un groupe d'exposant fini, J.
Algebra, 155 (1993), no. 2, 290-297.

[13] A. I. Kostrikin, The Burnside problem, Izv. Akad. Nauk SSSR, 23 (1959),
3-34.

[14] A. I. Kostrikin, Sandwiches in Lie algebras, Mat. Sb. 110 (152) (1979),
3-12.

[15] A. I. Kostrikin, Around Burnside, Ergebnisse der Mathematik und ihrer
Grenzgebiete, Springer-Verlag, Berlin, 1990.

[16] F. Levi and B. L. van der Waerden, Über eine besondere Klasse von
Gruppen, Abh. Math. Sem. Univ. Hamburg, 9 (1933), 154-158.

[17] A. J. Mann, On the orders of groups of exponent four, J. London Math.
Soc., 26 (1982), 64-76.

[18] B. H. Neumann, Varieties of groups, Bull. AMS, 73 (1967), 603-613.

[19] H. Neumann, Varieties of groups, Ergebnisse der Mathematik und ihrer
Grenzgebiete, 37, Berlin, Springer-Verlag (1967).

[20] P. S. Novikov and S. I. Adian, Infinite periodic groups I, Izv. Akad. Nauk
SSSR, 32 (1968), 212-244; Infinite periodic groups II, Izv. Akad. Nauk
SSSR, 32 (1968), 251-524; Infinite periodic groups III, Izv. Akad. Nauk
SSSR, 32 (1968), 709-731.

[21] Yu. P. Rasmyslov, On Engel Lie algebras, Algebra i Logica, 10 (1971),
33-44.

[22] S. Shelah, Primitive recursive bounds for van der Waerden numbers, J.
AMS (1988), 683-688.

[23] A. I. Shirshov, On rings with identical relations, Mat. Sb. 41 (1957),
277-283.

[24] M. Vaughan-Lee, The restricted Burnside problem, London Math. Soc. monographs, New series (5), Oxford, Clarendon Press (1990).

[25] M. Vaughan-Lee and E. Zelmanov, Upper bounds in the restricted Burnside problem, J. Algebra 162 (1993), 107-145.

[26] E. Zelmanov, The solution of the restricted Burnside problem for groups of odd exponents, Izv. Akad. Nauk SSSR, 54 (1990), no. 1, 42-59, 221.

[27] E. Zelmanov, The solution of the restricted Burnside problem for 2-groups, Mat. Sb. 182 (1991), no. 4, 568-592.

[28] E. Zelmanov, On additional laws in the Burnside problem on periodic groups, International J. Algebra and Computation 3 (1993) 583-600.

Department of Mathematics
University of Wisconsin-Madison
Madison, WI 53706
USA
email: zelmanov@math.wisc.edu

Problem Session

Problem 1 [I.M. Chiswell] A group G is n–*residually free* if, given any n non-trivial elements $g_1, \ldots, g_n$ of G, there exists a homomorphism $\phi : G \to F$ to a free group F such that $\phi(g_i) \neq 1$ for $i = 1, \ldots, n$. A group G is *fully residually free* if it is n–residually free for all n. Finitely generated surface groups are fully residually free. Does there exist a finitely generated fully residually free group which is not finitely presented?

Problem 2 [I.M. Chiswell] Let Λ be a totally ordered abelian group. A group G is Λ–*free* if it has a free action on some Λ–tree, and G is *tree-free* if it is Λ–free for some Λ. If G is a finitely generated tree-free group, does G act freely on some Λ–tree with Λ finitely generated?

This is true if G is fully residually free (see problem 1) (Remeslennikov). Not all tree-free groups are fully residually free: there are counterexamples due to D. Spellman.

Problem 3 [D.E. Cohen] For what meanings of "nice" is a graph product of nice groups nice?

Hermiller has shown that one "nice" property is that of having a finite, complete rewriting system. Baik, Howie and Pride (*J. Algebra* 162 (1993) 168-77) show that FP_3 is a "nice" property, whilst Harlander and Meinert have shown that FP_m is a "nice" property. Cohen himself has shown that FP_m, FP and having finite cohomological dimension are all "nice" properties.

Problem 4 [M.J. Dunwoody, attributed to W. Woess] Let X be a vertex-transitive locally finite graph. Is X quasi-isometric to a Cayley graph?

Note that every such X corresponds to a pair (G, H) where $H \leq G$ and $G = \langle H, g_1, \cdots, g_n \rangle$ with $G = \mathrm{Comm}(H)$, the commensurator of H.

Problem 5 [M.J. Dunwoody] Let G be a group and H an infinite cyclic subgroup of G such that G is the union of finitely many double cosets of H. Is H necessarily of finite index in G?

G.A. Niblo remarks that the answer is yes if G is residually finite.

Problem 6 [R.I. Grigorchuk] Let G be a finitely generated one-relator group, and set $\gamma(n) = |\{g \in G : |g| \leq n\}|$. Let $\Gamma(t) = \sum_{n=0}^{\infty} \gamma(n) t^n$. Is $\Gamma(t)$ rational?

Certain cases in which $\Gamma(t)$ is rational have been investigated by Edjvet and Johnson.

Problem 7 [R.I. Grigorchuk] A group G is called a Tychonoff group if, for every action of G by continuous affine transformations on a convex subcone with a compact base in a locally convex linear topological space, there is an invariant ray. Is every Tychonoff group virtually nilpotent?

Problem 8 [R.I. Grigorchuk] Trauber's theorem states that the bounded cohomology of an amenable group vanishes. Is the converse true?

Problem 9 [M. Hagelberg] Give conditions on a 3–dimensional orbifold $\mathcal{O}$ that ensure that $\pi_1\mathcal{O}$ is a generalized triangle group. Conversely, give conditions on a generalized triangle group that ensure it is the fundamental group of an orbifold.

Problem 10 [M. Hagelberg] A *generalized tetrahedron group* T is a group with a presentation of the form

$$T = \langle a, b, c | a^k = b^l = c^m = R_1^{n_1}(a, b) = R_2^{n_2}(b, c) = R_3^{n_3}(c, a) = 1 \rangle.$$

Explore the relations between such groups and orbifolds. Which generalized tetrahedron groups are infinite? Results of Baumslag, Morgan and Shalen on triangle groups have been adapted to tetrahedron groups by Vinberg.

Problem 11 [O. Kharlampovich] Does there exist a group G that is relatively finitely presented in $\mathcal{B}_n$, $(n \gg 1)$ and that has unsolvable word problem?

Problem 12 [O. Kharlampovich] Describe the varieties of soluble groups with the property that every relatively finitely presented group in the variety is residually finite.

Problem 13 [O. Kharlampovich] Is there a variety of algebras with solvable word problem and unsolvable isomorphism problem?

Problem 14 [O. Kharlampovich, attributed to J. Rhodes] Does there exist a finitely based variety of groups $\mathcal{W}$ such that there is no algorithm to decide for a given word $w \in F(x_1, \ldots, x_n)$ whether w is the identity on all finite groups in $\mathcal{W}$?

Problem 15 [D. Krammer] Let G act discretely and cocompactly on either a CAT(-1) piecewise hyperbolic complex or a CAT(0) piecewise euclidean complex. What is the complexity of the word problem and conjugacy problem for G?

Problem 16 [D. Krammer] Let G be a finitely generated discrete subgroup of $GL_n(\mathbb{Q})$. Is the conjugacy problem for G solvable?

Problem 17 [M.P. Latiolais] A $\mathbb{Z}G$–lattice M satisfies Eichler's condition if, for every simple ideal S in $\mathbb{Q}G$ such that $\text{End}(S)$ is a totally definite quaternion algebra, then either S is not a direct summand of $\mathbb{Q}M$, or $S \oplus S$ is a direct summand. Classify all groups which do or do not satisfy Eichler's condition.

Problem 18 [M.P. Latiolais] Let G be a finite group and $N = \sum_{g \in G} g \in \mathbb{Z}G$. Then G satisfies *weak cancellation* if, for all integers k coprime to the order of G, the submodule $\langle k, N \rangle \subseteq \mathbb{Z}G$ is free whenever it is stably free (Swan, Dyer). Classify all groups which satisfy weak cancellation.

Problem 19 [M. Lustig] Do all Cockcroft properties of a 2–complex K depend only on $\pi_1 K$ and $\chi(K)$?

Problem 20 [M. Lustig] Let $G = \langle x_1, \ldots, x_n | R \rangle$ with $d(G) = n$. Conjecture: the 2-sided ideal generated by $\{\partial R/\partial x_1, \cdots, \partial R/\partial x_n\}$ is properly contained in $\mathbb{Z}G$.

Problem 21 [M. Lustig] Let F_r be a free group of rank r and $\phi \in \mathrm{Aut} F_r$. Consider $F(r, n, \phi) := F_r \rtimes_{\phi^n} \langle t \rangle / \ll t \gg$. What can we say about $F(r, n, \phi)$? In particular, is $F(r, n, \phi)$ finite or infinite?

The isomorphism type of $F(r, n, \phi)$ depends only on the conjugacy class of ϕ in $\mathrm{Aut} F_r$. If $\phi : a_i \mapsto a_{i+1}$ $(1 \leq i \leq r - 1)$ and $\phi : a_r \mapsto a_1 a_2 \cdots a_r$ then $F(r, n, \phi)$ is the Fibonacci group $F(r, n)$. If ϕ is induced by a pseudo-Anosov homeomorphism of a surface with one boundary component then $F(r, n, \phi)$ is infinite for sufficiently large n.

Problem 22 [I.G. Lysionok] Does there exist a hyperbolic group that is not residually finite? Does there exist a hyperbolic group having no torsion-free subgroup of finite index?

Gersten remarks that there is a synchronously bounded combable group that is not residually finite.

Problem 23 [H. Meinert] Let $G = F_1 \times \cdots \times F_k$ be the direct product of k free groups of rank ≥ 2. Let H be a subgroup of G of type FP_k. Is H a finite extension of a direct product of k free groups of finite rank?

This is trivial for $k = 1$, and true for $k = 2$ by a result of Baumslag and Roseblade. Meinert has shown that the result is also true if $H \geq G'$.

Problem 24 [S.J. Pride] Is there a finitely presented small cancellation group which is not residually finite or does not have a proper subgroup of finite index? Is there a hyperbolic group that does not have a proper subgroup of finite index? (Compare problem 22.)

A possible example for the first question is a one-relator group G with relator of the form $tA = Bl$ with A and B words not involving t. If $|A| = |B|$ then G also has a $C(3), T(6)$ presentation (El-Mosalamy and Pride).

Problem 25 [G. Rosenberger] Let $G = \langle a, b | a^p = b^q = R^m(a, b) = 1 \rangle$, where $2 \leq p, q, m$. Is G either solvable-by-finite or SQ universal? Suppose that G has a faithful representation $\rho : G \to \mathrm{PSL}(2, \mathbb{C})$ with $\rho(G)$ discrete and of finite covolume. Is $1/p + 1/q + 1/m \geq 1$? This is true if $m \geq 3$. When is G arithmetic?

Problem 26 [G. Rosenberger] Let $G = \langle a, b | [a, b]^m = 1 \rangle$. Is G a non-trivial free product with amalgamation? This is true if m is not a power of 2.

Problem 27 [G. Rosenberger] Let F be a free group with basis $\{a_1, \ldots, a_n\}, n \geq 2$. Let $u, v \in F^p F'$ with $p \geq 2$ and u, v not proper powers, cyclically reduced and involving all of the $a_1, \ldots, a_n$. Let K be the HNN extension $\langle F, t | tu^\alpha t^{-1} = v^\beta \rangle, \alpha, \beta \geq 1$. Let X be a minimal generating system for K. Does K have a one-relator presentation on X? Do there exist only finitely many Nielsen equivalence classes of minimal generating systems for K? Is the isomorphism problem for K solvable in the class of one-relator groups? Is $\mathrm{Aut}(K)$ finitely presented? Is K hopfian? All these questions have affirmative answers if $\alpha = 1 = \beta$.

Problem 28 [M. Shapiro] Let Γ be a Cayley graph for a group G. The *metric cone* $C(g)$ at $g \in G$ is the set of all paths $p(t)$ with $p(0) = g$ such that the distance $d(1, p(t))$ increases at constant unit speed. Note that the paths in $C(g)$ need not be edge paths: they may terminate in subsegments of edges. Elements g, g' have the same metric cone type if translation by $g'g^{-1}$ carries $C(g)$ to $C(g')$. Let $WC(g)$ be the set of edge paths in $C(g)$: there corresponds a notion of word cone type for g. Note that the metric cone type determines the word cone type.

It is known that Γ has finitely many word cone types if and only if the language of geodesics is regular. Can Γ have finitely many word cone types and yet G not be finitely presented? Can Γ have finitely many metric cone types and yet G not be finitely presented? If Γ has finitely many word cone types, must Γ have finitely many metric cone types? If Γ has finitely many word cone types and G is finitely presented, must Γ have finitely many metric cone types?

Printed in the United States
By Bookmasters